THE PATHOLOGY AND IMAGING OF CORONAVIRUS PNEUMONIA

THE PATHOLOGY AND IMAGING OF CORONAVIRUS PNEUMONIA

Edited by

HONGJUN LI

Beijing You An Hospital, Capital Medical University, China

CHUNMING ZHANG

Henan University, Kaifeng, Henan, China

YINGMIN MA

Beijing You An Hospital Affiliated to Capital Medical University, Beijing, China

SERGEY TERNOVOY

Sechenov First Moscow State Medical University, Moscow, Russia

ELSEVIER

ACADEMIC PRESS

An imprint of Elsevier

Academic Press is an imprint of Elsevier
125 London Wall, London EC2Y 5AS, United Kingdom
525 B Street, Suite 1650, San Diego, CA 92101, United States
50 Hampshire Street, 5th Floor, Cambridge, MA 02139, United States

ISBN: 978-0-443-32930-2

For information on all Academic Press publications visit our website at https://www.elsevier.com/books-and-journals

Publisher: Stacy Masucci
Acquisitions Editor: Glyn Jones
Editorial Project Manager: Naomi Robertson
Production Project Manager: Gomathi Sugumar
Cover Designer: Mark Rogers

Typeset by TNQ Technologies

Contents

The Wisdom Shop

Thank you for ordering with us!

We hope that you are happy with your order! If there are any issues, we would be love for the opportunity to better your experience. Please reach out to us and we will be glad to assist you. If you are happy with your purchase, we would be grateful to receive feedback for our store. To leave seller feedback:

- Find your order and select the options that best reflect your experience.
- Click Submit Feedback.

To view our policies, and to purchase with us in the future, follow the link or scan the QR code below:

www.amazon.com/shops/The_Wisdom_Shop

We are a small business that hopes your experience with our store was one that was positive and memorable. Thank you for your time and valuable feedback!

Contributors

Liu Bailu
The Second Affiliated Hospital of Harbin Medical University, Harbin, China

Xu Bingren
Shenzhen Qianhai Shekou Free Trade Zone Hospital, Shenzhen, China

Fu Chaohong
The Sixth People Hospital of Zhengzhou, Zhengzhou, China

Xu Chuanjun
Affiliated Nanjing Hospital of Nanjing University of Chinese Medicine, Nanjing, China; Department of Radiology, The Second Hospital of Nanjing University of Chinese Medicine, Nanjing, China

Zhang Chunming
The Sixth People Hospital of Zhengzhou, Shanghai, China

Hou Dailun
Beijing Chest Hospital, Capital Medical University, Beijing, China

Zhao Dawei
Beijing YouAn Hospital, Capital Medical University, Beijing, China

Liu Dechun
Anhui Bengbu Medical University, Bengbu, China

Li Di
Beijing Children's Hospital, Capital Medical University, Beijing, China

Xu Dong
Beijing Chest Hospital, Capital Medical University, Beijing, China

Yu Dong
Shenzhen Center for Prehospital Care, Shenzhen, China

Zheng Guangping
The Third People's Hospital of Shenzhen, Shenzhen, China

Qiao Guoqing
The First Hospital of China Medical University, Shenyang, China

Wang Haibo
The Second Affiliated Hospital of Harbin Medical University, Harbin, China

Chen Haixia
The Central Hospital Affiliated to Chongqing University of Technology, Chongqing, China

Liu Heng
Affiliated Hospital of Zunyi Medical University, Zunyi, China

Lu Hong
The Central Hospital Affiliated to Chongqing University of Technology, Chongqing, China

Yuan Hong
The Third People's Hospital of Shenzhen, Shenzhen, China

Li Hongjun
Beijing YouAn Hospital, Capital Medical University, Beijing, China

Li Hongyan
The First Affiliated Hospital of Nanyang Medical College, Nanyang, China

Guo Hui
The Four Affiliated Hospital of Xinjiang Medical University, Ürümqi, China

Guo Jiachen
Beijing YouAn Hospital, Capital Medical University, Beijing, China

Xu Jianrong
Renji Hospital, Shanghai Jiaotong University School of Medicine, Shanghai, China

Liu Jiaping
Beijing Chest Hospital, Capital Medical University, Beijing, China

Liu Jing
Soochow University Affiliated Hospital of Infectious Diseases (Suzhou No. 5 People's Hospital), Jiangsu, China

Lou Jinli
Beijing YouAn Hospital, Capital Medical University, Beijing, China

Liu Jun
The Second Xiangya Hospital of Central South University, Changsha, China

Huang Kuanlong
Shenzhen Qianhai Shekou Free Trade Zone Hospital, Shenzhen, China

Song Lan
Chinese Academy of Medical Sciences Peking Union Medical College Hospital, Beijing, China

Li Li
Beijing YouAn Hospital, Capital Medical University, Beijing, China

Ai Li
The Central Hospital Affiliated to Chongqing University of Technology, Chongqing, China

Gong Lianggeng
The Second Affiliated Hospital of Nanchang University, Nanchang, China

Wu Lianming
Renji Hospital, Shanghai Jiaotong University School of Medicine, Shanghai, China

Xing Lihong
The Affiliated Hospital of Hebei University, Hebei, China

Sun Lijun
Beijing YouAn Hospital, Capital Medical University, Beijing, China

Kong Lili
Yantai Yeda Hospital, Yantai, China

Zhang Lina
Department of Radiology, The Fourth Affiliated Hospital of China Medical University, Liaoning, China

Qin Ling
Beijing YouAn Hospital, Capital Medical University, Beijing, China

Zhao Lingling
The Sixth People Hospital of Zhengzhou, Zhengzhou, China

Zeng Lingzi
Shenzhen Qianhai Shekou Free Trade Zone Hospital, Shenzhen, China

Yingmin Ma
Beijing YouAn Hospital, Capital Medical University, Beijing, China

Ren Meiji
Beijing YouAn Hospital, Capital Medical University, Beijing, China

Li Ping
The Second Affiliated Hospital of Harbin Medical University, Harbin, China

Lu Puxuan
Shenzhen Center for Chronic Disease Control, Shenzhen, China

Zheng Qiuting
Shenzhen Center for Chronic Disease Control, Shenzhen, China

Xu Qiuzhen
Department of Radiology, Zhongda Hospital Affiliated to Southeast University, Nanjing, China

Chen Tianwu
The Second Affiliated Hospital of Chongqing Medical University, Chongqing, China

Chen Tingting
The Second Affiliated Hospital of Harbin Medical University, Harbin, China

Li Tongzeng
Beijing YouAn Hospital, Capital Medical University, Beijing, China

Fang Wenchun
Comprehensive Clinic of Shenzhen Healthcare Committee Office, Shenzhen, China

Ma Xi
The Affiliated Hospital of Hebei University, Hebei, China

Tang Xiaoli
Shenzhen Qianhai Shekou Free Trade Zone Hospital, Shenzhen, China

Yin Xiaoping
The Affiliated Hospital of Hebei University, Hebei, China

Gong Xiaoming
Hubei Xianning Central Hospital, Xianyang, China

Zhang Yaming
Shenzhen Qianhai Shekou Free Trade Zone Hospital, Shenzhen, China

Dang Yan
Beijing YouAn Hospital, Capital Medical University, Beijing, China

Shi Yanbin
The Sixth People Hospital of Zhengzhou, Zhengzhou, China

Liu Yang
Beijing YouAn Hospital, Capital Medical University, Beijing, China

Lu Yibo
HIV/AIDS Clinical Treatment Center of Guangxi (Nanning), The Fourth People's Hospital of Nanning, Nanning, Guangxi Zhuang Autonomous Region, China; Nanning Fourth People's Hospital, Nanning, China

Guan Ying
The Second Affiliated Hospital of Harbin Medical University, Harbin, China

Zhang Yu
The Affiliated Hospital of Hebei University, Hebei, China

He Yulin
The First Affiliated Hospital of Nanchang University, Nanchang, China

Zhao Yunxia
Anhui Bengbu Medical University, Bengbu, China

Shi Yuxin
Shanghai Public Health Clinical Center, Shanghai, China

Lv Zhehao
The First Affiliated Hospital of Harbin Medical University, Harbin, China

Li Zhen
Peking University Health Science Center, Beijing, China

Zhou Zhen
Beijing Deepwise Science and Technology Co., Ltd., Beijing, China

Lu Zhiyan
Zhongnan Hospital of Wuhan University, Wuhan, China

About the editors

Dr. Hongjun Li, is a professor, chief physician, doctoral supervisor, postdoctoral supervisor and inheritance supervisor, expert enjoying a special government allowance of the State Council, national outstanding contribution expert, one of the first batch of "Ten-Hundred-Thousand Health Talents" in Beijing, discipline (backbone) leader of first batch of "215" High-level Health Talents in Beijing, chief medical imaging specialist of the National Clinical Research Center for Infectious Diseases, and medical imaging specialist in infectious diseases and founder of infectious disease imaging and innovative discipline system.

He has specialized in radiology diagnosis of infectious diseases, infection and inflammation, and inflammation-related tumors, advocating the concept of evidence-based medicine, and committed to noninvasive and accurate hierarchical diagnosis based on the fusion of imaging and multisource heterogeneous data.

Honors: he was honored as the candidate for "Beijing Scholar" in 2021, advanced individual in the Fight against COVID-19 in Beijing and advanced individual in the Fight against COVID-19 of the Chinese Peasant and Workers Democratic Party in 2020, and expert with the title of "National Renowned Doctor with Outstanding Achievements" in 2020. He was awarded the title of "People's Good Doctor" in 2019 and "Master

Teacher" from 2018 to 2020 by the Beijing Municipal Federation of Trade Union, China's capital health guardian title.

Current Positions: Director of Medical Imaging Center at Beijing YouAn Hospital, Capital Medical University, Deputy Director of the Department of Medical Imaging at Capital Medical University, the founding editor-in-chief of *Radiology of Infectious Diseases* (the managing journal of the National Health Commission), the founding editor-in-chief of *Radiology Science,* the associate editor of *BMC Neurology*, etc.

Academic Positions: the Chairman of the Professional Committee of Radiology in National Health Technology Promotion Inheritance and Application Project of National Health Commission, the Chairman of the Specialized Committee for Radiology of Infection and Inflammation of Chinese Research Hospital Association, the Chairman of the Infection (Infectious Disease) Imaging Working Committee of Chinese Association of STD and AIDS Prevention and Control, the Chairman of the Specialized Committee for Digital Diagnosis and Treatment of China Society for the Promotion of Science and Technology Commercialization, the Chairman of the Specialized Committee for Radiology of Infectious Diseases of Chinese Society of Radiology and the team leader of the Imaging Management Team for Infectious Diseases of Chinese Hospital Association Management Branch for Infectious Diseases, etc. Expert reviewer of the National Prize for Progress in Science and Technology, expert reviewer of the Chinese Medical Science and Technology Progress Award, expert reviewer of Major R&D Projects of the Ministry of Science and Technology, expert reviewer of National Natural Science Foundation projects, etc.

Scientific Research Experience: he was approved for two major research and development projects by the Ministry of Science and Technology, approved and hosting six key projects of the National Natural Science Foundation, surface projects and of the Beijing Natural Science Foundation and hosting more than 11 projects of the Beijing Major Scientific and Technological R&D Programs, etc. He has published more than 230 papers in Chinese and English and has won 10 provincial and ministerial awards including China Medical Science and Technology Award and 30 national patents for invention, intellectual property rights and software copyrights, etc. He has edited 50 monographs (including 16 in English, Springer, Elsevier and PMPH), five textbooks, guidelines and 26 standards. He has four sets of products by medicine and engineering combination and interdisciplinary integration. In 2023, he won the Springer Nature China New Development Award.

Prominent Contributions: he has been engaged in clinical imaging diagnosis and treatment for more than 30 years. He has followed the discipline construction idea of "International Vision, Patient Needs, Systematic Thinking and Overall Promotion" and international discipline construction thought of "Standardization of Medical Technology, Modernization of Technology and Equipment, Informatization of Medicine and Engineering and Specialization of Technical Team." He has conducted the systematic theory system and technical research on infectious disease imaging as well as pathogenic and pathological mechanisms and together with national famous medical imaging specialists, created a systematic and innovative theory system of global AIDS radiology, infectious disease radiology, infection and inflammation radiology and infectious inflammation-related tumor radiology and formed the discipline model for modern medical imaging information. He took the lead in building an international innovative integration discipline—modern imaging informatics for infectious diseases, he is known as the pioneer and founder of the imaging informatics for infectious diseases and promoted the development of treatment and prevention of infectious disease both at home and abroad.

Chunming Zhang, surgical chief physician, secondary professor, master tutor at Henan University, the affiliated infectious disease hospital of Henan province (sixth people's hospital of Zhengzhou city), vice secretary of party committee of the China medical doctor association of obesity and diabetes surgery committee, director of clinical medical research center, vice President of Henan province, director of the AIDS professional committee of the committee. He has been awarded the honors of "Academic and Technical Leader of Zhengzhou," "Top Professional and Technical Talents of Zhengzhou," "Young and Middle-aged Health Science and Technology Innovation Talents in Henan Province," Advanced Individual of Disease Prevention and Control in Henan Province, personal achievements of medical personnel in the new stage of epidemic prevention and control in Henan Province in 2022, and the ninth batch of "People's Good Doctors" in China. He has published 2 monographs, published more than 30 academic papers, and won 3 scientific and technological achievements awards at or above the municipal department level.

Yingmin Ma, Doctor of Medicine, Chief Physician, Professor, Doctoral Supervisor.

Educational background: He graduated from the Fourth Military Medical University of the PLA. Work experience: He has worked at the PLA General Hospital, Beijing Millennium Temple Hospital affiliated to Capital Medical University, and Beijing Chaoyang Hospital affiliated to Capital Medical University. Professional expertise: Diagnosis and treatment of various diseases in respiratory medicine. He is good at respiratory and critical medicine; clinical treatment and the research of respiratory failure, especially the clinical diagnosis and treatment of lung tumors and lung interstitial diseases; and tobacco control and cessation.

Scientific research experience: He has successively undertaken major projects of the Ministry of Science and Technology, the National Natural Fund, the Beijing Traditional Chinese Medicine Science and Technology Project, the Beijing Health System "Ten, Hundred, Thousand" Talent Funding Project, the first medical development fund project, and the Beijing Science and Technology Plan Project ("Capital characteristics" special project) National Science and Technology Support Plan Project. He has also undertaken the Beijing Medical Management Center sail plan key projects, the Beijing Science and Technology Commission of the capital characteristics and capital cultivation, and other topics of research work.

Awards: He has won two second prizes and one third prize for Military Science and Technology Progress. He has also won the second prize for National Science and Technology Progress.

Social part-time job: He is also the Chairman of the Branch of the Infectious Diseases Hospital of the Chinese Hospital Association; a Member of the Critical Care Medicine Group of the Ninth Committee of the Respiratory Society of the Chinese Medical Association; the Chairman of the Anti-infective Pharmacology Committee of the Beijing Pharmacological Society; the Executive Director of the Fourth Council of the Beijing Pharmacological Society; the Deputy Chairman of the Professional Committee of Drug Clinical Evaluation, Chinese Pharmaceutical Society; a Standing Member of the Therapeutic Drug Monitoring and Research Committee of the Chinese Pharmacological Society; the Executive Director of the Infectious Diseases Committee, Chinese Medical Education Association; the Deputy Chairman of the Ninth Committee of the Respiratory Branch of the Beijing Medical Association; a Standing Member of the Branch of Allergy (Allergy) Reaction, Beijing Medical Association; a Member of the Ninth Council of the China Anti-Cancer Association; and an Editorial Member of the *Chinese Journal of Tuberculosis and Respiratory Diseases*.

Professor Sergey Ternovoy, M.D. (MD) and Ph.D. (PhD), is the Head of the Department of Radiology at the National Centre for Cardiology Research and the Head of the Department of Radiology at the First Moscow State Medical University, and a member of the Russian Academy of Sciences.He graduated from the Odessa State Medical University with a degree in General Practitioner in 1972, and then completed a postgraduate course in Radiology in 1975, and received a Candidate of Medical Sciences degree in 1977, and a PhD in 1984. He completed a postgraduate programme in Radiology in 1975 and received his Candidate of Medical Sciences degree in 1977 and his PhD in 1984. Professor Ternovoy has gained extensive experience in the field of medical imaging and has held a number of key positions, including Head of the Department of Radiology at the Central Scientific Research Laboratory from 1989 and Head of the CT Department at the Cardiology Research Centre from 1990.

Professor Ternovoy is well known in academic circles and has received numerous honours, including the USSR State Prize for Medicine in 1989, the Government Prize of the Russian Federation in the field of education in 2011 and the title of Honorary Scientist in 2014. His scientific output is notable with over 650 scholarly articles, 54 monographs, over 30 patents and over 35 MD and PhD theses under his supervision. In addition, he serves as a member of the editorial boards of several medical journals, including the founder and deputy editor-in-chief of the Russian Journal of Electron Radiology. Professor Ternovoy's main areas of research include modern digital radiology techniques (CT, MRI, ultrasound and digital radiology) and hybrid imaging techniques, and he is committed to advancing the development and application of medical imaging.

Preface

At the end of 2019, the Coronavirus Disease 2019 (COVID-19) began to spread globally. The COVID-19 pandemic, with its long duration, widespread impact, the large-number infections, and the profound consequences, was considered an unprecedented event in nearly a century. On January 25, 2020, General Secretary Xi Jinping presided over the meeting of the Standing Committee of the Political Bureau of the CPC Central Committee and delivered an important speech. He emphasized that "Party committees and governments at all levels shall fully mobilize, deploy and strengthen their efforts in accordance with the decisions and arrangements of the CPC Central Committee, put the people's lives and their physical well-being as the top priority and take the epidemic prevention and control as the most important task at present." General Secretary Xi Jinping showed great concern for the development of the epidemic and the treatment of patients and issued important instructions on multiple occasions, emphasizing "putting the people's lives and their physical well-being as top priority."

On January 30, 2020, in accordance with the Notice of the General Office of the Ministry of Science and Technology on Matters Related to Strengthening the Management of Scientific and Technological Research Projects on COVID-19, it was proposed that all project units shall prioritize the epidemic prevention and control as the most important task at hand, organize researchers to concentrate their efforts, and collaborate closely with each other to ensure high-efficient and high-quality completion of scientific and technological research on COVID-19 prevention and control, adhere to put the interests of the state and the people first, "write the paper on the land of our country," and apply the research results to the epidemic prevention and control as soon as possible. The real-time release of the updated guidelines on diagnosis, treatment, and prevention by the National Health Commission clearly demonstrated the great importance attached by the CPC Central Committee and the State Council, and the determination and will of all government departments to unite and coordinate to fight the epidemic.

Saving lives is of paramount importance, and the prevention and control of the epidemic are our responsibilities. In order to effectively implement the decisions and arrangements of the CPC Central Committee and promote the public awareness of enhancing self-protection so as to decisively win the battle against the epidemic, the Infectious Disease Group of Chinese Society of Radiology, the Professional Committee for Infection Imaging of Chinese Medical Doctor Association Radiologist Branch, and other infectious disease imaging teams gave up the Chinese New Year holiday and worked day and night. On February 1, 2020, they timely released Guidelines for Imaging Diagnosis of Novel Coronavirus Pneumonia in both Chinese and English versions (first edition 2020) and compiled the second edition on February 26, 2020. Although our team has made continuous efforts in deep learning, active practice, and exploration, there is still a lot of follow-up work to achieve a comprehensive understanding of these newly emerging infectious diseases. Based on the imaging diagnosis experience from the outbreak of coronavirus-family pneumonia, including Severe Acute Respiratory Syndrome (SARS) in 2003 and Middle East Respiratory Syndrome (MERS) in 2015, as well as global epidemics such as the H1N1 and H7N9 avian influenza, our infectious disease imaging team has experienced 23 years (since 1998) of infectious disease imaging research and practice, accumulated extensive expertise, and made a profound academic influence both at home and abroad, all of which have played a crucial role in providing significant professional support for the successful compilation of this book. While newly emerging and recurring infectious diseases pose a serious threat to human life, health, and national security, the World Health Organization and the Chinese government have attached great importance to the prevention and control of infectious diseases. In the past 30 years, the rapid development of the visualized precision diagnosis and treatment technology has facilitated the emergency of new medical perspectives and theories and led the advancement of medical diagnosis and treatment technologies, representing the forefront of medical practice. The visualized precision diagnosis and treatment technology has played a vital role during the diagnostic and therapeutic process. By leveraging the clinical heterogeneous data in visualized precision diagnosis and treatment, we have made breakthroughs in establishing technical specifications for precision diagnosis and treatment of infectious diseases, exploring diagnosis and treatment pathways, formulating digital noninvasive diagnosis and treatment standards, and implementing artificial intelligence, not only providing evidence for the early diagnosis and assessment of treatment effect against newly emerging and epidemic infectious diseases but also offering technological support and innovative contributions to mitigate the threats posed by newly emerging and recurring infectious diseases to the health of the Chinese population.

This book has the following characteristics: ①Focus on the clinical and imaging features as well as distinguishing signs of coronavirus-family pneumonia, including SARS, MERS, and COVID-19; ②With the deepening of clinical practice and research, the data collection becomes more comprehensive, reliable and objectively truthful, and the experience becomes more mature, including longitudinal follow-ups and pathogenic pathology verification; it is authored by more than 60 experts holding a doctoral degree and/or the title of chief physician or associate chief physicians from more than 10 provinces, municipalities, and autonomous regions across the country, and reviewed by experienced industry experts for quality control and consensus. We sincerely hope that this book will become an essential reference for medical imaging professionals and clinical personnel in the diagnosis and treatment of coronavirus pneumonia and its family pneumonia, the identification and diagnosis of variant subtype virus pneumonias, as well as their clinical practice.

As scientific development is the process of continuous exploration, accumulation, and summarization of knowledge, both cognition and recognition require constant validation through practical application. It is inevitable that this book may be lacking in deep analysis or even contain errors. Your valuable insights and feedback are greatly appreciated.

Editor-in-chief Hongjun Li

Brief introduction

This book focuses on the study of the coronavirus family—related pulmonary diseases, specifically targeting the three subtypes of coronaviruses that have been confirmed to infect humans: SARS, MERS, and COVID-19, to provide a comprehensive description and introduction based on heterogeneous data from imaging and clinical sources, and particularly emphasize the imaging-based staging diagnostic model rooted in the pathogenic and pathological aspects of clinical staging, aiming to better align with the requirements of evidence-based medicine in clinical diagnosis and treatment. This book provides a comprehensive overview of the pathology and imaging manifestations of the coronavirus pneumonia, covering the integrated theories and practical achievements of etiology, epidemiology, pathology, clinical diagnosis, treatment, radiology, cleaning, disinfection, and protection measures commonly used in Imaging Department, as well as prevention and prognosis of SARS, MERS, and COVID-19.

Featuring substantial contents, comprehensive information, vivid illustrations, typical cases and high practicality, this book is an invaluable medical monograph reflecting the latest research findings, which is suitable for medical professionals and students from various clinical departments.

List of editorial board members

Editor-in-chief

Hongjun Li
Chunming Zhang
Yingmin Ma
Sergey Ternovoy

Associate Editor-in-chief

Lu Puxuan
Lu Hong
Li Li
Yin Xiaoping
Gong Lianggeng
Chen Tianwu
Zhao Lingling
Fu Chaohong
Shi Yanbin
Li Hongyan

Editors (in order of pinyin of last name)

Ai Li	Affiliated Hospital of Chongqing University of Technology (Chongqing Seventh People's Hospital)
Chen Haixia	Affiliated Hospital of Chongqing University of Technology (Chongqing Seventh People's Hospital)
Chen Tianwu	The Second Affiliated Hospital of Chongqing Medical University
Chen Tingting	The Second Affiliated Hospital of Harbin Medical University
Dang Yan	Beijing YouAn Hospital, Capital Medical University
Fang Wenchun	Comprehensive Clinic of Shenzhen Healthcare Committee Office
Fu Chaohong	The Sixth People Hospital of Zhengzhou
Gong Lianggeng	The Second Affiliated Hospital of Nanchang University
Gong Xiaoming	Hubei Xianning Central Hospital
Guan Ying	The Second Affiliated Hospital of Harbin Medical University
Guo Jiachen	Beijing YouAn Hospital, Capital Medical University
Guo Hui	The First Affiliated Hospital of Xinjiang Medical University
He Yulin	The First Affiliated Hospital of Nanchang University
Hou Dailun	Beijing Chest Hospital, Capital Medical University
Huang Kuanlong	Shenzhen Qianhai Shekou Free Trade Zone Hospital
Kong Lili	Yantai Yeda Hospital

Li Di	Beijing Children's Hospital, Capital Medical University
Li Li	Beijing YouAn Hospital, Capital Medical University
Li Ping	The Second Affiliated Hospital of Harbin Medical University
Li Zhen	Peking University Health Science Center
Li Hongjun	Beijing YouAn Hospital, Capital Medical University
Li Hongyan	The First Affiliated Hospital of Nanyang Medical College
Li Tongzeng	Beijing YouAn Hospital, Capital Medical University
Liu Heng	Affiliated Hospital of Zunyi Medical University
Liu Jing	Soochow University Affiliated Hospital of Infectious Diseases (Suzhou No. 5 People's Hospital)
Liu Jun	The Second Xiangya Hospital of Central South University
Liu Yang	Beijing YouAn Hospital, Capital Medical University
Liu Bailu	The Second Affiliated Hospital of Harbin Medical University
Liu Dechun	Anhui Bengbu Medical College
Liu Jiaping	Beijing Chest Hospital, Capital Medical University
Lou Jinli	Beijing YouAn Hospital, Capital Medical University
Lu Yibo	Nanning Fourth People's Hospital
Lu Hong	Affiliated Hospital of Chongqing University of Technology (Chongqing Seventh People's Hospital)
Lu Zhiyan	Zhongnan Hospital of Wuhan University
Lu Puxuan	Shenzhen Center for Chronic Disease Control
Lv Zhehao	The First Affiliated Hospital of Harbin Medical University
Ma Xi	The Affiliated Hospital of Hebei University
Yingmin Ma	Beijing YouAn Hospital, Capital Medical University
Qiao Guoqing	The First Hospital of China Medical University
Qin Ling	Beijing YouAn Hospital, Capital Medical University
Ren Meiji	Beijing YouAn Hospital, Capital Medical University
Shi Yanbin	The Sixth People Hospital of Zhengzhou
Shi Yuxin	Shanghai Public Health Clinical Center
Song Lan	Chinese Academy of Medical Sciences Peking Union Medical College Hospital
Sun Lijun	Beijing YouAn Hospital, Capital Medical University
Tang Xiaoli	Shenzhen Qianhai Shekou Free Trade Zone Hospital
Wang Haibo	The Second Affiliated Hospital of Harbin Medical University
Wu Lianming	Renji Hospital, Shanghai Jiaotong University School of Medicine
Xing Lihong	The Affiliated Hospital of Hebei University

CHAPTER 1

Overview

Guo Hui[1] and Zhang Chunming[2]
[1]The Four Affiliated Hospital of Xinjiang Medical University, Ürümqi, China; [2]The Sixth People Hospital of Zhengzhou, Shanghai, China

1.1 The discovery and types of coronaviruses

Coronavirus (CoV) is an enveloped, positive-sense single-stranded RNA virus with the genome length of 27.32 kb. As the largest RNA virus currently known, it is classified into two genera: torovirus and coronavirus. The coronavirus genus is composed of multiple subtypes that infect mammals and poultry, including humans, mice, pigs, cats, dogs and avian vertebrate, etc. The International Committee on Taxonomy of Viruses (ICTV) has classified coronaviruses into four categories: α-coronavirus, β-coronavirus, γ-coronavirus and δ-coronavirus. Among them, β-coronavirus is further divided into four lineages: A, B, C and D. So far, there are seven coronaviruses known to infect humans, namely: α-HCoV-229E and HCoV-NL63, β-HCoV-OC43 and HCoV-HKUl in lineage A, SARS-CoV and SARS-CoV-2 in lineage B and MERS-CoV in lineage C.

Most studies believe that coronaviruses are divided into four groups, of which Group 1, 2 and 4 are mammalian viruses and Group 3 is avian virus. Group 1 includes Human Coronavirus 229E (HCoV-229E), Porcine Transmissible Gastroenteritis Virus (TGEV), Porcine Epidemic Diarrhea Virus (PEDV), Canine Coronavirus (CCoV) and Feline Infections Peritonitis Virus (FCoV); Group 2 includes Human Coronavirus OC43 (HCoV-OC43), Bovine Coronavirus (BCoV), Mouse Hepatitis Virus (MHV), Rat Sialodacryoadenitis Coronavirus (SDAV), Swine Hemagglutinating Encephalomyelitis Virus (HEV) and Rat Coronavirus (RtCoV); Group 3 includes Avian Infectious Bronchitis Virus (IBV), Human Enteric Coronavirus (HECV), Rabbit Coronavirus (RBCV), Monkey Coronavirus (MCV), Cheetah Coronavirus (CHCV) and Turkey Coronavirus (TCoV). Due to low homology of their genomes with the sequence of existing coronaviruses, Group 4 coronaviruses are composed of severe acute

The Pathology and Imaging of Coronavirus Pneumonia
ISBN 978-0-443-32930-2
https://doi.org/10.1016/B978-0-443-32930-2.00002-7

respiratory syndrome coronavirus (SARS-CoV) and severe acute respiratory syndrome coronavirus 2 (SARS-CoV-2). According to research, classic human coronaviruses consist of two serotypes: HCoV-229E and HCoV-OC43. Currently, the coronaviruses known to infect humans including HCoV-229E and HCoV-OC43 discovered in 1960s, SARS-CoV emerged in 2003, HCoV-NL63 discovered in 2004, HCoV-HKU discovered in 2005 and MERS-CoV, a novel human coronavirus emerged in the Middle East in July 2012; In late December 2019, a patient with pneumonia of unknown cause was found to be infected with a novel coronavirus, SARS-CoV-2.

SARS-CoV is the pathogen of severe acute respiratory syndrome (SARS), characterized primarily by symptoms such as fever and cough, etc. The genome of SARS-CoV belongs to the typical coronavirus genotype and possesses the general characteristics of coronaviruses. But it is neither a mutant nor recombinant of any known coronavirus. At present, the virus has disappeared from the population. People infected with MERS-CoV may develop acute and severe respiratory illness with symptoms such as fever, cough, shortness of breath and respiratory distress, and in some cases kidney failure and even death. MERS- CoV is more closely related to bat-CoV-HKU4 and bat-CoV-HKU5 with a genomic similarity of 70.1%. In comparison, its genomic similarity with the SARS virus is 54.9%. Current research has found that MERS-CoV can infect human cells as well as cells of various bats and pigs, making animals a continuous source of infection. The virus has the potential to spread among multiple species. The genome of SARS-CoV-2 is single RNA with the length of approximately 30 kb, and expected to encode 16 nonstructural proteins (nsp1-16), 4 structural proteins, and 8 accessory proteins. The spike glycoprotein mediates viral entry by binding to human angiotensin-converting enzyme 2 (ACE2), which is followed by proteolytic hydrolysis. It is triggered by trans-membrane serine protease 2 (TMPRSS2), furin protease and other possible pulmonary proteases, leading to the fusion of the virus and cell membrane. The spike glycoprotein is also the primary target of neutralizing antibodies (nAbs) in the host.

1.2 Types of human coronavirus

The coronavirus is the largest positive-sense RNA virus currently known, with the genome length ranging from 26,000 to 32,000 bp. Mature coronavirus has a diameter of 60—220 nm and exhibits a corona-like or crown-like appearance under the electron microscopy, it is hence named as "coronavirus". As early as 1931, Schalk and Hawn reported on the disease caused by coronaviruses and studied how to distinguish avian infectious bronchitis from other diseases of respiratory system. In 1937, coronavirus was isolated from chickens. In 1965, Tyrrell and other researchers isolated a virus from the nasal wash of patients suffered from common cold and named it B814 virus. In 1966, a similar virus was isolated from human embryonic kidney cells by Hamer et al. in the United States and the representative strain was named 229E virus. In 1967, Mclntosh et al. cultured and isolated a group of viruses from human embryonic trachea of patients suffered from common cold and the representative strain was named OC43 strain. Almeida et al. conducted morphological studies on these viruses and observed them under the electron microscope. They discovered corona-like spikes on the envelopes of these viruses and hence named "coronavirus". In 1975, the International Committee on Taxonomy of Viruses officially named the Coronaviridae. In recent years, it has been found that the Coronaviridae primarily contains five new coronaviruses that infect humans.

(1) SARS-CoV: In November 2002, a severe viral pneumonia broke out and rapidly spread to more than 30 countries worldwide. Research has found that the intermediate host was the *Paguma larvata*, and the typical coronavirus genotype was SARS-CoV gene. In 2003, the International Committee on Taxonomy of Viruses named this novel coronavirus as SARS-CoV. SARS-CoV belongs to β-coronavirus Group B and exhibits typical characteristics of coronaviruses.

(2) HCoV-NL63: HCoV-NL63 was isolated from a 7-month-old infant with bronchiolitis and conjunctivitis in 2004, and is considered as the second coronavirus discovered to be infectious to humans after SARS-CoV. HCoV-NL63 belongs to α-coronavirus with the genome length of 27.5 kb. Compared with other coronaviruses, it has a relatively lower proportion of guanine and cytosine (GC contents) ranging from 37% to 42%. HCoV-NL63 utilizes ACE2 as its receptor. Cases of HCoV-NL63 infection have been reported globally with the epidemic interval lasting for approximately 2 years.

(3) HCoV-HKU1: The coronavirus HKU1 was isolated from the nasopharyngeal aspirate of a 71-year-old male patient with pneumonia in 2005. It is considered as the third coronavirus confirmed to be pathogenic to humans. HCoV-HKU1 belongs to

β-coronavirus Group A and can not be cultured in continuous culture and lines currently. It consists of three genotypes, namely A, B and C with the genome length of 29.9 kb and the GC content of 32%, which is the lowest among known coronaviruses. In 2012, it was reported that HCoV-HKU1 was detected in the nasopharyngeal aspirate of a 2-year-7-month-old child with acute pharyngitis-induced asthma attack.

(4) MERS-CoV: In September 2012, a novel coronavirus strain was isolated for the first time from a male patient in Saudi Arabia. Because the symptoms caused by this virus are similar to that of SARS virus, it was initially called a SARS-like virus. Subsequently, the United Kingdom reported to the World Health Organization (WHO) that a 49-year-old male who had traveled to Saudi Arabia and Qatar exhibited clinical symptoms of the novel coronavirus infection, and named the isolated virus London Novel CoV. In November 2012, the Erasmus Medical Center in the Netherlands confirmed that the virus infected to one patient belonged to the Novel B Coronavirus HCoV-EMC. Since then, the novel coronavirus infection began to spread sporadically. In February 2013, the WHO initially considered the risk of sustained human-to-human transmission of the disease to be low, but by April and May of the same year, a total of 19 new cases were reported, primarily concentrated in the Middle East region or among travelers who had been to such region, most of whom had a history of contact with infected people. On May 15, 2013, the Coronavirus Study Group of the International Committee on Taxonomy of Viruses decided to designate this novel coronavirus as the Middle East respiratory syndrome coronavirus (MERS-CoV). On May 23, 2013, the WHO officially replaced the term "Novel Coronavirus" with "MERS-CoV" upon epidemic notification.

(5) SARS-CoV-2: In late December 2019, patients were found to be infected with pneumonia caused by a novel coronavirus, which is an acute infectious pneumonia; its pathogen is a new strain of coronavirus not previously found in humans, known as the 2019 novel coronavirus (2019-nCoV). On January 30, 2020, the WHO declared the outbreak of the novel coronavirus pneumonia as a Public Health Emergency of International Concern (PHEIC). Through the research on comparative genomic data, especially the analysis on the mutations in conserved proteins involved in viral

replication, it has been discovered that this novel coronavirus belongs to the same genus of coronavirus family as SARS-CoV; due to its similarity to SARS-CoV, this virus was subsequently named SARS-CoV-2, a member of Sarbecovirus subspecies (Beta-CoV lineage B). On February 7, 2020, the National Health Commission decided to temporarily name "Novel Coronavirus-infected Pneumonia" as "Novel Coronavirus Pneumonia" (NCP for short). On February 11, 2020, the International Committee on Taxonomy of Viruses named this virus severe acute respiratory syndrome coronavirus 2 (SARS-CoV-2) and the WHO named the pneumonia caused by SARS-CoV-2 "COVID-19". At present, it is experiencing a global pandemic and has not been effectively controlled.

1.3 The hazards of coronaviruses

Coronaviruses are currently classified into two genera: Torovirus and coronavirus. So far, it has been found that coronaviruses exclusively infect vertebrates, particularly bats, where they are widely present. However, coronaviruses have also been detected in other animals, including cats, birds, pigs, horses, dogs, mice, whales and humans, which can cause infections in the respiratory tract, intestinal tract, liver, cardiovascular and nervous systems, with the severity of the disease varying among different animal species.

The host range of coronaviruses is directly related to the virus-specific cellular receptors. The virions are adsorbed on the cell membrane of sensitive cells through receptor mediation, bound to glycoprotein receptors on the cell membrane via Hemaglutinin-esterase (HE) or spike glycoprotein (S-protein), and then invade the cells through membrane fusion or endocytosis. After oral or nasal infection, the virus exhibits cytotoxic effect on lymphocytes, reticuloendothelial cells, epithelial cells and parenchymal cells, thereby causing damage to various organs in animals and humans. After acute infection, coronavirus may also result in persistent infection; the virus spreads chronically from cell to cell, leading to cell death and pathological changes in organs. Other coronaviruses, such as avian infectious bronchitis virus and human coronavirus, can also lead to persistent infection. Human coronavirus is one of the viruses that cause the occurrence of common cold and pharyngitis in humans; infected patients typically present with symptoms of acute upper respiratory tract infection, such as rhinitis, nasal congestion, sneezing and pharyngitis, etc.; SARS-CoV, MERS-CoV and SARS-CoV-2, on the other hand, cause severe symptoms in the lungs; while coronaviruses possess numerous distinctive characteristics, the clinical symptoms caused are remarkably similar to those of common cold caused by rhinovirus or of influenza caused by influenza virus.

From 2002 to 2003, the severe acute respiratory syndrome (SARS) caused by SARS-CoV affected more than 30 countries and regions all over the world, with a fatality rate of approximately 10%. The epidemic of SARS mainly occurred in winter and spring, with its primary symptoms including fever, cough, progressive respiratory distress and hypoxemia. In severe cases, it can develop to the acute respiratory distress syndrome (ARDS). Diabetes and heart diseases are important risk factors for mortality of SARS patients. It typically represents with reduced lymphocyte count and abnormal liver function in the laboratory test. Patients may die from rapidly progressive respiratory failure due to alveolar damage. Compared with adult patients, children infected with SARS-CoV generally feature milder clinical symptoms and lower probability of developing SARS. Studies have shown that children suffered from SARS exhibit milder symptoms and signs, better prognosis and no significant long-term severe complications than that of adults.

In 2004, HCoV-NL63 infection was reported worldwide, with its peak season occurring in winter. However, there were regional variations, such as in Chongqing of China, where the peak season of infection was reported to be in summer and autumn. HCoV-NL63 is most commonly discovered in infants and has a detection rate of 1.0%—9.3% in children with respiratory tract infection. HCoV-NL63 is associated with both upper and lower respiratory tract infections, and its clinical symptoms are similar to those caused by HCoV-OC43 and HCoV-229E. However severe lower respiratory tract infections are also frequently observed in infants, patients with underlying diseases and the elderly. The primary manifestations of HCoV-NL63-related lower respiratory tract infections include croup and bronchiolitis, mainly occurred in infants, among which bronchiolitis is considered the most concerning condition associated with HCoV-NL63-related diseases. Studies have shown that HCoV-NL63 positive children are more prone to developing lower respiratory tract infections and exhibiting more severe clinical symptoms compared with HCoV-OC43 positive patients; it has also been found that, compared with those infected with Respiratory Syncytial Virus, children with HCoV-NL63-infected respiratory diseases feature shorter duration of fever and hospitalization. Some researchers believe that HCoV-NL63 infection is related to young age, croup, febrile convulsion, acute gastroenteritis, etc. It is common for HCoV-NL63 to be coinfected with other respiratory viruses, including other subtypes of human coronaviruses, respiratory syncytial viruses, parainfluenza viruses, influenza A and B viruses, etc.

There were reports of HCoV-HKU1 infection in 2005 with its peak incidence in winters and associated with both upper and lower respiratory tract infections. Moreover, the majority of cases exhibited self-limiting symptoms. HCoV-HKU1-related upper respiratory tract infection are mainly manifested as fever, running nose and cough, while the main symptoms of HCoV-HKU1-related lower respiratory tract infection include fever, cough, expectoration and difficulty in breathing. HCoV-HKU1 infection has been reported throughout the world, with an incidence ranging from 0% to 4.4% with the median of 0.9%, showing no significant difference compared with other human coronaviruses. A study has reported that one child suffered from severe combined immunodeficiency with HCoV-HKU1 infection-related pneumonia resulted in fatal respiratory distress syndrome. Severe conditions such as pneumonia, bronchiolitis and acute exacerbation of asthma have also been reported, mainly occurred in individuals with underlying medical conditions. HCoV-HKU1 infection in young children is also associated with the high incidence of febrile convulsion.

In September 2012, the new coronavirus was detected for the first time in Saudi Arabia of the Middle East region, known as MERS-CoV. The infection was quickly spread to 27 countries, mainly in the Middle East region and South Korea, with the fatality rate of up to 36%, surpassing that of SARS-CoV. MERS-CoV infection may not only cause typical cold-like symptoms such as headache, fever, cough, sore throat and difficulty in breathing, but also severe respiratory symptoms including fever, cough, chest pain and shortness of breath, accompanied by a higher incidence of acute kidney failure and even mortality. It is typically manifested as elevated lactate dehydrogenase (LDH), increased glutamic-oxaloacetic transaminase (GOT) as well as reduced blood platelet count and lymphocyte count in the laboratory test. Due to its clinical symptoms being similar to that of the SARS in 2003, it was initially called "SARS-like virus". Chest X-ray examinations showed that abnormal findings were observed in almost all confirmed MERS-CoV cases. Both unilateral and bilateral lung lesions, such as vacuolation and ground glass fibrosis, can be seen, but such abnormalities in the chest X-ray examination are not specific and similar focal changes can be seen in lung diseases caused by other pathogens. In addition to causing respiratory symptoms and abnormalities on chest X-ray and CT scans, MERS-CoV infection can also lead to pathological changes in other system. The most severe and primary fatal complication is renal failure. The majority of the severe clinical cases reported earlier have eventually caused respiratory failure with renal failure and even death. And these cases often necessitate treatment in the Intensive Care Unit. In addition to respiratory symptoms, MERS-CoV may also cause abdominal symptoms. As a clinical analysis on a severe case in Saudi Arabia showed, the rate of occurrence of diarrhea, vomiting and abdominal pain was 26%, 21% and 17%, respectively. In terms of blood routine examination, patients may exhibit either decreased or increased white blood cell count, along with normal neutrophil count and decreased platelet count. MERS-CoV is prone to causing serious illness in patients with underlying diseases. According to research reports, the mortality rate of MERS-CoV infection increases with age, and 96% of patients have underlying diseases, including 68% with diabetes, 34% with hypertension, 28% with chronic heart diseases and 49% with chronic kidney diseases. Research has found that children accounted for 3.54% of cases with the majority exhibiting mild or asymptomatic clinical manifestations and good prognosis. Among the deceased pediatric cases, there were individuals who had underlying chronic lung diseases and combined nephrotic syndrome.

The clinical symptoms caused by SARS-CoV-2 infection primarily include fever and cough, as well as muscle pain, fatigue and neurological symptoms. As the disease progresses, it can cause ARDS, shock and even death; besides, SARS-CoV-2 also presents characteristic laboratory test results and lung CT abnormalities. Compared with SARS-CoV, MERS-CoV and seasonal influenza, the novel coronavirus exhibits different tropism. Research suggests that the proportion of patients with SARS-CoV-2 who do not experience fever is higher than those infected with SARS-CoV (1%) and MERS-CoV (2%). Therefore, if the case definition emphasizes the monitoring of fever, then patients who do not have a fever may be missed or misdiagnosed. The most common complications among hospitalized patients with SARS-CoV-2 are pneumonia (79.10%), ARDS (3.37%) and shock (1.00%). The incidence of complications is significantly higher in critical patients compared with noncritical cases. SARS-CoV-2 mainly infects the respiratory tract, with a small proportion manifesting as bilateral lower lobe pneumonia and diffuse alveolar damage (DAD), and may progress to ARDS, especially in the elderly and those with complications. Compared with MERS-CoV and SARS-CoV infections,

SARS-CoV-2 infection presents with a wide range of clinical symptoms and shows greater variability. The differences in viral transmission capacity and viral detachment suggest that the intracellular replication sites and/or replication efficiency of SARS-CoV-2 are significantly different from those of SARS-CoV.

Bibliography

[1] Adams MJ, Carstens EB. Ratification vote on taxonomic proposals to the International Committee on Taxonomy of Viruses. Arch Virol 2012;157:1411−22.
[2] van Boheemen S, de Graaf M, Lauber C, et al. Genomic characterization of a newly discovered coronavirus associated with acute respiratory distress syndrome in humans. mBio 2012;3(6):e00473.
[3] Woo PC, Lau SK, Li KS, et al. Genetic relatedness of the novel human group C betacoronavirus to *Tylonycteris* bat coronavirus HKU4 and *Pipistrellus* bat coronavirus HKU5. Emerg Microbes Infect 2012;1(11):e35.
[4] Zhu CY, Shen YZ, Lu HZ. The prevalence, transmission and prevention of middle east respiratory syndrome coronavirus infection. Chin J Infect Chemother 2014;14(4):353−6.
[5] Sun S, Wang Y, Liu L, et al. An overview of the coronaviruses. China Animal Health Inspect 2013;30(6):68−71.
[6] Ksiazek TG, Erdman DC, Cynthia S, et al. Anoval coronavirus associated with severe acute respiratory syndrome. N Engl J Med 2003;348:1947−58.
[7] Rota PA, Oberste MS, Monroe SS, et al. Characterization of a novel coronavirus associated with severe acute respiratory syndrome. Science 2003;300(5624):1394−9.
[8] Lan X, Liu J. Progress of molecular biological study on coronavirus. Basic Clin Med 2005;25(12):1095−100.
[9] Pengfei H, Wang X, Shouwen Du, et al. Coronavirus origin, receptor and recent advances in novel coronavirus detection and vaccine. Electron J Emerg Infect Dis 2020;5(2):74−8.
[10] Geng H, Tan W. Progress of study on newly discovered coronavirus. Chin J Virol 2013;29(1):65−9.
[11] Leung TF, Chan PK, Wong WK, et al. Human coronavirus NL63 in children: epidemiology, disease spectrum, and genetic diversity. Hong Kong Med J 2012;18(Suppl. 112):27−30.
[12] Woo PC, Lau SK, Yip CC, et al. More and more coronaviruses: human coronavirus HKU1. Viruses 2009;1(1):57−71.
[13] Lee SS, Wong NS. Probable transmission chains of Middle East respiratory syndrome coronavirus and the multiple generations of secondary infection in South Korea. Int J Infect Dis 2015;38:65−7.
[14] Assiri A, Mcgeer A, Perl TM, et al. Hospital outbreak of Middle East respiratory syndrome coronavirus. N Engl J Med 2013;369(5):407−16.
[15] Mao L, Jin H, Wang M, et al. Neurological manifestations of hospitalized patients with COVID-19 in Wuhan, China: a retrospective case series study. JAMA Neurol 2020;77(6):683−90.
[16] Chan PK, Chan MC. Tracing the SARS-coronavirus. J Thorac Dis 2013;2:118−21.
[17] Fielding BC. Human coronavirus NL63: a clinically important virus? Future Microbiol 2011;6(2):153−9.

[18] Amini R, Jahanshiri F, Amini Y, et al. Detection of human coronavirus strain HKU1 in a 2 years old girl with asthma exacerbation caused by acute pharyngitis. Virol J 2012;9:142.

[19] Zaki AM, Vall Boheemen S, Bestebroer TM, et al. Isolation of a novel coronavirus from a man with pneumonia in Saudi Arabia. N Engl J Med 2012;367(19):1814–20.

[20] DE Groot RJ, Baker SC, Baric RS, et al. Middle east respiratory syndrome coronavirus (MERS-CoV): announcement of the coronavirus study group. J Viml 2013;87(14):7790–2.

[21] Ong SWX, Tan YK, Chia PY, et al. Air, surface environmental, and personal protective equipment contamination by severe acute respiratory syndrome coronavirus 2 (SARS-CoV-2) from a symptomatic patient. JAMA 2020;323(16):1610–2.

[22] Wang D, Hu B, Hu C, et al. Clinical characteristics of 138 hospitalized patients with 2019 novel coronavirus-infected pneumonia in Wuhan, China. JAMA 2020;323(11):1061–9.

CHAPTER 2

Etiology

Qin Ling
Beijing YouAn Hospital, Capital Medical University, Beijing, China

2.1 Morphological structure and physicochemical properties of coronaviruses

2.1.1 Morphological structure

As a positive-sense single-stranded RNA virus with envelope, SARS-CoV-2 belongs to the β genus, and is the seventh member of coronaviridae that infect humans following the MERS coronavirus and SARS coronavirus. They are generally round or oval, often exhibit polymorphism and have a diameter ranging from 60 to 140 nm.

The viral envelope structure is divided into three layers (Fig. 2.1): ① The inner layer is the viral mucleocapsid protein (N) enveloped by a segment of viral RNA sequence; SARS-Cov-2 RNA, containing 26,000 to 32,000 bases, is considered as one of the largest RNA viruses in the known genome; the N protein is a kind of highly immunogenic phosphoprotein featuring high conservation; ② The middle layer is the viral envelope, composed of Envelope Protein (E) and Membrane Protein (M), with the latter involving in the formation and budding of the viral envelope; ③ The outer layer is composed of two different glycoproteins, including the HE

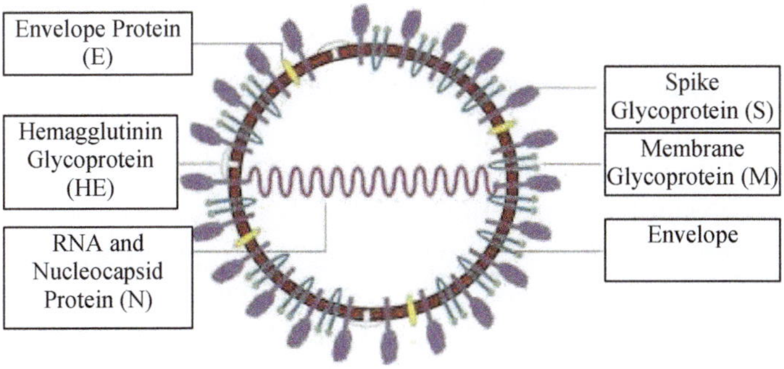

Figure 2.1 Diagram for virus envelope pattern.

The Pathology and Imaging of Coronavirus Pneumonia
ISBN 978-0-443-32930-2
https://doi.org/10.1016/B978-0-443-32930-2.00007-6

protein and S protein, which form prominent spikes with the length of 9–12 nm. The HE protein may be associated with the early adsorption of the virus; the S protein is key for the virus to enter cells and can be divided into two functional subunits, S1 and S2; S1 is to facilitate viral infection to host receptor with the C-terminal RNA Binding Domain (RBD) responsible for recognizing the receptor ACE2 on the surface of respiratory cells and entering the cells to infect humans. S2 contains the essential components required for membrane fusion process, and mainly mediates virus-cell and cell-cell membrane fusion. Additionally, the S protein also exhibits antigenicity.

2.1.2 Physicochemical properties

At present, the understanding of physical and chemical properties of coronaviruses largely stems from research on SARS-CoV and MERS-CoV. The virus is sensitive to ultraviolet light and heat. It can be effectively inactivated by exposure to a 56°C water bath for 30 min, as well as by treatment with ether, 75% ethanol, chlorine-based disinfectants, peracetic acid, chloroform and other lipid solvents. However, chlorhexidine is not effective in inactivating the virus.

2.2 Isolation and culture of coronavirus

2.2.1 Collection of specimens

2.2.1.1 Collection objects

Suspected cases, clinically diagnosed cases and clustered cases of novel coronavirus pneumonia, as well as individuals requiring infection diagnosis or differential diagnosis of novel coronavirus, or other environment or biological materials (such as traceability analysis) that require further screening and testing.

2.2.1.2 Requirements for specimen collection

(1) Technicians engaged in the collection of specimens for novel coronavirus testing shall obtain biosafety training (with qualified certification) and possess relevant experiment skill.

(2) Sampling personnel shall wear professional protective equipment, including N95 or higher-grade masks, goggles, protective jump suit, double-layer latex gloves and waterproof boot covers; after contacting with the patient's blood, body fluids, secretions or excretions, it is required to promptly replace the outer layer of the latex gloves.

2.2.1.3 *Type of specimen collection*

(1) Upper respiratory tract specimens: Include throat swabs, nasal swabs and nasopharyngeal aspirates, etc.

(2) Lower respiratory tract specimens: Include deep cough sputum, respiratory tract aspirates, bronchoalveolar lavage fluid, pulmonary lavage fluid, and biopsy specimen of lung tissue, etc.

(3) Blood specimens: Anticoagulated blood specimen in the acute stage within 7 days after the onset of the disease. The collection volume is 5 mL, preferably in a fasting state. It is advised to use vacuum blood collection tubes containing Ethylene Diamine Tetraacetic Acid (EDTA) anticoagulant for blood collection.

(4) Serum specimens: It is recommended to collect paired serum specimens during the acute stage and convalescent stage if possible. The first serum sample shall be collected as early as possible (preferably within 7 days after onset of illness), and the second serum sample shall be collected in the third to fourth week after onset of illness. The collection volume is 5 mL and a vacuum blood collection tube without anticoagulant is advised. Serum samples are mainly used for antibody determination, and the infection status of the case is confirmed from the serum antibody level. Serum specimens are not subjected to nucleic acid testing.

(5) Eye conjunctiva specimens: Cases with symptoms of eye infection require the collection of conjunctival swab specimens.

(6) Stool specimens: Patients with symptoms of diarrhea require the collection of stool specimens.

(7) Anal swab: It can increase the detection rate, effective reduce the probability of false negatives and minimize missed diagnoses.

2.2.1.4 *Method of specimen collection*

(1) Throat Swab: Use two plastic stick swabs with polypropylene fiber heads to swab the bilateral tonsils and posterior pharyngeal wall simultaneously. Immerse the swab heads in a tube containing 3 mL of sampling fluid, discard the tails and tighten the tube cap.

(2) Nasal Swab: Gently insert a plastic swab with a polypropylene fiber tip into the nasal passage, until it reaches the area near the nasopalatine, leave it in place for a moment and then slowly rotate and withdraw it. Take another plastic stick with a polypropylene fiber head and collect a sample from the other nostril using the same method. Immerse both swabs into the same tube containing 3 mL sampling fluid, discard the tails and tighten the tube cap.

(3) Nasopharyngeal aspirates or respiratory tract aspirates: Use a collector connected to a negative pressure pump to extract mucus from the nasopharynx or respiratory secretions from the trachea. Insert the head of the collector into the nasal cavity or trachea, connect the negative pressure, rotate the head of the collector and slowly withdraw it, then collect the extracted mucus and rinse the collector with 3 mL sampling fluid once (a pediatric urinary catheter can also be attached to a 50 mL syringe instead of the collector).

(4) Deep Cough Sputum: Instruct the patient to cough deeply and collect the coughed-up sputum into a 50 mL screwed plastic tube containing 3 mL sampling fluid.

(5) Bronchoalveolar Lavage Fluid: Insert the head of the collector through the nostril or tracheal opening into the trachea (approximately 30 cm deep), inject 5 mL saline solution, connect the negative pressure, rotate the head of the collector and slowly withdraw it. Collect the extracted mucus and rinse the collector with sampling fluid once (a pediatric urinary catheter can also be attached to a 50 mL syringe instead of the collector).

(6) Pulmonary Lavage Fluid: After local anesthesia, insert a bronchoscope through the mouth or nose, passing through the pharynx and into the bronchus of the middle lobe of the right lung or the lingular segment of the left lung, engage the tip of the bronchoscope into the opening of the bronchial branch, and slowly add sterile saline solution through the biopsy channel of the bronchoscope. Infuse 30–50 mL of saline solution each time with a total volume of 100–250 mL (not exceeding 300 mL). Then collect the lavage fluid in 3s with a recovery rate of not less than 40%.

(7) Blood Specimens: It is recommended to collect 5 mL blood specimen using a vacuum blood collection tube containing anticoagulants, let the tube stand at room temperature for 30 min, then centrifuge at a speed of 1500–2000 r/min for 10 min, and carefully collect the plasma and blood cells in separate sterile screwed plastic tubes.

(8) Serum Specimens: Collect a 5 mL blood specimen using a vacuum negative-pressure blood collection tube, let the tube stand at room temperature for 30 min, then centrifuge at a speed of 1500–2000 r/min for 10 min, and carefully collect the serum in a sterile screwed plastic tube.

(9) Stool Specimens: If the patient exhibits symptoms of diarrhea in the early stage of the disease, collect 3–5 mL of a stool specimen;

(10) Anal Swab Specimen: There are three steps for sampling. Insert a disinfectant cotton swab into the anus for 3–5 cm; Rotate the cotton swab inside the anus and then pull it out; Place the swab in the sampling tube.

(11) Eye Conjunctiva Swab Specimen: After gently wiping the surface of the eye conjunctiva with a swab, insert the swab head into the sampling tube, discard the tail and tighten the tube cap.

(12) Other Materials: Collect according to the design requirements.

2.2.1.5 Specimen packaging

After collection, the specimens are subpackaged in the biosafety cabinet of the biosafety level II laboratory.

(1) All specimens shall be placed in a sample collection tube featuring a suitable screw cap (with a gasket inside) and being resistant to freezing, and then tightened. Indicate the sample number, type, name and sampling date on the outside of the container.

(2) Place the sealed specimen in a suitable plastic bag and seal it properly, with each specimen in one bag.

2.2.1.6 Specimen storage

Specimens used for virus isolation and nucleic acid detection shall be detected as soon as possible, and specimens that can be detected within 24 h can be stored at 4°C; Specimens that cannot be detected within 24 h shall be stored at −70°C or below (if there is no condition for storage at −70°C, they shall be stored in −20°C refrigerator temporarily). Serum can be stored at 4°C for 3 days or frozen at −20°C for a long time. A special library or counter shall be set up to keep specimens separately. Repeated freezing and thawing shall be avoided during the specimen transportation.

2.2.1.7 Specimen submission

Samples shall be sent to the laboratory as soon as possible after collection. If long-distance transportation is required, it is recommended to use dry ice and other refrigeration methods for preservation.

2.2.2 Virus isolation and culture

Virus culture generally contains two methods: Cell Culture and Embryonated Egg Culture (9–11-day-old embryonated egg). The identification of viruses isolated and cultured from respiratory specimens is often considered the gold standard for laboratory diagnosis of respiratory pathogens, with both its sensitivity and specificity of more than 90%, however, due to the long cultivation period, high technical difficulty, complex operations and stringent biosafety requirements, it is generally not suitable for clinical detection. The isolation, cultivation and identification of the virus are not included in the etiological detection items in the published guidelines for the prevention and treatment of SARS-CoV-2.

2.3 Genetic structure and replication of coronavirus

2.3.1 Gene structure

The genome of SARS-CoV-2 is composed of a single-stranded RNA, with a length of approximately 30 kb and containing 26,000 to 32,000 bases. The genome of SARS-CoV-2 contains six open reading frames (ORF), and ORF1a and ORF1b are primarily responsible for encoding enzymes involved in virus replication and transcription; the function of ORF3a is not clear currently; ORF7a is a single transmembrane protein sequence, which is relatively conserved and may be the basic functional gene of coronavirus; the function of ORF8 is unclear but its protein sequence varies greatly; ORF10 is a nucleocapsid phosphoprotein that packages the viral genome into the ribonucleoprotein (RNP) and plays a fundamental role in viral self-assembly. The genome of SARS-CoV-2 can encode structural protein, nonstructural protein and accessory protein. The structural proteins, including the S protein, E protein, M protein and N protein, are essential components that maintain the structure of the viral particles; the nonstructural proteins, primarily composed of ORF1a and ORF1b, are responsible for encoding enzymes related to virus replication and transcriptions, these enzymes include papain-like protease (PLpro), chymotrypsin-like protease (3CLpro), methyltransferase and RNA-dependent RNA polymerase, etc. These proteins serve as the genetic information for synthesizing progeny viruses and also transcribe the mRNA required for the structural proteins of the progeny viruses; the accessory proteins do not directly participate in the physiological processes of SARS-CoV-2 but play a role in evading the hose cell's antiviral immune system. Currently, nucleic acid detection methods primarily target the ORF1a/ORF1b and N protein of the SARS-CoV-2 genome.

2.3.2 Virus replication process

Like other viruses, SARS-CoV-2 shall rely on host cells for replication and undergo the process of adsorption, penetration, replication, assembly and release.

2.3.2.1 *Adsorption and penetration*

When the S protein in the outermost layer of SARS-CoV-2 binds to the ACE2 molecule of the receptor protein on the surface of the host cell, it initiates the entrance of the virus into the host cell. Featuring protease activity, ACE2 can cleave and activate the S protein, thereby inducing fusion between the virus and the host cell membrane and entering the hose cell through endocytosis or binding with the TMPRSS2 receptor on the host cell membrane.

2.3.2.2 *Uncoating and replication*

After entering the cell, the virus releases the SARS-CoV-2 genome into the cytoplasm through an uncoating process, and the viral RNA, having a 5′-methylated terminal and 3′-polyadenylated tail, can attach to the ribosomes of the host cell. Using the genetic material RNA as a template, a nonstructural protein with a length of 4405 amino acids is directly translated under the action of the ribosome of the host cell. This nonstructural protein features the activity of papaya-like protease and 3C-like protease, allowing it to undergo self-cleavage into 15 to 16 proteins and perform distinct functions. Of utmost importance among these proteins is the RNA-dependent RNA polymerase, which can transcribe the negative strand complementary to the genomic RNA, and utilize the negative-strand RNA as the template to synthesize more positive-strand viral RNAs, thereby amplifying the viral genome; meanwhile, the negative-strand RNA template can also transcribe other subgenomic RNAs used for translating structural proteins necessary for viral assembly, such as the M protein, N protein and S protein, etc. See Fig. 2.2.

2.3.2.3 *Assembly and release*

Once different proteins are produced, the N protein, M protein and S protein assemble the newly synthesized progeny viral genome RNA into viral particles within the endoplasmic reticulum lumen of the host cell, and transport them to the extracellular space through the Golgi vesicles and eventually released via exocytosis (Fig. 2.2). The new viruses will continue to utilize the same mechanism to infect more host cells.

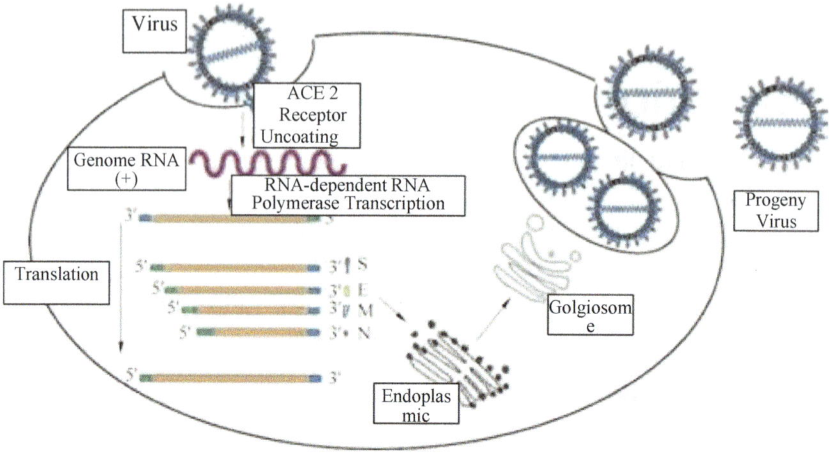

Figure 2.2 Diagram for SARS-CoV-2 infection in human cells.

Bibliography

[1] Adams MJ, Carstens EB. Ratification vote on taxonomic proposals to the international committee on taxonomy of viruses. Arch Virol 2012;157:1411–22.

[2] General Office of National Health Committee, State Administration of Traditional Chinese Medicine Office. The COVID-19 diagnosis and treatment program (Trial 6th ed.) [EB/OL]. http://www.nhc.gov.cn/yzygj/s7653p/202002/8334a8326dd94d329df351d7da8aefc2/files/b218cfeb1bc54639af227f922bf6b817.

[3] National Health Commission. Guidelines for detection technology of COVID-19 laboratory (2nd ed.) [EB/OL]. http://www.nhc.gov.cn/jkj/s3577/202001/c67cfe29ecf1470e8c7fc47d3b751e88/files/7db05db9e315401389bc8b69252c25ef.docx.

[4] Chinese Society of Imaging Technology. Expert consensus on radiological examination scheme and infection prevention and control of COVID-19. Electron J Emerg Infect Dis 2020;5(2):65–73 (trial first edition).

[5] Chen HX, Ai L, Lu H, et al. Clinical and imaging features of COVID-19. Radiol Infect Dis 2020;7(2):43–50.

[6] National Health Commission. Guidelines for detection technology of COVID-19 laboratory. 4th ed. 2020 [EB/OL] http://www.nhc.gov.cn/.

[7] Zheng Q, Lu Y, Tan L, et al. Advances in clinical and imaging studies of novel coronavirus pneumonia. Electron J Emerg Infect Dis 2020;5(2):140–4.

[8] Zhu X, Liu Q, Du L, et al. Receptor-binding domain as a target for developing SARS vaccines. J Thorac Dis 2013;5(2):142–8.

CHAPTER 3

Epidemiology

Li Hongjun[1], Liu Jing[2] and Li Zhen[3]
[1]Beijing YouAn Hospital, Capital Medical University, Beijing, China; [2]Soochow University Affiliated Hospital of Infectious Diseases (Suzhou No. 5 People's Hospital), Jiangsu, China; [3]Peking University Health Science Center, Beijing, China

3.1 Epidemic profile

3.1.1 Prevalence of severe acute respiratory syndrome (SARS)

SARS is an infectious disease that has erupted globally and is also known as atypical pneumonia in China. On February 21, 2003, a Chinese doctor who had taken care of SARS patients went to Hongkong, China to visit relatives and stayed at the M Hotel, which was followed by the SARS outbreak in multiple regions. The SARS broke out with high incidence in Guangdong Province and Beijing of mainland China, with the peak period from February to May 2003. As of July 31, 2003, a total of 8096 cases were reported to the World Health Organization worldwide, including 744 deaths.

3.1.2 Prevalence of middle east respiratory syndrome (MERS)

MERS is a viral respiratory disease that has predominantly existed in sporadic form over the past years. In 2012, MERS-CoV infection emerged in the Kingdom of Saudi Arabia (KSA) and subsequently spread to 26 countries, with 80% of the cases occurring in KSA. In 2015, the largest outbreak outside the KSA occurred in South Korea. As of November 2019, a total of 2494 laboratory-confirmed MERS cases, including 858 deaths, have been reported to the World Health Organization, with the case fatality rate (CFR) of 34.4%.

3.1.3 Prevalence of coronavirus disease 2019 (COVID-19)

On January 21, 2020, the National Health Commission of China categorized the novel coronavirus pneumonia as a newly emerging Class B infectious disease and implemented prevention and control measures equivalent to Class A infectious disease. Provinces and cities have also launched the first-level

The Pathology and Imaging of Coronavirus Pneumonia
ISBN 978-0-443-32930-2
https://doi.org/10.1016/B978-0-443-32930-2.00017-9

response for public health incidents. On February 11, 2020, the World Health Organization named it Coronavirus Disease 2019 (COVID-19). On March 11, 2020, the World Health Organization declared a global pandemic of COVID-19. As of July 2, 2020, China has reported a total of approximately 85,000 confirmed cases with 4648 deaths; the global cumulative confirmed cases have exceeded 20.54 million with a death toll of over 730,000. The United States, Brazil, Russia, India and the United Kingdom account for the top five countries in terms of total confirmed cases (https://www.who.int/emergencies/diseases/novel-coronavirus-2019).

3.2 Source of infection

3.2.1 Infection source of SARS

The natural host of SARS-CoV is the *Rhinolophus sinicus*, and it is believed that the *Paguma larvata* may have transmitted the virus to humans as an intermediate host. SARS-CoV can survive on dry surfaces for up to 24 h. At room temperature, SARS-CoV can survive stably in feces and urine for at least 1−2 day(s); in the feces of diarrhea patients (with higher pH value compared with normal feces), SARS-CoV can survive for up to 4 days (https://www.who.int/csr/sars/survival_2003_05_04/en/).

3.2.2 Infection source of MERS

The origin of MERS-CoV is not clear. MERS belongs to Coronavirus β genus lineage C, it is a zoonotic virus, meaning that it can be transmitted between animals and humans. MERS-CoV is the first confirmed virus in this lineage that can be transmitted from animals to humans. Its infection sources include MERS patients, MERS-CoV carriers and infected animals, with the possibility of hidden sources of transmission. Some studies have speculated that bats may be the animal source of MERS-CoV transmission and then infect the virus to the dromedary at some point. MERS-CoV has been found in dromedary camels in several countries in the Middle East, Africa and South Asia, and the virus was already circulating in dromedary camels before it started to infect humans. Studies have shown that MERS virus in genetically sequenced camels is highly similar to the fragments detected in human bodies. Humans are infected through direct or indirect contact with infected dromedary camels. Therefore, camels may be the direct source of infection and an intermediate host [https://www.who.int/en/news-room/fact-sheets/detail/middle-east-respiratory-syndrome-coronavirus-(mers-cov)].

3.2.3 Infection source of COVID-19

The zoonotic source of this virus is still unclear currently. The genetic sequencing results of SARS-CoV-2 show that it belongs to β-coronavirus with the pathogen of SARS-CoV-2. The most likely ecological host of SARS-CoV-2 is bats, probably rhinolophus sinicus. It is highly likely that the transmission of the virus to human occurred through an intermediate host (domestic animals, wildlife, or yet unidentified domesticated wild animals). Researches have showed that SARS-CoV-2 can survive for up to 72 h on plastic and stainless surfaces, but no more than 4 h on copper products and no more than 24 h on cardboard surfaces.

3.3 Route of transmission

3.3.1 Transmission route of SARS

SARS is mainly transmitted through close-range respiratory droplets and close contact, as well as pollutants and aerosols. SARS is infected through close contact within households, visitation and medical nursing as well as enclosed public spaces. It features human-to-human transmission and clear clustering in hospitals and households. Till now, no transmission between neighbors or localized outbreak within communities have been found yet.

3.3.2 Transmission route of MERS

As a zoonotic virus, MERS-CoV can persist in the human body for a prolonged period, featuring the capability of animal—human transmission and limited and nonsustained human-to-human transmission. Close contact may cluster and result in outbreak within households and hospitals. It is speculated that the transmission route of MERS-CoV may be bats—dromedary camels—humans. However, the exact role and route in virus transmission are still unclear, which may include airborne droplet transmission, contact transmission, fecal—oral transmission and secretion transmission.

3.3.3 Transmission route of COVID-19

At present, it has not been conclusively determined, but recent studies have found the potential for both animal-to-human and human-to-human transmission. The virus can spread through respiratory droplets, close contact and contact with contaminated objects and surfaces. Feces and urine can cause aerosol or contact transmission, and prolonged exposure in an enclosed environment has the potential for aerosol transmission. And there is also a risk of mother-to-child transmission.

3.4 Susceptible population

3.4.1 Susceptible population of SARS

People are generally susceptible to SARS with young adults (20–40 years old) being a key demographic for infection. It is estimated to be due to the increased mobility, larger range of activities and more opportunities of contact with infectious sources. Many medical staff who have treated and cared for SARS patients, as well as personnel who have been in contact with SARS patients are considered high-risk groups.

3.4.2 Susceptible population of MERS

People are generally susceptible to MERS. Individuals who have contact with camels are considered high-risk groups and medical staff also account for a significant proportion of infections. In global MERS cases, there is a higher incidence among males compared with females, with an average age of 49. And the majority of cases fall within the age range of 50–59 years old. WHO has classified patients with diabetes, kidney failure, chronic lung diseases and immune deficiencies as high-risk individuals for MERS-CoV infection. MERS cases that are older in age or undergo underlying health conditions have a higher mortality rate.

3.4.3 Susceptible population of COVID-19

People are generally susceptible to COVID-19. Currently, there are not enough data available to classify by age and gender, primarily for the age group of 30–70 years old. A history of travel or residency in relevant or other areas with ongoing local transmission of cases within 14 days prior to the onset of symptoms; a history of contact with patients with fever or respiratory symptoms from areas with ongoing transmission of cases within 14 days prior to the onset of symptoms; prone to clustering onset.

3.5 Epidemiologic characteristics

3.5.1 Epidemiologic characteristics of SARS

Countries and regions with high incidence of SARS include Mainland China, Hong Kong, Singapore, Canada, Vietnam and the Philippines, etc.

It initially emerged in the winter and spring seasons, with a clustering pattern of occurrence primarily concentrated in areas with high population mobility. It has a moderate level of infectivity and involves both asymptomatic infected persons and super spreaders. The incubation period of SARS is 2—10 days, with an average period of 6.4 days. The average tome from onset of symptoms to hospital admission is 3—5 days. Easy symptoms are similar to those of flu, such as fever, muscle pain and headache; almost all patients exhibit symptoms of fever, typically with hyperpyrexia, occasionally accompanied by chills; it is reported that as many as 25% of patients suffer from diarrhea. Common laboratory features include reduced lymphocyte count, decreased platelet count, as well as elevated glutamic-pyruvic transaminase, lactic dehydrogenase and creatine kinase. The SARS virus is sensitive to ultraviolet light and heat. It can be effectively inactivated by exposure to a 56°C water bath for 30 min, as well as by using ether, 75% ethanol, chlorine-containing disinfectants, peracetic acid, chloroform and other lipid solvents.

3.5.2 Epidemiologic characteristics of MERS

Since its emergence in 2012, MERS has consistently exhibited non-sustained transmission. The Middle East countries are the primarily affected areas. Other 25 countries and regions have reported initial cases with a travel history to the Middle East or contact history with individuals from the Middle East, and there are super spreaders and asymptomatic infected persons. Fever and gastrointestinal symptoms serve as prodromal symptoms, followed by the development of more severe respiratory syndrome. Research has determined that the average incubation period of MERS is 5—6 days. In patients with progressive disease, the median time to death is 11—13 days. Currently, there is no effective vaccine or specific treatment method available.

3.5.3 Epidemiologic characteristics of COVID-19

Since December 2019, COVID-19 has gone through three stages: local outbreak, epidemic spread and epidemic expansion. It initially occurred in winter and spring, but after the global pandemic, the distinction in seasonal occurrence is no longer evident. According to The COVID-19 Diagnosis and Treatment Program (Trial Seventh Edition) issued by National Health Commission, the incubation period of COVID-19 is 1—14 day(s), with most patients of 3—7 days; the most common symptoms in patients are fever, dry cough and fatigue. COVID-19 is highly contagious, and prolonged unprotected close contact is a high-risk factor for infection. Therefore, family members, relatives and friends of patients are at a high risk of infection. There are both asymptomatic infected persons and super spreaders of COVID-19. This disease features high incidence, rapid transmission and low overall crude death rate. SARS-CoV-2 is sensitive to ultraviolet light and heat. It can be effectively inactivated by exposure to a 56°C water bath for 30 min, as well as by using ether, 75% ethanol, chlorine-containing disinfectants, peracetic acid, chloroform and other lipid solvents.

Bibliography

[1] Sampathkumar P, Temesgen Z, Smith TF, et al. SARS: epidemiology, clinical presentation, management, and infection control measures. Mayo Clin Proc 2003;78(7):882–90.
[2] Liu X, Zhuang H. Advances in epidemiologic study on SARS. Chin J Prev Med 2003;37(4):283–5.
[3] Chen X, Chughtai AA, Dyda A, et al. Comparative epidemiology of Middle East respiratory syndrome coronavirus (MERS-CoV) in Saudi Arabia and South Korea. Emerg Microbes Infect 2017;6(6):51.
[4] Ran L, Gao Z. Clinical and epidemiological characteristics of the Middle East respiratory syndrome coronavirus (MERS-CoV) infection. Chin J Tuberc Respir Dis 2013;36(9):684–6.
[5] Yali W, Qiao Z, Liu W, et al. Guidelines for imaging diagnosis of novel coronavirus pneumonia (2020 second edition, simple version). J Capit Med Univ 2020;41(2):168–73.
[6] Mcmichael TM, Currie DW, Clark S, et al. Epidemiology of COVID-19 in a long-term care facility in King County, Washington. N Engl J Med 2020;382(21):2005–11.
[7] Epidemiology Working Group for NCIP Epidemic Response. The epidemiological characteristics of an outbreak of 2019 novel coronavirus diseases (COVID-19) in China. Chin J Epidemiol 2020;41(2):145–51.
[8] Szmuda T, Ali S, Hetzger TV, et al. Are online searches for the novel coronavirus (COVID-19) related to media or epidemiology? A cross-sectional study. Int J Infect Dis 2020:386–90.
[9] Committee of the Infectious Diseases Radiology Group of Chinese Society of Radiology; Committee of the Infectious Diseases Radiology Section of Chinese Medical Doctor Association; Infectious Disease Imaging Group, Infectious Disease Branch, Chinese Research Hospital Association, etc. Guideline for Imaging Diagnosis of novel coronavirus (2019-nCoV) infected pneumonia (1st edition 2020). New Med 2020;30(1):22–34.
[10] Liu G, Tan W. Epidemiological advance of the Middle East respiratory syndrome coronavirus infection. Chin J Microbiol Immunol 2016;36(2):81–7.
[11] Tong Z, Wang J, Wu H, et al. A review on the epidemiology of Middle East respiratory syndrome. Chin J Epidemiol 2015;36(7):765–8.
[12] Van Boheemen S, de Graaf M, Lauber C, et al. Genomic characterization of a newly discovered coronavirus associated with acute respiratory distress syndrome in humans. mBio 2012;3(6):e00473.

[13] Hongjun L. Guideline for medical imaging in auxiliary diagnosis of coronavirus disease 2019. Chin J Med Imaging Technol 2020;36(3):321–31.
[14] Qun G, Miao L, Zhuang Y, et al. Epidemiological investigation of a family clustering of COVID-19. Chin J Epidemiol 2020;5:629–33.
[15] Jiang R, Liu J, Zhang L, et al. CT manifestations and outcome with family clustering coronavirus disease 2019 in 8 families. Electron J Emerg Infect Dis 2020;5(2):87–90.
[16] Peng G, He J, Lin J, et al. Epidemiological study on severe acute respiratory syndrome in Guangdong Province. Chin J Epidemiol 2003;5:20–2.
[17] Khan S, Morabet RE, Khan RA, et al. Where we missed? Middle East respiratory syndrome (MERS-CoV) epidemiology in Saudi Arabia 2012–2019. Sci Total Environ 2020;747:141369.
[18] Hui DS, Chan MC, Wu AK, et al. Severe acute respiratory syndrome (SARS): epidemiology and clinical features. Postgrad Med J 2004;80(945):373–81.
[19] Mackay IM, Arden KE. MERS coronavirus: diagnostics, epidemiology and transmission. Virol J 2015;12:222.
[20] Xueqiu L, Wenfeng C, Lifen H, et al. Comparison of epidemic characteristics between SARS in 2003 and COVID-19 in 2020 in Guangzhou. Chin J Epidemiol 2020;(5):634–7.
[21] Zheng Q, Lu Y, Tan L, et al. Advances in clinical and imaging studies of novel coronavirus pneumonia. Electron J Emerg Infect Dis 2020;5(2):140–4.
[22] Yin X, Xing L, Zhang Y, et al. Interpretation of the imaging diagnostic guidelines of the novel coronavirus pneumonia. Med Res Educ 2020;37(2):1–12.

CHAPTER 4

Pathogenesis and pathophysiology

Liu Dechun and Zhao Yunxia
Anhui Bengbu Medical University, Bengbu, China

4.1 Section 1 Pathogenesis

Coronavirus is an enveloped, positive-stranded RNA virus. At present, seven types of coronaviruses have been identified that being capable of infecting humans. The coronaviruses that pose a more significant threat to human health are SARS-CoV and MERS-CoV, which have been identified since 2003, as well as SARS-CoV-2/2019-nCoV discovered in 2019. They share similar biological characteristics and have comparable pathogenicity, primarily manifesting as severe lower respiratory tract infections. Besides, there are also many similarities in clinical, radiological and pathological manifestations. Therefore, they are expounded collectively. The term "Coronavirus" as described below primarily refers to these three newly-identified coronaviruses, excluding the other four coronaviruses that cause common cold or mild upper respiratory tract infections.

Coronaviruses have the ability to transmit between animals and humans, and the primary source of infection in population is individuals infected with the virus. It can be transmitted not only through respiratory droplets and aerosols but also through contact, primarily causing respiratory tract infections. The autopsy pathology study has revealed that coronavirus particles can be found in various tissues (respiratory tract, gastrointestinal tract, etc.), body fluid and excreta (feces, urine, sweat), indicating the possibility of multiple transmission routes, such as digestive tract transmission and transmission through contact with skin and various body fluids. People are generally susceptible, and in the case of SARS-CoV-2 infection, the clustering transmission is more prominent. The elderly, obese individuals and those with underlying diseases (such as diabetes, coronary heart diseases, hypertension, chronic hepatitis, etc.) tend to experience

The Pathology and Imaging of Coronavirus Pneumonia
ISBN 978-0-443-32930-2
https://doi.org/10.1016/B978-0-443-32930-2.00005-2

more severe illness after infection. Children and infants can also be infected. Extensive researches have been conducted on how coronaviruses, specifically SARS-CoV and MERS-CoV, cause diseases after invading the human body. Preliminary studies on SARS-CoV-2 suggest that its pathogenicity is also similar to that of SARS-CoV. At present, the following viewpoints are primarily presented in the literature.

4.1.1 Receptor binding theory

The binding of the coronavirus S protein to ACE2 receptor leads to the invasion of the virus into human cells. Research has found that all coronaviruses carry S protein, which has affinity to the respiratory mucosal epithelium. The S protein of coronaviruses is prone to adhere to the epithelial cells in the nasal cavity, pharynx, trachea and bronchi, which are rick in ACE2 receptors; after binding to the ACE2 receptor on the surface of epithelial cells, the virus penetrates the cell membrane and enters the target cells, then replicates and products large number of new viral particles. These particles are then released from the cells as budding and continue to invade neighboring epithelial cells, or transmitted through the excretion of the body through sputum. During the virus infection process, S protein can also induce the production of neutralizing antibody, T-cell immune response and protective immunity in the host. Both bronchial and alveolar epithelial cells carry ACE2. ACE2 serves as the receptor of coronaviruses and consists of 805 amino acids, which is a Type I transmembrane glycoprotein with single extracellular catalytic domain. Typically located on the luminal surface of epithelial cells, ACE2 is susceptible to respiratory infections. ACE2 is a homolog of carboxypeptidase ACE. Angiotensin II (Ang II), generated by carboxypeptidase, is the primary active peptide of the Renin-angiotensin-system (RAS). ACE2 is reported to be internalized as an intact molecule and/or its transmembrane region, along with the SARS-CoV envelope during infection, and this endocytosis is crucial for virus infection. Even when the recombinant SARS-CoV surface ligand S protein interacts with ACE2, internalization can still occur. When SARS-CoV is infected by expressing ACE2 on their luminal surface, its infectivity can be enhanced by 10 times. ACE2-mediated Ang II degradation also plays an important role in protecting the lungs from SARS-CoV-induced damage. Current research suggests that SARS-CoV-2 also has similar mechanism. At the genomic level, SARS-CoV-2 shows a high degree of sequence similarity to SARS-CoV. The S protein of SARS-CoV-2, just like that of

SARS-CoV, utilizes the same cellular receptor ACE2 to enter cells, indicating that the pathogenic mechanism of SARS-CoV-2 and SARS-CoV are similar. Research has demonstrated through the analysis of the extracellular domain structure of SARS-CoV-2 that its S protein has a higher affinity for binding to the ACE2 receptor on cells. Other research suggests that the affinity of SARS-CoV-2 for binding to the ACE2 receptor cells is 10–20 times higher than that of SARS-CoV, indicating that SARS-CoV-2 may have a higher infectivity potential compared with the SARS-CoV virus. Domestic research reveals that the structure of SARS-CoV-2 is highly similar to that of SARS-CoV, its difference lies in four key amino acids in the S protein, which results in a stronger binding capacity to ACE2 and makes it easier for the virus to spread and infect.

However, the receptor of MERS-CoV is dipeptidyl peptidase-4 (DPP4), which is also present in the alveolar epithelial cells. It can bind to the S protein of MERS-CoV, thereby facilitating the invasion of the virus. DPP4 is a serine protease expressed on the cell surface, highly conserved in human cells and widely distributed in organs such as the lungs, gastrointestinal tract, liver, pancreas and thymus, etc. In the respiratory system, DPP4 is mainly expressed in alveolar epithelial cells but is rarely found within alveolar macrophages. Given that DPP4 is a key factor in activating T cells and providing costimulatory signals in T cell immune response, indicating its potential involvement in regulating the host immune system.

Recent studies have found that, in addition to ACE acceptors, coronaviruses may also invade host cells by binding to them through sialic acid receptor, transmembrane protease serine 2 (TMPRSS2) or extracellular matrix metalloprotease inducer CD_{147} (also called basal immunoglobulin). Blocking the binding of the virus's S protein to receptors on the target cell membrane is also a current approach in developing antiviral drugs.

4.1.2 Immune damage theory

Coronavirus invasion into the body can also trigger the immune response. Based on the early onset of viremia in patients, and the decrease of lymphocytes in the blood during the course of the disease, especially the significant reduction of CD_4 and CD_8 T cells, clinical treatment with glucocorticoid can improve lung inflammation and clinical symptoms, indicating that this disease may be an immune-mediated damage caused by the body's immune response. Pathological examination showed that the splenic capsule was wrinkled and the volume was obviously reduced. The lymphocytes in the white pulp of the spleen and lymph nodes are significantly reduced. Both the spleen and lymph node tissues show abundant expression of ACE2, especially in macrophages. The pathology of these immune organs may be caused by the direct attack of virus and indirect immune damage. Cytokines produced during infection can also induce large number of lymphocytes (especially T cells), leading to damage to the immune system. Flow cytometry of a COVID-19 deceased patient in Beijing showed that CD_4 and CD_8 lymphocytes in peripheral blood of the patient were significantly reduced, the concentration of the highly proinflammatory Th17 subset ($CCR_4^+CCR_6^+$) was elevated in the CD_4 T lymphocytes, and CD_8 T cells exhibited higher concentration of cytotoxic granules with 31.6% being positive for perforin, 64.2% being positive for granulysin and 30.5% being double positive. The overexpression of Th17 and CD_8 T cells both indicated that the patient was in a state of severe immune impairment.

4.1.3 Cytokine storm theory

The body mobilizes neutrophils, monocytes/macrophages and lymphocytes through fever. Mononuclear macrophages not only have phagocytic function but can also process the antigen information of the virus, and present it to lymphocytes, produce specific antibodies and cytokines, and thus activate inflammatory and immune response. During the progressive stage of the disease, the body's own excessive immune response to virus-infected cells and the activation of mononuclear macrophages can not only phagocytize normal blood cells, including red blood cells, white blood cells, lymphocytes or platelets, but also produce large number of cytokines (such as tumor necrosis factor, interleukin, chemokines, etc.), and even release a massive amount of cytokines or inflammatory mediators through a cytokine storm or cascade of inflammatory mediators, causing severe inflammation and tissue damage, which may be a crucial mechanism in the development of DAD. The inflammatory mediators nonspecifically activate mononuclear macrophages and lymphocytes, which in turn release large number of cytokines and form a positive feedback loop. Cytokines are small molecules released into the blood by the affected cells through autocrine and paracrine activities, which can promote the migration and concentration of immune cells to the infection site, triggering an inflammatory response and leading to systemic fever and local pain. Infection with microorganisms causes various cytokines in body fluids such as tumor necrosis factor-α (TNF-α), interleukin (IL-1β, IL-6, IL-8, IL-12), α interferon, β interferon, interferon, monocyte chemoattractant protein-1 (MCP-1), transforming growth factor (TGF)-β_1 and other proinflammatory cytokines, which is an crucial cause of acute respiratory distress and multiorgan failure. Most patients with SARS and COVID-19 present mild symptoms at the onset of the disease, and in some cases, the condition suddenly worsens, which may be related to the body's excessive response to the cytokines produced after the onset of the disease, leading to a phenomenon known as "cytokine storm". The cytokine storm is actually a distress signal aimed at triggering a rapid and intense immune response in the body and killing the virus with a suicide attack. But it also leads to the damage to blood vessels, organ tissues and cells, increased vascular permeability and inflammatory exudation. Besides, the Cytokine Storm can also trigger the massive release of nitric oxide, which further dilutes the blood and damages to blood vessels. When all these factors are integrated, the blood pressure will be lowered to dangerous levels, leading to tissue hypoxia, hypertension, multiorgan dysfunction and disseminated intravascular coagulation (DIC).

4.1.4 Direct damage theory

The virus may directly damage target cells (such as alveolar epithelial cells, etc.) and cause acute lung injury. Virus particles may propagate in the Golgi apparatus and endoplasmic reticulum of infected cells and eventually lead to degeneration, necrosis and exfoliation of type II alveolar epithelial cells into the alveolar space. Virus infection can also trigger the apoptosis process, causing apoptosis and detachment of alveolar cells. Ding Yanqing et al. used the apoptosis and immunohistochemistry dual staining method in SARS pathological tissues and discovered a significant increase in apoptosis in the spleen, lung tissue and lymph nodes. The apoptotic cells in the affected lung tissues are primarily alveolar epithelial cells, bronchial epithelial cells and mononuclear macrophages. In the spleen, the apoptotic cells are mainly located in the red pulp. T and B lymphocytes in the lymph nodes also exhibit significant apoptosis. There is also a significant infiltration of mononuclear macrophages in the affected organs, lymph nodes and spleen, indicating that apoptosis and mononuclear macrophages infiltration may play an important role in the pathogenesis of SARS. The virus may also enter the bloodstream, causing damage to the vascular endothelial cells, leading to increased vascular permeability, extravasation of plasma and red blood cells and pulmonary inflammation. As entering other organs along with the bloodstream, the virus may cause varying degrees of damage to multiple organs. This has been confirmed in both SARS and MERS, and in autopsy data from COVID-19 cases, damage and inflammatory reactions can also be observed in organs such as the liver, heart, kidneys and gastrointestinal tract, indicating that coronaviruses can directly damage various tissues. Ding Yanqing's team found that cells with positive SARS virus detection and varying degrees of positive ACE2 expression include alveolar epithelial cells, bronchial epithelial cells and serous cells, lymphocytes, mononuclear macrophages, vascular endothelial cells, nerve cells, esophageal squamous epithelial cells, gastrointestinal epithelial cells, hepatocytes, myocardial cells, epithelial cells of distal convoluted renal tubule, adrenocortical cells, pancreatic islet cells, parathyroid acidophilic cells, pituitary acidophilic cells and sweat gland epithelial cells. Similar phenomena have also been found in autopsy studies of COVID-19 patients, suggesting that these cells can be damaged as target cells of coronaviruses and that the damaged organs are closely related to the distribution of SARS-CoV receptor ACE2 in the human body.

4.1.5 Endothelial injury theory

Clinical and pathological studies have found that many COVID-19 patients experienced varying degrees of thrombosis, thromboembolism, pulmonary infarction and other related phenomena. Zhongnan Hospital of Wuhan University conducted lower limb venous thrombosis screening in 48 patients with severe COVID-19, and found a high incidence rate of 85.4% (41/48), including 36 cases (75%) located in the distal veins and 5 cases (10.4%) located in the proximal veins. In the autopsy and biopsy pathology studies of COVID-19 patients in China, vascular diseases have been found both inside and outside the lungs, including vasculitis, mixed thrombus and hyaline thrombus formation, thromboembolism, pulmonary infarction and other lesions. Lower extremity thrombosis is also the pathological basis for the occurrence of the so-called "COVID-19 foot" in some patients. According to the studies, the main cause of thrombosis is severe SARS-CoV-2 infection and inflammation that damages the vascular endothelial cells. ACE2 receptors are widely present in the vascular endothelial cells of many tissues, such as lungs, kidneys, livers and intestines, which are susceptible to SARS-CoV-2 infection and damage, and thus induce apoptosis or pyroptosis of endothelial cells. Researchers have also confirmed that the presence of viral components in endothelial cells, and inflammatory cell filtration can be observed in the vascular wall under the electron microscopy. Endothelial damage and vasculitis are important conditions for thrombosis, and in critical patients, bed rest and venous catheterization are also significant triggering factors for thrombosis. Microthrombosis in capillaries can exacerbate local tissue congestion and edema. The formation of microthrombus consumes large number of platelets and cellulose, and can also promote disseminated intravascular coagulation to induce bleeding. Venous endothelial damage can lead to thrombosis in lower limb veins, prostatic veins, pulmonary artery embolism and local hemorrhagic infarction. These lesions are often observed in the lungs and also important mechanisms of pulmonary function impairment.

In conclusion, after entering the lungs through the respiratory tract, coronavirus invades the epithelial cells of bronchus, bronchioles and alveoli through the mediated action of S protein on the surface of the virus and ACE2 receptors on the surface of the target cells, and replicates rapidly within these cells, which can not only directly cause epithelial cell degeneration, necrosis, apoptosis and shedding, but also lead to lymphocytes, immune system dysregulation or suppression and elevated levels of cytokines, resulting in tissue congestion, edema, increased vascular permeability and exudation of fluids, cellulose and white blood cell, causing DAD, increased edema fluid and protein exudation from alveolar cavity, even forming transparent membrane-like structure and leading to thickened pulmonary interstitium, and even impairing gas flow and exchange function in severe cases. As a result, patients may experience coughing, asthma, and in severe case dyspnea, hypoxemia and even cardiopulmonary failure, posing a life-threatening situation.

4.2 Section 2 Pathophysiology

Clinical studies indicate that a series of pathophysiological changes also occur in patients with coronavirus pneumonia, mainly manifested as fever and hypoxia. A few severe patients may develop acute respiratory distress syndrome (ARDS), septic shock and acid-base disturbance, etc.

4.2.1 Fever

Fever is one of the common clinical symptoms of virus infection, and also the fundamental pathological process common to many infectious diseases. All patients with coronavirus pneumonia experience fever with a body temperature of 38°C or higher. The testing indicator for COVID-19 patients is set at 37.3°C for isolation observation, but there are also many patients without obvious fever symptoms. According to the theory of body temperature regulating set-point, fever is a regulated increase in body temperature caused by the upward shift of the set-point of the body's temperature-regulating center under the action of pyrogens. As an exogenous pyrogen, the coronavirus activates endogenous pyrogenic cells within the body, producing and releasing a large amount of endogenous pyrogens, such as Interleukin-1 (IL-1), Tumor Necrosis Factor (TNF), Interferon (IFN) and Interleukin-6 (IL-6), etc. The information of endogenous pyrogens enters the brain, resulting in the production of prostaglandin E2 (PGE2), cyclic adenosine monophosphate (cAMP), arginine vasopressin (AVP) and melanocyte stimulating hormone (MSH), etc. These mediators act on the body's temperature-regulating center to shift the set point upward, increase the heat production, reduce the heat loss, increase the body temperature and lead to fever. Fever helps activate the body's immune system to fight viral infection.

4.2.2 Hypoxia

Hypoxia is a pathological process characterized by metabolism, function and morphology abnormalities in cells due to the reduced oxygen supply or impaired oxygen utilization. The impact of hypoxia on the body varies with the cause, rate, degree of hypoxia and the patient's reactivity. For patients with coronavirus pneumonia, hypoxia may cause headache, chest tightness, cyanosis, shortness of breath (RR $\geq$ 30 times/min), dyspnea and even confusion, etc. Hypoxia may affect various systems in the body, and severe hypoxia can impair the myocardial systolic and diastolic function, and cause arrhythmia. Hypoxia can be divided into hypoxic hypoxia, hemic hypoxia, circulatory hypoxia and histogenous hypoxia based on different obstacles in oxygen supply and utilization. Hypoxic hypoxia refers to the reduction of arterial oxygen content and arterial partial pressure of oxygen (PaO_2) due to reduced oxygen partial pressure in inhaled air, external respiratory dysfunction or venous blood shunting into the arterial circulation and reduced oxygen uptake from lungs, which is called hypotonic hypoxemia. Patients with coronavirus pneumonia suffer from impaired lung functions and external respiratory dysfunction caused by conditions such as pulmonary edema, pulmonary consolidation, hyaline membrane formation, DAD and infiltration of interstitial inflammatory cells, leading to decreased alveolar ventilation, difficulty of blood oxygen exchange, reduction of dissolved oxygen in blood and decreased PaO_2 and thus causing hypoxic hypoxia. When hypoxic hypoxia occurs, the content of oxygenated hemoglobin in the blood decreases and the deoxygenated hemoglobin increases. When the average concentration of deoxygenated hemoglobin in the capillary blood exceeds 5 g/dL, the skin mucosa appears bluish purple, which is called cyanosis.

4.2.3 ARDS

Acute respiratory distress syndrome (ARDS) is a form of acute respiratory failure caused by acute lung injury. Patients present with progressive respiratory distress and uncorrectable hypoxemia, which is one of the main clinical manifestations and causes of death in severe cases of COVID-19. There are various causes of acute lung injury, including coronavirus infections, but its exact mechanism has not been fully elucidated yet. Possible mechanisms of acute lung injury caused by coronavirus: ① Directly act on alveolar epithelium, causing cell degeneration, necrosis and shedding and even DAD; during the inflammatory process, large amount of exudates accumulates in the alveolar cavity, causing consolidation of lung tissues and widening of alveolar septa, thereby impairing the gas exchange. ② Activate white blood cells, macrophages and platelets to release substantial cytokines and thus indirectly cause lung injury. The activated cells aggregate in the lungs and adhere to the capillary endothelium to release free radicals, proteases and inflammatory mediators, which may damage the alveolar epithelial cells and capillary endothelial cells, enhance the vascular permeability and cause pulmonary edema and formation of a hyaline membrane, thereby impairing the lung diffusion function. The pulmonary vasoconstriction and disseminated intravascular coagulation caused by inflammation mediators lead to inadequate blood flow in some alveoli, resulting in ineffective cavity ventilation. ③ The damage to the alveolar type II epithelial cells results in a decrease in the production of pulmonary surfactant. The dilution of edema fluid and compensatory hyperventilation may consume the surfactant, resulting in the increase in the surface tension of alveoli, decrease in lung compliance and formation of pulmonary atelectasis. Certain inflammatory mediators may cause bronchospasm and inadequate ventilation in some alveoli, thereby resulting in functional shunt and even increased anatomical shunt. The ARDS-induced respiratory failure is mainly caused by pulmonary diffusion impairment, intrapulmonary shunt and ineffective cavity ventilation; therefore, the changes in the blood oxygen are manifested by the decrease of arterial partial pressure of oxygen, classified as Type I Respiratory Failure (Hypoxemic Respiratory Failure). When the lesions in the lungs are extensive and extremely severe, the total ventilation of the lungs is decreased, leading to an increase in partial pressure of carbon dioxide in artery, which can result in Type II Respiratory Failure.

4.2.4 Shock

Shock refers to the systemic pathological process that causes histiocytic ischemia and hypoxia, as well as functional, metabolic and structural damage to vital organs due to sharply decreased effective circulating blood volume and severely deficiency in tissue blood perfusion under the influence of severe blood loss, fluid loss, infection, trauma and other pathogenic factors. The clinical manifestations include pale or cyanotic complexion, cold and clammy extremities, rapid and weak pulse, reduced pulse pressure, decreased urine volume and haziness of spirit-mind, etc. Multiple organ dysfunction or even failure may occur due to cell damage during the process of shock. The multiple organ dysfunction syndrome (MODS) refers to a condition in which patients without preexisting organ dysfunction experience the simultaneous or sequential dysfunction of two or more organs in a short period of time during severe trauma, infection or shock. Patients with coronavirus pneumonia, suffered from shock and respiratory failure, are considered critical and need to be warded in the intensive care unit (ICU). A small number of patients may die from MODS. The shock caused by coronavirus infection is known as septic shock, the mechanism of which is still unclear, and it may involve both microcirculation and cellular mechanisms simultaneously. Coronaviruses can directly damage target cells and also cause cell damage by producing and releasing a variety of cytokines, inflammatory mediators and free radicals through the activation of white blood cells, mononuclear macrophages, platelets and endothelial cells. The changes of cytokines in body fluid can cause microcirculation disorder and abnormal biochemical indices. Excessive activation of inflammatory cells leads to uncontrolled self-sustained amplification and self-destructive inflammation, known as Systemic Inflammatory Response

Syndrome (SIRS), which is characterized by widespread activation of inflammatory cells and massive release of inflammatory mediators into the blood stream (inflammatory cytokine cascade). This may cause a systemic inflammatory response in distant organs and damage the parenchymal cells of various organs. SIRS is also a major mechanism contributing to MODS during the shock, and there is a close relationship between viral load and the progression of the disease.

4.2.5 Acid-base disturbance

In certain diseases or pathological progresses, the body can experience disruptions in acid-base balance due to excessive or severely insufficient acid-base load or impaired regulatory mechanism. The acid-base disturbance is often a secondary change in certain diseases or pathological progresses, but once it occurs, it may complicate and exacerbate the disease conditions, therefore, it is crucial to timely detect and correctly treat the disease. Patients with severe coronavirus pneumonia may experience mixed acid-base disturbance due to the presence of various pathological processes. Severe hypoxia may lead to enhanced anaerobic glycolysis, lactic acids and other metabolites, as well as respiratory failure and even shock, resulting in renal insufficiency; the acid expelling and base retaining dysfunction of renal tubule may cause metabolic acidosis. In type I respiratory failure, respiratory alkalosis may occur due to excessive ventilation of the lungs caused by hypoxia. In type II respiratory failure, the retention of large amount of carbon dioxide can lead to respiratory acidosis. The acid-base disturbance in the body may often cause electrolyte metabolism disorder and further exacerbate the disease.

Bibliography

[1] Yin Y, Wunderink RG. MERS, SARS and other coronaviruses as causes of pneumonia. Respirology 2018;23:130−7.
[2] Cui J, Li F, Shi ZL. Origin and evolution of pathogenic coronaviruses. Nat Rev Microbiol 2019;17:181−92.
[3] Li MY, Xu ZK. Medical microbiology. 3rd ed. Beijing: People's Medical Publishing House, 2015.
[4] Chan JF, Yuan S, Kok KH, et al. A familial cluster of pneumonia associated with the 2019 novel coronavirus indicating person-to-person transmission: a study of a family cluster. Lancet 2020;395:514−23.
[5] Weiss SR, Leibowitz JL. Coronavirus pathogenesis. Adv Virus Res 2011;81:85−164.
[6] Wang X, Ding YQ. From SARS to COVID-19: pathogens, receptor, pathogenesis and principles of the treatment. Chin J Pathol 2020;49(6):647−52.

[7] To KF, Tong JH, Chan PK, et al. Tissue and cellular tropism of the coronavirus associated with severe acute respiratory syndrome: an in-situ hybridization study of fatal cases. J Pathol 2004;202:157−63.
[8] Li W, Moore MJ, Vasilieva N, et al. Angiotensin-converting enzyme 2 is a functional receptor for the SARS coronavirus. Nature 2003;426(6965):450−4.
[9] Chan JFW, Kok KH, Zhu Z, et al. Genomic characterization of the 2019 novel human-pathogenic coronavirus isolated from a patient with atypical pneumonia after visiting Wuhan. Emerg Microb Infect 2020;9(1):221−36.
[10] Liu M, Feng RE, Li Q, et al. Comparison of pathological changes and pathogenic mechanisms caused by H1N1 influenza virus, highly pathogenic H5N1 avian influenza virus, SARS-CoV, MERS-CoV and 2019-nCoV. Chin J Pathol 2020;49(5):511−6.
[11] Meyerholz DK, Lambertz AM, Mccray PB. Dipeptidyl peptidase 4 distribution in the human respiratory tract: implications for the Middle East Respiratory Syndrome. Am J Pathol 2016;186(1):78−86.
[12] Tripathy S, Dassarma B, Roy S, et al. A review on possible modes of action of chloroquine/hydroxychloroquine: repurposing against SAR-CoV-2 (COVID-19) pandemic. Int J Antimicrob Agents 2020;56:1−7.
[13] Gu J, Gong EC, Zhang B, et al. Multiple organ infection and the pathogenesis of SARS. J Exp Med 2005;202(3):415−24.
[14] Xu Z, Shi L, Wang Y, et al. Pathological findings of COVID-19 associated with acute respiratory distress syndrome. Lancet Respir Med 2020;8(4):420−2.
[15] Huang KJ, Su IJ, Theron M, et al. An interferon-gamma-related cytokine storm in SARS patients. J Med Virol 2005;75(2):185−94.
[16] Zhou J, Chu H, Li C, et al. Active replication of Middle East Respiratory Syndrome coronavirus and aberrant induction of inflammatory cytokines and chemokines in human macrophages: implications for pathogenesis. J Infect Dis 2014;209(9):1331−42.
[17] Wong CK, Lam CW, Wu AK, et al. Plasma inflammatory cytokines and chemokines in severe acute respiratory syndrome. Clin Exp Immunol 2004;136(1):95−103.
[18] He L, Ding Y, Zhang Q, et al. Expression of elevated levels of pro-inflammatory cytokines in SARS-CoV- infected ACE2 cells in SARS patients: relation to the acute lung injury and pathogenesis of SARS. J Pathol 2006;210(3):288−97.
[19] Chu H, Zhou J, Wong BH, et al. Middle East Respiratory Syndrome coronavirus efficiently infects human primary T lymphocytes and activates the extrinsic and intrinsic apoptosis pathways. J Infect Dis 2016;213(6):904−14.
[20] Ding YQ, He L, Zhang QL, et al. Organ distribution of severe acute respiratory syndrome (SARS) associated coronavirus (SARS-CoV) in SARS patients: implications for pathogenesis and virus transmission pathways. J Pathol 2004;203:622−30.
[21] Bian XW. The COVID-19 pathology team. Autopsy of COVID-19 victims in China. Natl Sci Rev 2020;7(9):1414−8.
[22] Wang D, Hu B, Hong C, et al. Clinical characteristics of 138 hospitalized patients with 2019 novel coronavirus- infected pneumonia in Wuhan, China. JAMA 2020;323(11):1061−9.
[23] Ding YQ, Wang HJ, Shen H, et al. The clinical pathology of severe acute respiratory syndrome (SARS): a report from China. J Pathol 2003;200(3):282−9.

CHAPTER 5

Pathological manifestations

Liu Dechun and Zhao Yunxia
Anhui Bengbu Medical University, Bengbu, China

5.1 Pathological manifestations of SARS

The coronavirus can infect a variety of hosts, including humans, and in humans, it mainly causes respiratory system diseases, but may also lead to multiple organ diseases affecting lymphatic tissue, gastrointestinal tract, heart, liver, kidney, brain and secretory glands.

In terms of pathology, studies on SARS, MERS and COVID-19 have found that their pathological features are quite similar. Lungs, blood vessels and immune system are the main target organs of coronaviruses. The pathological manifestations of lung tissue in autopsy are all DAD. DAD is a pathological change caused by severe diseases inside and outside the lungs (including viral infection), which is based on diffuse injury of capillary endothelium and alveolar epithelium as well as enhanced vascular permeability, and mainly manifested by pulmonary edema and formation of hyaline membranes. It is an acute respiratory distress syndrome characterized by progressive respiratory distress and refractory hypoxemia. The course of the disease is generally divided into exudative stage and proliferative stage. Generally, it appears within 7—10 days after the onset of the disease, mainly manifested as pulmonary edema and the formation of hyaline membrane, and then enters the proliferative stage, characterized by the proliferation of type II alveolar epithelial cells, as well as proliferation and fibrosis of interstitial fibroblast. In a few cases, pulmonary interstitial fibrosis or the formation of honeycomb lung may occur.

Since 2003, a large number of autopsy cases of SARS have been reported in mainland China, Hong Kong, Singapore, Canada and other countries and regions, providing comprehensive descriptions on pathological changes of SARS. Based on these autopsy pathological studies, the pathological changes of SARS can be summarized into four aspects: pulmonary lesions, immune organ damage, vascular lesions and systemic toxicity.

The Pathology and Imaging of Coronavirus Pneumonia
ISBN 978-0-443-32930-2
https://doi.org/10.1016/B978-0-443-32930-2.00011-8

5.1.1 Pulmonary lesions

In the autopsy of SARS patients, it is found that both lungs were significantly swollen, distended, increased in volume and weight due to congestion, bleeding and inflammatory changes. Both the surface and section of the lungs appear gray-red, purplish-red or dark red. Plaque-like or extensive consolidation can be observed with a relatively firm texture. Additionally, there may be patchy or scattered hemorrhagic foci, and dark red fluid may be flowing out.

The primary manifestation of pulmonary inflammation under the microscope is DAD, which can be roughly divided into the exudative stage (acute stage) and proliferative stage (organizing stage). ① Exudative Stage: Lung tissue congestion, hemorrhage, edema, fibrin deposition and inflammatory cell infiltration, formation of pulmonary hyaline membranes (Fig. 5.1), and type II alveolar epithelial hyperplasia are commonly observed. In all cases, focally distributed alveolar epithelial cell lesions can be observed, characterized by increased cell volume, presence of bicolor eosinophilic cytoplasmic granules, enlarged nuclei, and irregular and prominent eosinophilic nucleoli. In some alveolar cavities, there are abundant desquamated alveolar epithelial cells (desquamative alveolar

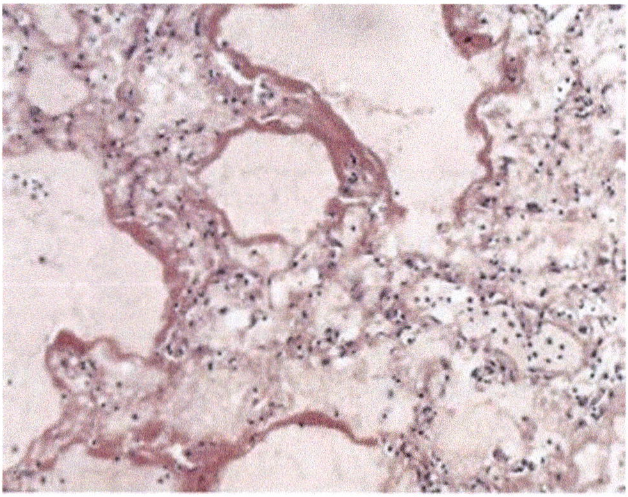

Figure 5.1 Pathological image of pulmonary hyaline membranes. The pulmonary hyaline membrane is a layer of reddish materials that is formed by proteinaceous substances, and tightly adhered to the alveolar wall, impairing the gas exchange. *(Images are provided by Professor Ding Yanqing from Southern Medical University.)*

catarrh), and syncytial multinucleated giant cells formed by the fusion of macrophages or alveolar epithelial cells (Fig. 5.2), which could be confirmed by immunohistochemistry, respectively. The exudate contains a large number of monocytes, lymphocytes and plasma cells. A small number of mononuclear macrophages show signs of apoptosis. ② Proliferative Stage: Partial small airways and alveolar spaces undergo the organization of exudates, accompanied by interstitial inflammatory cell infiltration, fibrosis and the formation of honeycomb lungs. Additionally, vasculitis and the formation of intravascular fibrinous microthrombus may also be observed. The above pathological changes are more typical in the subpleural regions. The characteristic manifestations include the presence of viral inclusion bodies in alveolar cells (Fig. 5.3). Under electron microscopy, coronavirus-like particles can be observed (Fig. 5.4). In-situ hybridization reveals the presence of viral nucleic acid, and immunohistochemical markers can demonstrate the viral antigens, proving viral infection. The autopsy has found that the staging of DAD is significantly correlated with the course of the disease. Cases with a disease course of less than 14 days are mostly in the exudative stage of DAD, while that with a disease course exceeding 14 days are manifested as the proliferative stage of DAD, accompanied by

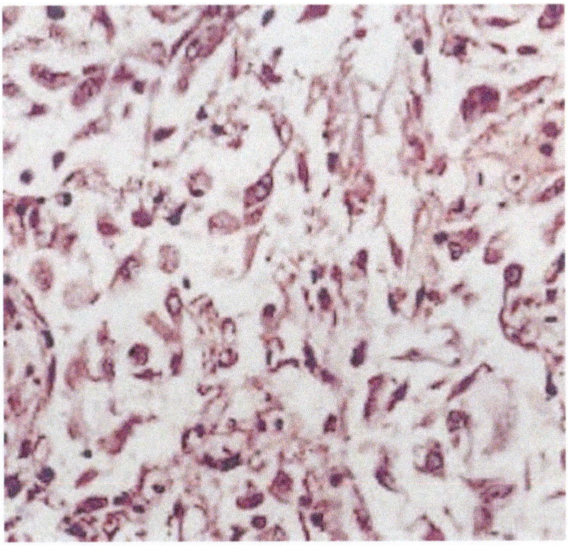

Figure 5.2 Pathological image of alveolar exfoliative cells. The alveolar cavity is filled with desquamated alveolar epithelial cells and infiltrated mononuclear macrophages, which could be confirmed using immunohistochemistry, respectively. *(Images are provided by Professor Ding Yanqing from Southern Medical University.)*

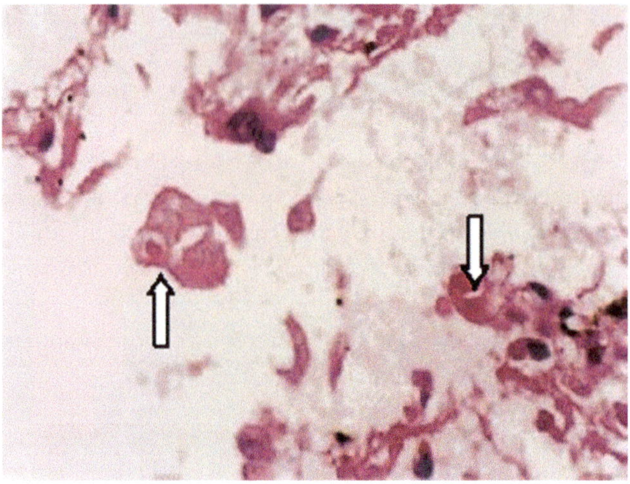

Figure 5.3 Pathological image of viral inclusion bodies. Viral inclusion bodies can be observed in the cytoplasm of alveolar epithelial cells, pink in color and surrounded by a halo (*indicated by arrows*), which can be confirmed by immunohistochemistry and special staining methods. *(Images are provided by Professor Ding Yanqing from Southern Medical University.)*

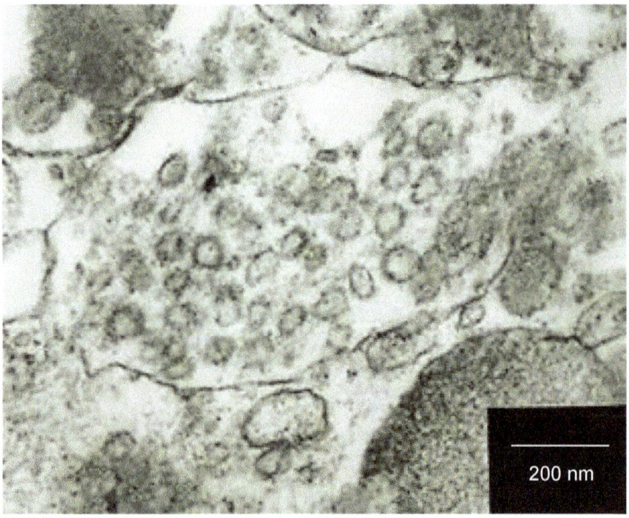

Figure 5.4 Pathological image of virus-like particles. Transmission electron microscopy: numerous virus-like particles can be observed (locally magnified) in the expanded endoplasmic reticulum in the cytoplasm of the alveolar epithelial cells, with a diameter ranging from 70 to 90 nm. *(Images are provided by Professor Ding Yanqing from Southern Medical University.)*

organization and fibrosis. In a few autopsy cases, concurrent bacterial pneumonia, pulmonary aspergillosis or cytomegalovirus infection are also identified.

5.1.2 Damage to immune organs

In the autopsy of SARS patients, it is found that lymph nodes and the spleen also frequently experience immune-related lesions. ① The envelope of the spleen is wrinkled with its volume being reduced obviously; the lymphoid tissues of the white pulp and marginal sinus exhibit extensive necrosis, and the intrinsic lymphoid tissue undergoes atrophy with a significant decrease in lymphocytes and proliferation of mononuclear macrophages; the structure of perivascular lymphoid sheath is disrupted; macrophages are proliferated in the red pulp area with phagocytosis of red blood cells and lymphocytes being observed. And focal hemorrhage and patchy necrosis can be observed in the red pulp area. Due to abnormal coagulation mechanism and vasculitic lesions, splenic infarction with anemia are commonly observed. ② Multifocal hemorrhagic necrosis is present in lymph nodes, with atrophy or disappearance of lymphoid follicles, loss of germinal center and a significant decrease in lymphocytes. Immunohistochemical staining reveals a decrease in CD_4 T cells and CD_8 T cells within the spleen and lymph nodes, as well as a reduction in B lymphocytes. Conversely, there is a significant increase in proliferation of mononuclear macrophages and cellular activity. Both the spleen and lymph node tissues show abundant expression of ACE2, especially in macrophages. The lesions in these immune organs may be caused by direct viral attack and indirect immune damage, suggesting that SARS-CoV can lead to immune injury.

5.1.3 Vasculopathy

The invasion of the virus into blood vessels can cause various organs and tissues (such as the lungs, heart, kidneys, adrenal glands, brain and skeletal muscle) to experience proliferation, swelling, degeneration and apoptosis of venular endothelial cells; it can also lead to vascular wall edema, vascuclar wall damage and diffuse infiltration of monocytes, lymphocytes and neutrophils (vasculitis). In the lungs, fibrinoid necrosis and intravascular thrombus can be observed, while microthrombi formation, inflammatory cell infiltration, and partial vascular wall fibrinoid necrosis or hyaline thrombus formation and thromboembolism may occur in the lungs and adrenal glands.

5.1.4 Systemic toxic changes

The coronavirus can invade pulmonary blood vessels and lymphatic vessels, enter other organs through the blood stream (viremia) and cause parenchymal cell degeneration, focal necrosis and hemorrhage in various organs such as the lungs, liver, heart, brain, kidneys and adrenal glands, as well as mononuclear cell infiltration of interstitial lymphocytes (interstitial inflammation), swelling and generation of striated muscle, degeneration and swelling of brain nerve cells, segmental bleeding in the intestinal tract, reduction of hematopoietic cells in the bone marrow and appearance of hemophagocytes, etc. These pathological changes may be related to the systemic toxic reaction caused by viremia from the coronavirus, but they could also be the results of the coronavirus infection itself, because strong positive expressions of SARS-CoV nucleocapsid protein antibodies and SARS-CoV RNA polymerase are observed in respiratory epithelial cell, gastrointestinal epithelial cells, distal convoluted renal tubule epithelial cells and skin sweat gland cells.

Given the aforementioned pathological changes, researchers such as Ding Yanqing and his team believe that SARS is a systemic multiorgan injury disease caused by the virus, with the lungs being the main target organ of injury.

5.2 Pathological manifestations of MERS

In the autopsy study of MERS death cases, the lesions are found to be similar to those of SARS patients with the most common one of DAD, and the organs outside the lungs may also be affected. The pulmonary lesions are manifested as acute pulmonary edema, alveolitis, pulmonary consolidation and interstitial inflammation, with the most characteristic ones including the formation of hyaline membrane, massive infiltration of inflammatory cells in the alveolar cavity, fibrin exudation and necrosis and exfoliation of alveolar epithelium, causing pulmonary consolidation, type II alveolar epithelial hyperplasia and cytopathy, formation of syncytial multinuclear cell and infiltration of interstitial lymphocytes, etc. In addition, focal pulmonary hemorrhage, infarction and pleural effusion can also be observed. In the late stage, organizing pneumonia and interstitial fibrosis may occur. Neutrophil and viral inclusion bodies are rarely seen in the lungs.

Extrapulmonary manifestations include acute kidney injury, mild portal vein and lobular hepatitis, necrosis and atrophy of skeletal muscle fiber; disappearance of lymphatic follicles and reactive hyperplasia of lymphocytes in the interfollicular area may occur for lymph node lesions. In severe cases, the incidence of acute kidney injury can be as high as 75%. No significant histopathological changes are found in the brain and heart.

5.3 Pathological manifestations of COVID-19

For COVID-19 patients 37 cases of systemic autopsy and 54 cases of minimally invasive autopsy have been performed in China. There are also some reports from European and American countries. Based on the currently collected biopsy and autopsy data, the pathological characteristics of SARS-CoV-2 infection are similar to those of SARS-CoV infection, but there are still some differences. Typically, the main lesions are in the lungs, but also in many organs and tissues outside the lungs. The primary lesions are infectious in nature, but it can also be accompanied by lesions related to various underlying diseases or complications.

5.3.1 Pulmonary lesions

It can be seen that the lung tissues are swollen with reduced elasticity, increased mass and toughened texture, showed varying degrees of consolidation and lost inherent sponge texture. Due to diffuse congestion and bleeding, the surface appears purplish red or dark red and locally purplish brown, the bleeding focus is patchy or spot-like, and large local bleeding can also be observed. The pleura is thickened, may be rough or adhere to lung tissues. The section shows grayish white patchy solid lesion and purplish red bleeding area. The grayish white consolidation area and purplish red bleeding area may appear alternately. The consolidation area can extrude viscous liquid due to more exudation in the alveolar cavity, and there is bloody fluid outflow in the bleeding area. The mucosa of tracheobronchial cavity is congestive with increased secretion, in which white and foamy liquid is observed. There may be a small amount of pleural effusion. The visual findings at autopsy are consistent with the imaging changes, and the ground-glass opacity seen on imaging corresponds to the grayish white solid lesions in the lungs by naked eyes, which are more common in the pleura, suggesting that COVID-19 mainly causes deep airway and alveolar damage.

The main pathological changes of the lungs are concentrated in the alveoli and bronchus of all levels, and the histological changes are mainly manifested as DAD, with acute diffuse inflammatory changes of alveoli and pulmonary interstitium as the main changes. It can also be divided into the exudative stage (acute stage) and proliferative stage (organizing stage).

5.3.1.1 Lesions in exudative stage

The exudative phase corresponds to the early clinical stage and progressive stage. In the early stage of COVID-19, some patients do not develop pneumonia, but are admitted for surgery for other diseases such as lung cancer, colon cancer, brain tumor, ovarian tumor, etc., which have provided an opportunity to understand the lung lesions in the early stage of infection. The pathology departments from three hospitals in Wuhan examined the lung lesions of four patients with SARS-CoV-2 who were admitted for lung cancer. In addition to lung cancer, the following plaque-like or diffuse lesions are seen: ① The alveolar wall and interstitial blood vessels are dilated and congested with increased permeability, resulting in serous exudation from the alveolar cavity and even pulmonary edema, which contains protein, cellulose-like or globular proteinaceous exudates that can form mucoid substances or transparent membrane along the inner side of the alveolar cavity (Figs. 5.5 and 5.6). ② With the progression of the disease, under the influence of vascular permeability and inflammatory

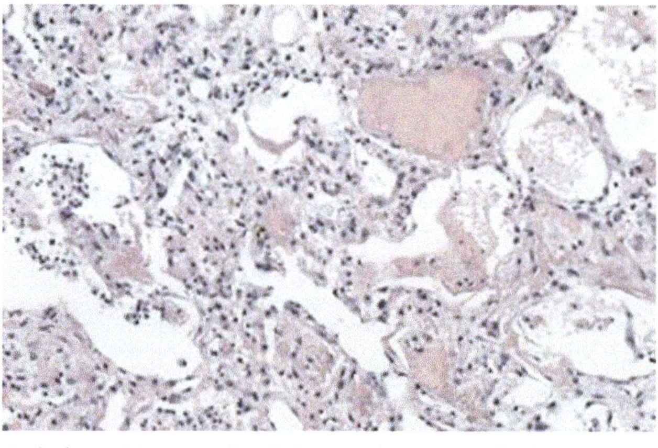

Figure 5.5 Pathological images of early lesion of COVID-19 (I). The patient's pulmonary blood vessels are diluted and congested and there are varying amounts of proteinaceous fluid exudates in the alveolar spaces, resulting in pulmonary edema and infiltration of lymphocytes in the interstitium. *(Photo provided by Director Li Hongjun of Beijing YouAn Hospital, Capital Medical University.)*

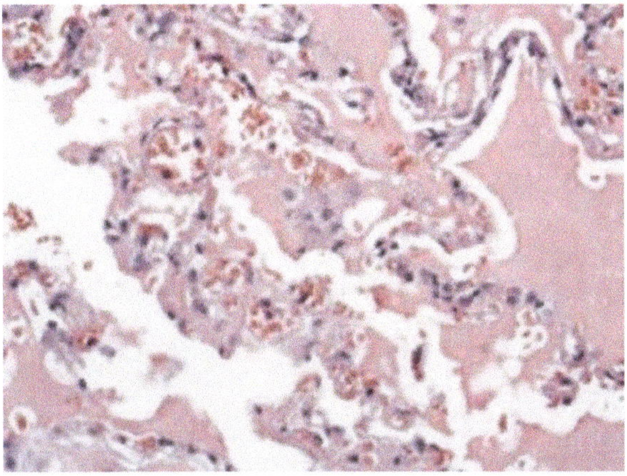

Figure 5.6 Pathological images of early lesion of COVID-19 (II). The patient's pulmonary blood vessels are diluted and congested and there are large amount of proteinaceous fluid exudates in the alveolar spaces, resulting in pulmonary edema and containing leaked red blood cells. There is a tendency for the formation of local hyaline membranes. *(Photo provided by Director Yuan Jingping of Wuhan University Renmin Hospital.)*

chemokines, the alveolar cavity is filled with a large number of inflammatory exudates, mainly consisting of serous fluid, cellulose and a large number of mononuclear macrophages, a small quantity of lymphocytes and multinucleated giant cells, and exfoliated alveolar epithelium could be observed, leading to lung consolidation, and sometimes neutrophilic exudation could be seen in varying amounts. The exudation of a large number of mononuclear macrophages in the alveolar cavity is the active response of the body to resist the pathogen. ③ The alveolar epithelial cells are the target cells of SARS-CoV-2, so there is obvious damage, which may cause degeneration, necrosis and exfoliation, similar to desquamative pneumonia; the alveolar epithelial cells can also be swollen and increased, mainly focal or extensive proliferation of type II alveolar epithelial cells, which may be erected as "shoe studs" (Figs. 5.7 and 5.8); besides, the fusion of alveolar epithelial cells can also be seen to form multinuclear giant cells, which generally have two to three nuclei, but can be as many as 10 to 15. ④ Intracytoplasmic or intranuclear inclusions can be observed in Type II alveolar epithelial cells (Figs. 5.9 and 5.10); viral inclusions are formed by the massive proliferation and accumulation of viral proteins and particles inside the host cells. Viral inclusions are present as single or multiple structure(s) in the cytoplasm of proliferative type II alveolar epithelial cells

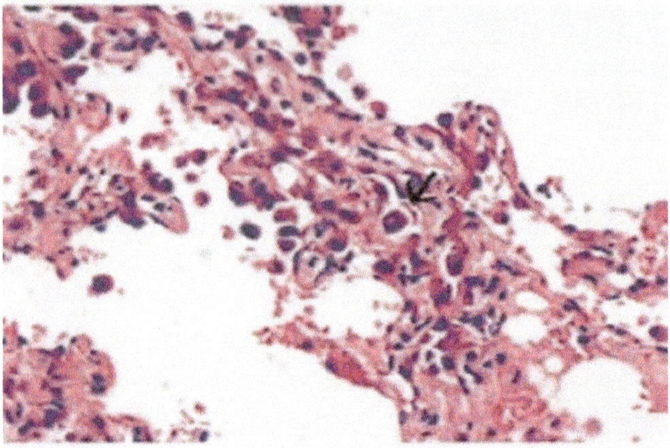

Figure 5.7 Alveolar epithelial hyperplasia (I). COVID-19 patients with type II alveolar epithelial hyperplasia, increased volume, enlarged and deeply stained cell nuclei, some of which exhibit suspicious viral inclusions (*indicated by arrows*). *(Photo cited with the consent of Professor Tian Sufang and Xiao Shuyuan from Zhongnan Hospital of Wuhan University.)*

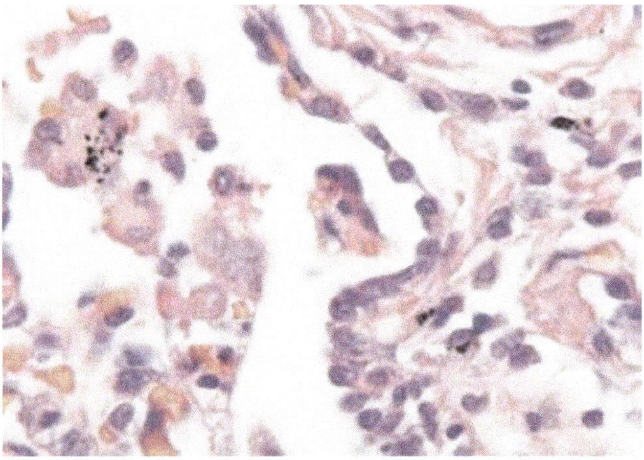

Figure 5.8 Alveolar epithelial hyperplasia (II). COVID-19 patients with type II alveolar epithelial hyperplasia, increased volume, and exfoliated alveolar epithelial cells and exudated inflammatory cells seen in the alveolar space. Lymphocyte infiltration is observed in the interstitium. *(Photo provided by Director Yuan Jingping of Wuhan University Renmin Hospital.)*

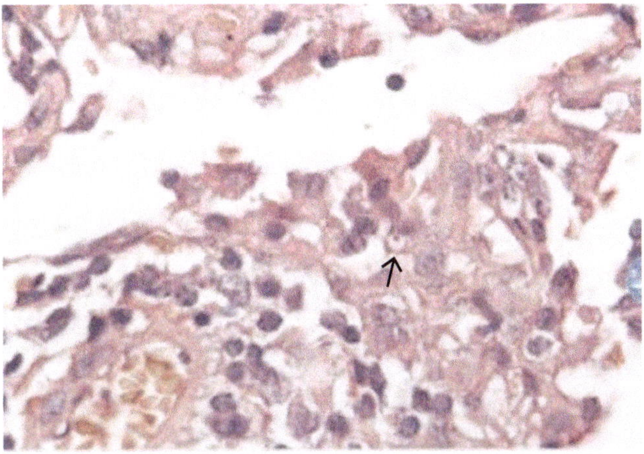

Figure 5.9 Viral inclusions (I). Scattered or clustered epithelial cells and macrophages can be seen in the alveolar space. Pink viral inclusions can be seen in the individual cytoplasm (*indicated by arrow*), surrounded by hollow halo. *(Photo provided by Director Yuan Jingping of Wuhan University Renmin Hospital.)*

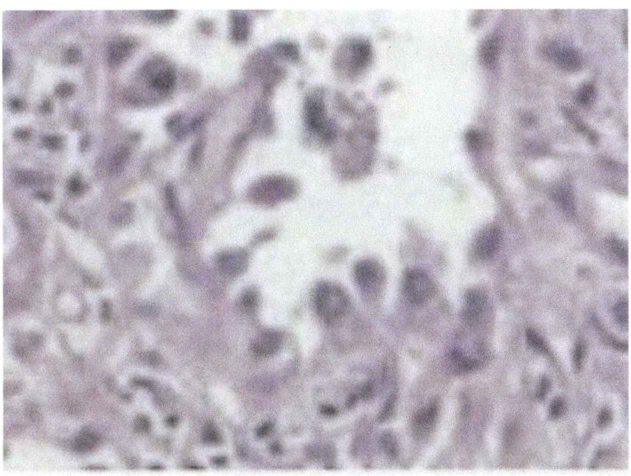

Figure 5.10 Viral inclusions (II). Purplish red inclusions can be seen in the nuclei of alveolar epithelial cells with no halo surrounded. Lymphocyte infiltration is observed in the pulmonary interstitium. *(Photo provided by Director Yuan Jingping of Wuhan University Renmin Hospital.)*

and macrophages; they exist as round or oval structures, which are eosinophilic and surrounded by a halo. Under the microscope, viral inclusions hold significant diagnostic value for viral infections. Viral nucleic acid sequences have also been found in the alveolar epithelial cells, providing further evidence supporting viral infection. ⑤ There are varying numbers of mononuclear macrophages and lymphocytes infiltrating the pulmonary interstitium (Figs. 5.5 and 5.8), leading to the formation of interstitial pneumonia; the pulmonary interstitium is widened, and the infiltrating lymphocytes are predominantly CD_4 T cells, with some CD_8 T cells and B cells. Plasmocytes are rare. These findings suggest the involvement of cellular immune response. Lymphocytopenia in peripheral blood may also be related to the exudation of T cells. Microthrombosis and small-vessel vasculitis can be observed in interstitial substances, with T cells mainly infiltrating around the blood vessels. ⑥ The bronchial mucosa is contested and edematous, with some areas showing epithelial cell detachment. The exfoliated epithelial cells mix with necrotic materials, exudates and mucus, forming bronchial plugs (mucous plugs), which are mainly concentrated in bronchioles and terminal bronchioles. These lesions in the airways (especially in small airways) worsen the ventilation obstruction. If the contents in the airways and alveoli are liquefied, absorbed or coughed out and disappear, and the lesions regress, the condition can be turned to recovery.

5.3.1.2 Lesions of proliferative stage

The proliferative stage roughly corresponds to the clinical severe and late stage. In cases with a prolonged disease duration, the lesions are distributed in a patchy pattern of varying sizes. The above-mentioned lesions are more pronounced and accompanied by organizing lesions. Pathological manifestation: ① Pulmonary hemorrhage, showing scattered and multifocal distribution with varying sizes. The hemorrhagic areas appear purplish red. Under the microscope, numerous red blood cells are observed, with only the outline of alveolar structures visible, indicative of necrosis or hemorrhagic infarction. Vasculitis manifestations can also be observed in the interstitium, with the presence of small blood vessels possibly showing the formation of microthrombi (hyaline thrombus/microthrombus and mixed thrombus). Thromboembolism can also been seen. Some patients in the acute stage may also experience pulmonary hemorrhage. ② In the lesion areas, there is a more pronounced proteinaceous inflammatory exudates in the alveolar space, containing abundant protein-rich exudates, red blood cells, fibrin and dispersed proteinaceous microspheres. Hyaline membranes are still formed at the edges of the alveolar spaces. ③ The inflammatory cells in the alveolar spaces are more prominent, leading to the formation of alveolitis, mainly containing mononuclear macrophages or multinucleated giant cells, lymphocytes and plasma cells. Foam cells are often formed, or there are neutrophil exudates. Some researchers have also observed granulomatous nodules composed of fibrin, inflammatory cells and multinucleated giant cells (Fig. 5.11). ④ The alveolar epithelial cells show proliferation, swelling, deformation or degenerative necrosis, resulting in cytopathic changes, with the characteristics of granular cytoplasm, enlarged and deeply stained nuclei with prominent nucleoli. Additionally, binucleated or multinucleated syncytial giant cells may be observed due to the fusion of alveolar epithelial cells, some of which may contain viral inclusions. ⑤ The lung tissues become solid and organized gradually. Spindle-shaped fibroblasts can be observed undergoing proliferation, forming clusters that fill in the alveolar spaces, resulting in organizing pneumonia and fibrous occlusion of the alveoli. The lung tissue consolidation, parenchymal change and organization are the main pathological basis for the patient's respiratory distress, subsequent hypoxia and respiratory failure. The broader the extent of consolidation and organization is, the more pronounced the symptoms of hypoxia will become. ⑥ The alveolar walls and interstitium show congested edema and inflammatory infiltration, leading to interstitial pneumonia mainly characterized by lymphocytic infiltration (Fig. 5.12). Macrophages may also be present, along with proliferation of interstitial fibroblast and increased collagen fibers, resulting in diffuse thickening of the interstitium. Additionally, the fusion of organizing focus can be observed, further expanding or aggravating pulmonary interstitial fibrosis (Figs. 5.13 and 5.14). Masson Staining can clearly demonstrate the proliferated fibrous tissues. A few alveoli are excessively

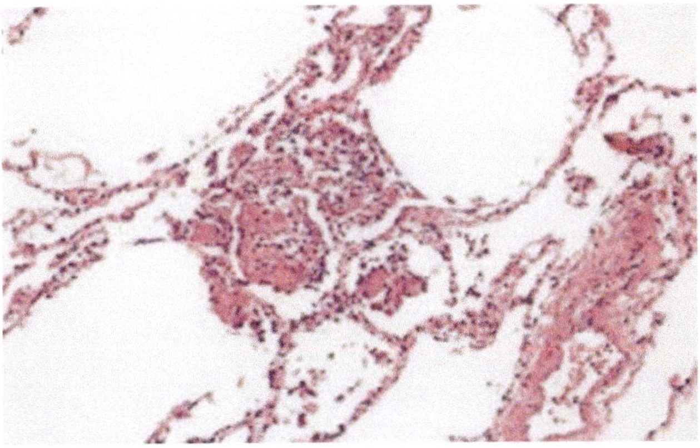

Figure 5.11 Granulomatous nodules. They are composed of fibrin, inflammatory cells and multinucleated giant cells. *(Photo cited with the consent of Professor Tian Sufang and Xiao Shuyuan from Zhongnan Hospital of Wuhan University.)*

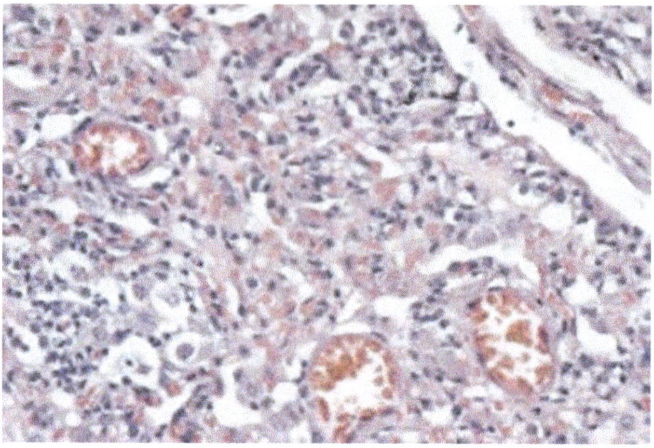

Figure 5.12 Interstitial pneumonia. The illustration shows pulmonary interstitial congestion with leakage of red blood cells. Mononuclear macrophages and lymphocytes infiltrate the interstitium, leading to its widening. *(Photo provided by Director Yuan Jingping of Wuhan University Renmin Hospital.)*

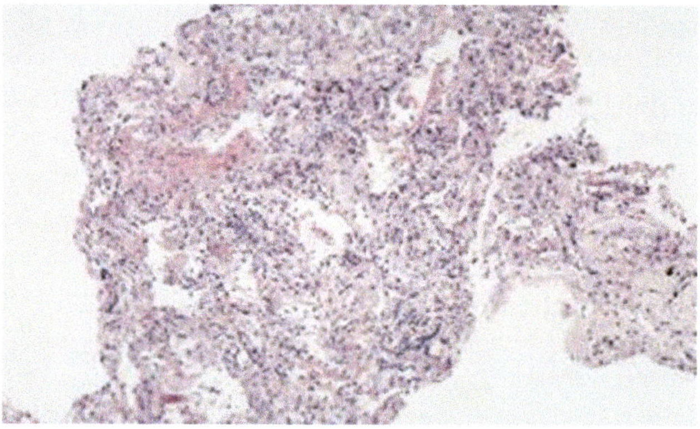

Figure 5.13 Pulmonary interstitial fibrosis (I). Pulmonary interstitial fibrosis is characterized by interstitial widening, proliferation of fibrous tissues and chronic inflammatory cell infiltration, with some areas exhibiting a glassy appearance. *(Photo provided by Director Li Hongjun of Beijing YouAn Hospital, Capital Medical University.)*

overinflated, leading to the rupture of alveolar septa or the formation of cystic space. Fibrous thickening may also occur in the pleura. ⑦ Chronic inflammation of the bronchial mucosa results in an increase in mucus cells and enhanced mucus secretion in the bronchial mucosa. This combines with exfoliated cells to form a thick, sticky liquid or mucus plug. Epithelial squamous metaplasia of the bronchus can also be observed.

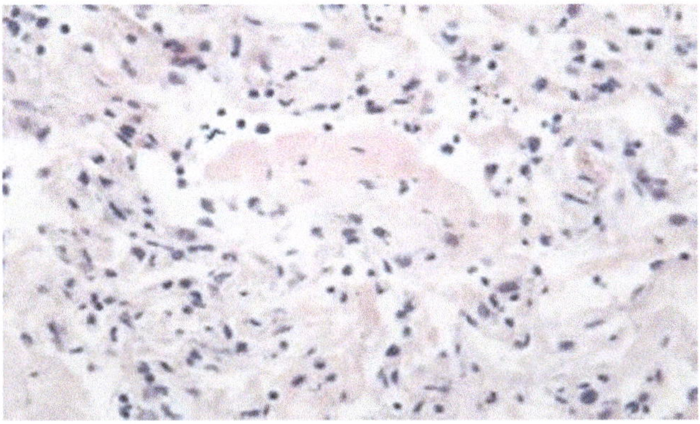

Figure 5.14 Pulmonary interstitial fibrosis (II). At high magnification, fibroblast proliferation can be observed, along with localized collagenization and glassy degeneration appearance. A small amount of lymphocyte infiltration may also be observed. *(Photo provided by Director Li Hongjun of Beijing YouAn Hospital, Capital Medical University.)*

In fact, in the course of disease progression, the mentioned inflammatory and fibrotic lesions represent a continuous process and can not be distinctly separated. In autopsy cases, upon careful examination of different regions of the lung tissues, all the aforementioned lesions can be observed. The diversity of lesions indicates that the progression of pulmonary pathology is not synchronous.

5.3.2 Extrapulmonary lesions

Extrapulmonary lesions with varying degrees have been found in the autopsy of COVID-19 patients. ① Lymphohematopoietic tissue lesion: The spleen shows significant shrinkage with focal hemorrhage and necrosis. Hemorrhage and anemic infarction are commonly observed in the spleen. Macrophages and phagocytosis are observed. Focal necrosis can also been seen in the lymph nodes; Immunohistochemistry reveals a decrease in CD_4 and CD_8 positive T cells in the spleen and lymph nodes, indicating impairment of cellular immune function. In most cases, there is a reduction in the erythroid, myeloid and megakaryocyte series in the bone marrow, while few cases show proliferation. ② Hepatobiliary lesion: The liver is enlarged with tense capsule and a dark red color. Hepatocytes show fatty degeneration, and in severe cases, there may be spotty, patchy, bridging or massive necrosis. Infiltration of mononuclear cells and neutrophils can be observed both inside and around the areas of necrosis; the hepatic sinuses are congested and there is a proliferation in Kupffer cells; there is a mild inflammation in the portal tracts with infiltration of lymphocytes and monocytes. Microthrombus formation can also be observed. The gallbladder is distended and enlarged, and its mucosa shows chronic inflammatory lesions. ③ Cardiovascular Lesion: Cardiomyocytes show degenerative swelling, scattered small focal necrosis, and interstitial edema. In the necrotic areas, there is a mild infiltration of mononuclear cells, lymphocytes, and/or neutrophils. Some blood vessels undergo endothelial cell exfoliation, with formation of intimitis and thrombus. A small amount of inflammatory cell infiltration is observed in the epicardium. Partial organs (such as lungs and heart) exhibit endarteritis and panvasculitis with endothelial cell exfoliation and thrombus formation in the blood vessels. Microthrombus can be observed in some organs. Thrombosis can also be seen in the blood vessels such as the prostate and lower limb veins. ④ Kidney Lesion: The renal glomerular capillary network is congested, with segmental proliferation or necrosis. The glomerular capsules show proteinaceous exudate in a flocculent pattern. The proximal convoluted tubules show epithelial cell swelling, degeneration, focal necrosis and exfoliation; hyalinous tubules are commonly formed in the distal convoluted tubules. The renal interstitium shows vascular congestion, mild inflammatory cell infiltration and focal fibrosis. Microthrombus can be observed in some small blood vessels. ⑤ Brain meninges and parenchyma show congestion and edema with partial neuronal degeneration. In some cases, there are manifestations of neuronophagia and satellitism, as well as neuronal ischemic changes with inflammatory cell infiltrated around blood vessels, focal cerebral infarction, and even cerebral hernia in a few cases. Currently, there is no pathological evidence of viral encephalitis. ⑥ Other Tissues: The esophageal, gastric and intestinal mucosal epithelial cells exhibit varying degrees of degeneration, necrosis and exfoliation. Adrenal cortex shows degeneration, focal hemorrhagic necrosis, and infiltration of mononuclear cells, lymphocytes and neutrophils around and within the necrotic foci. In most cases, the bladder is highly distended, and the mucosal epithelial cells are exfoliated. Occasionally, degeneration and dissolution of islet cells or small focal necrosis are detected. The testis shows varying degrees of reduction and even damage to spermatogenic cells. ACE2 receptors are found in testicular tissues, and viral nucleic acid is detected in semen. These lesions are generally consistent with interstitial inflammation caused by viral infection or with the influence of drugs.

5.3.3 Underlying lesions

The subjects of the autopsy are mostly the elderly, hence usually accompanied by underlying or concomitant diseases. ① Diseases of respiratory system, such as chronic bronchitis, bronchiectasis, pulmonary fibrosis, emphysema, and pulmonary bulla, etc.; ② Cardiovascular diseases, such as atherosclerosis of large and medium-sized arteries, aneurysm rupture, heart hypertrophy, senile myocardial infarction and arteriolosclerosis, etc.; ③ Diseases of digestive system, such as viral hepatitis, liver cirrhosis, chronic hepatitis and calculous cholecystitis, etc.; ④ Diseases of genitourinary system, such as chronic glomerulonephritis, renal cysts, chronic cystitis and benign prostatic hyperplasia, etc.; ⑤ Bone marrow fatty degeneration and fibrosis, etc.; ⑥ Brain lesions, such as senile cerebral infarction, lipofuscin accumulation and amyloid deposition, etc. These lesions, combined with acute injuries caused by SARS-CoV-2, constitute the pathological basis of multiple organ dysfunction syndrome, with the lung, heart, kidney and liver lesions leading to more severe consequences.

5.3.4 Cause of death

Autopsy cases in China have shown that, the main cause of death from COVID-19 is the multiple organ dysfunction syndrome, especially acute respiratory distress syndrome. Some autopsy cases have died from complications or underlying diseases, including severe lung infections (bacterial, fungal or multiple infections), pulmonary embolism, infarction or hemorrhagic shock, etc. However, in COVID-19 autopsy cases in Europe, thrombosis and its related bleeding, embolism and infarction are found to be relatively common. Some patients in Europe have died of pulmonary or cerebral embolism due to the exfoliation of venous thrombus. According to the autopsy reports from Italy, diffuse intravascular coagulation is the main lesion and the main cause of death from COVID-19 is the formation of venous thrombosis. In Germany, some patients have also been found to die of pulmonary embolism caused by thrombosis. Studies in France and the Netherlands have found that approximately 30% of patients experience disseminated intravascular coagulation. Therefore, they advocate anticoagulant therapy. As suggested in a large-scale autopsy study in Germany, patients' underlying diseases such as cardiovascular diseases, liver or kidney diseases, are significant causes of death, and SARS-CoV-2 infection only induced or promoted the deterioration of the diseases and the pulmonary lesions are not the direct cause of death. In severe COVID-19 patients, up to 85% are reported to suffer from deep vein thrombosis. In conclusion, severe hypoxia or respiratory distress, thrombosis and embolism, bleeding, shock or disseminated intravascular coagulation, as well as secondary opportunistic infections and multiorgan failure are all possible causes of death in patients.

5.3.5 Pathological diagnosis and differential diagnosis

In terms of pathological diagnosis, COVID-19 lacks significant specific lesions, and requires differentiation from other known viral pneumonias such as influenza virus, parainfluenza virus, adenovirus, cytomegalovirus, respiratory syncytial virus, rhinovirus and human metapneumovirus, as well as from nonviral diseases such as mycoplasma pneumoniae pneumonia, chlamydial pneumonia, bacterial pneumonia, etc. In addition, it also needs to be differentiated from noninfectious diseases, such as vasculitis, sarcoidosis and other pulmonary lesions and organizing pneumonia. The key is to detect viral components in situ in the pulmonary lesions. Viral inclusions are important diagnostic clues for viral pneumonia, but there are many viruses that can form inclusions, such as adenovirus, cytomegalovirus, herpes simplex virus, measles virus and respiratory syncytial virus, etc., which still need to be differentiated.

The use of electron microscopy to examine viral particles, immunohistochemistry to detect viral antigens, in-situ hybridization to examine viral nucleic acid and RT-PRC to detect viral gene sequences have been extensively applied in the diagnostic researches for SARS and MERS. In pathological specimens of COVID-19, the coronavirus particles can be observed using the electron microscopy in the cytoplasm of tracheal and bronchial mucosal epithelial cells as well as type II alveolar epithelial cells, and immunohistochemistry staining shows that some tracheal and bronchial mucosal epithelial cells, type II alveolar epithelial cells and infiltrating macrophages expressing ACE2 exhibit a positive reaction to SARS-CoV-2 proteins. SARS-CoV-2 nucleic acid has been detected as positive in the trachea and pulmonary tissues through qRT-PCR-based viral nucleic acid testing. SARS-CoV-2 RNA and viral particles have also been detected in the bronchial and alveolar epithelial cells, pulmonary hilar lymph nodes, spleen, heart, liver, gallbladder, kidneys, stomach, breast, testis, skin, nasopharynx and oral mucosa, which is sufficient to confirm the etiology and also suggests that SARS-CoV-2 may have a wider range of target organs.

Bibliography

[1] Liu J, Zheng X, Tong QX, et al. Overlapping and discrete aspects of the pathology and pathogenesis of the emerging human pathogenic coronaviruses SARS-CoV, MERS-CoV, and 2019-nCoV. J Med Virol 2020;92:491−4.
[2] Yin Y, Wunderink RG. MERS, SARS and other coronaviruses as causes of pneumonia. Respirology 2018;23:130−7.
[3] Wang X, Ding YQ. From SARS to COVID-19: pathogens, receptor, pathogenesis and principles of the treatment. Chin J Pathol 2020;49(6):647−52.
[4] Liu M, Feng RE, Li Q, et al. Comparison of pathological changes and pathogenic mechanisms caused by H1N1 influenza virus, highly pathogenic H5N1 avian influenza virus, SARS-CoV, MERS-CoV and 2019-nCoV. Chin J Pathol 2020;49(5):511−6.
[5] Ding YQ, Wang HJ, Shen H, et al. The clinical pathology of severe acute respiratory syndrome (SARS): a report from China. J Pathol 2003;200(3):282−9.
[6] Ding YQ, Bian XW. Analysis of coronavirus disease-19 (COVID-19) based on SARS autopsy. Chin J Pathol 2020;49(4):291−3.
[7] Liu T. Diagnostic pathology. 3rd ed. Beijing: People's Medical Publishing House; 2014. p. 224−8.
[8] Ding YQ, He L, Zhang QL, et al. Organ distribution of severe acute respiratory syndrome (SARS) associated coronavirus (SARS-CoV) in SARS patients: implications for pathogenesis and virus transmission pathways. J Pathol 2004;203:622−30.
[9] Chen J, Zhang H, Xie YQ, et al. Pathological changes of severe acute respiratory syndrome. Chin J Pathol 2003;32(6):516−20.
[10] Pei F, Zheng J, Gao ZF, et al. Lung pathology and pathogenesis of severe acute respiratory syndrome: a report of autopsies. Chin J Pathol 2005;34(10):656−60.
[11] Franks TJ, Chong PY, Chui P, et al. Lung pathology of severe acute respiratory syndrome (SARS): a study of 8 autopsy cases from Singapore. Hum Pathol 2003;34:743−8.
[12] Tse GMK, To KF, Chan PKS, et al. Pulmonary pathological features in corona- virus associated severe acute respiratory syndrome (SARS). J Clin Pathol 2004;57:260−5.
[13] Hwang DM, Chamberlain DW, Poutanen SM, et al. Pulmonary pathology of severe acute respiratory syndrome in Toronto. Mod Pathol 2005;18:1−10.
[14] Lang Z, Zhang L, Zhang S, et al. Pathological study on severe acute respiratory syndrome. Chin Med J 2003;116:976−80.
[15] Mazzulli T, Farcas GA, Poutanen SM, et al. Severe acute respiratory syndrome-associated coronavirus in lung tissue. Emerg Infect Dis 2004;10:20−4.
[16] Nicholls JM, Poon LLM, Lee KC, et al. Lung pathology of fatal severe acute respiratory syndrome. Lancet 2003;361(9371):1773−8.
[17] Cheung OY, Chan JW, Ng CK, et al. The spectrum of pathological changes in severe acute respiratory syndrome (SARS). Histopathology 2004;45(2):119−24.
[18] Farcas GA, Poutanen SM, Mazzulli T, et al. Fatal severe acute respiratory syndrome is associated with multiorgan involvement by coronavirus. J Infect Dis 2005;191:193−7.
[19] Wang H, Ding Y, Li X, et al. Fatal aspergillosis in a patient with SARS who was treated with corticosteroids. N Engl J Med 2003;349:507−8.
[20] Gu J, Gong EC, Zhang B, et al. Multiple organ infection and the pathogenesis of SARS. J Exp Med 2005;202(3):415−24.
[21] To KF, Tong JH, Chan PK, et al. Tissue and cellular tropism of the coronavirus associated with severe acute respiratory syndrome: an in-situ hybridization study of fatal cases. J Pathol 2004;202:157−63.
[22] Chu H, Zhou J, Wong BH, et al. Middle East respiratory syndrome coronavirus efficiently infects human primary T lymphocytes and activates the extrinsic and intrinsic apoptosis pathways. J Infect Dis 2016;213(6):904−14.

[23] Ng DL, Al Hosani F, Keating MK, et al. Clinicopathologic, immunohistochemical and ultrastructural findings of a fatal case of Middle East respiratory syndrome coronavirus infection in the United Arab Emirates. Am J Pathol 2016;186(3):652–8.
[24] Alsaad KO, Hajeer AH, Balwi M, et al. Histopathology of Middle East respiratory syndrome coronovirus (MERS-CoV) infection-clinico-pathological and ultra- structural study. Histopathology 2018;72(3):516–24.
[25] Walker DH. Value of autopsy emphasized in the case report of a single patient with Middle East respiratory syndrome. Am J Pathol 2016;186(3):507–10.
[26] Xu Z, Shi L, Wang Y, et al. Pathological findings of COVID-19 associated with acute respiratory distress syndrome. Lancet Respir Med 2020;8(4):420–2.
[27] National Health Committee, State Administration of Traditional Chinese Medicine. The COVID-19 diagnosis and treatment program. Infectious Disease Information 2020;33(4):289–96. Trial Eighth Edition.
[28] Bian XW, The COVID-19 Pathology Team. Autopsy of COVID-19 victims in China. Natl Sci Rev 2020;7(9):1414–8.
[29] Huang CL, Wang YM, Li XW, et al. Clinical features of patients infected with 2019 novel coronavirus in Wuhan, China. Lancet 2020;395(10223):497–506.
[30] Liu Q, Wang RS, Qu GQ, et al. Gross examination report of a COVID-19 death autopsy. Chin J Forensic Med 2020;36(1):1–3.
[31] Kuang D, Xu SP, Hu Y, et al. Pathological changes with novel coronavirus infection in lung cancer surgical specimen. Chin J Pathol 2020;49(5):471–3.
[32] Tian S, Hu W, Niu L, et al. Pulmonary pathology of early phase 2019 novel coronavirus (COVID-19) pneumonia in two patients with lung cancer. J Thorac Oncol 2020;15:700–4.
[33] Xu SP, Kuang D, Hu Y, et al. Detection of 2019-nCoV in the pathological paraffin embedded tissue. Chin J Pathol 2020;49(4):354–7.
[34] Wu JH, Li X, Huang B, et al. Pathological changes of fatal coronavirus disease 2019 (COVID-19) in the lungs: report of 10 cases by postmortem needle autopsy. Chin J Pathol 2020;49(6):568–75.
[35] Yao XH, Li TY, He ZC, et al. A pathological report of three COVID-19 cases by minimal invasive autopsies. Chin J Pathol 2020;49(5):411–7.
[36] Xu X, Chang XN, Pan HX, et al. Pathological changes of the spleen in ten patients with coronavirus disease 2019 (COVID-19) by postmortem needle autopsy. Chin J Pathol 2020;49(6):576–82.

CHAPTER 6

Clinical manifestations

Zhao Dawei[1], Yuan Hong[2], He Yulin[3], Lu Hong[4], Ai Li[4] and Chen Haixia[4]
[1]Beijing YouAn Hospital, Capital Medical University, Beijing, China; [2]The Third People's Hospital of Shenzhen, Shenzhen, China; [3]The First Affiliated Hospital of Nanchang University, Nanchang, China; [4]The Central Hospital Affiliated to Chongqing University of Technology, Chongqing, China

6.1 Clinical manifestations of SARS

The incubation period of SARS is 1—10 day(s) (average 5 days) and the fever period usually lasts for 3 days, followed by dry cough, shortness of breath and hypoxemia, with 10%—40% of cases requiring mechanical ventilation support. The most common clinical symptoms and signs are fever, cough, difficulty in breathing, muscle pain, fatigue and moist rales. Common laboratory abnormalities include a mild decrease in white blood cell count, a decrease in lymphocyte count and an increase in Aspartate Aminotransferase, Alanine Transaminase, Lactic Dehydrogenase and Creatine Kinase. The treatment of SARS includes supportive measures and empirical use of ribavirin. Laboratory etiological examination is required for a confirmed diagnosis. Imaging examinations play a crucial role in the early diagnosis and differential diagnosis of this disease and are also utilized to monitor the dynamic changes of the lesions to understand the clinical efficacy.

6.2 Clinical manifestations of MERS

The incubation period of MERS is 2—14 days, typically around 5—6 days. The clinical manifestations of MERS-CoV infection range from asymptomatic or mild respiratory symptoms to sever acute respiratory illness and even death. However, most confirmed cases of MERS-CoV suffer from severe acute respiratory illness, primarily characterized by fever, cough and shortness of breath. The onset is sudden with high fever, and the body temperature can reach 39—40°C. It may also be accompanied by fear of cold,

The Pathology and Imaging of Coronavirus Pneumonia
ISBN 978-0-443-32930-2
https://doi.org/10.1016/B978-0-443-32930-2.00015-5

chills, chest pain, headache, generalized muscle and joint aches, fatigue and loss of appetite, etc. On the basis of pneumonia, the clinical conditions progress rapidly and quickly develop into respiratory failure, acute respiratory distress syndrome or multiple organ failure, especially acute renal failure, and even be life-threatening. Atypical clinical manifestations such as diarrhea may be present in some individual cases (such as patients with immune deficiency). Although the majority of cases present with severe clinical symptoms of both respiratory and gastrointestinal infections, the clinical manifestations are more severe in the elderly, immunocompromised individuals, and chronic patients with cancer, chronic lung diseases and diabetes.

6.3 Clinical manifestations of COVID-19

6.3.1 Main manifestations and course of disease

Based on current epidemiological investigations, the incubation period of this disease ranges from 1 to 14 day(s), with most cases falling between 3 and 7 days. Literature reports suggest that the median incubation period is 3 days. According to the disease progression, it is clinically divided into three stages.

Early Stage (Day 1 to Day 7 during the Disease Course): In most patients, the clinical symptoms mainly include fever, dry cough and fatigue. The fever is typically low or moderate, and a few patients may experience additional symptoms such as nasal congestion, running nose, sore throat, muscle pain and diarrhea, etc. According to the literature reports, the most common symptoms are fever (87.9%) and cough (67.7%), while diarrhea (3.7%) and vomiting (5.0%) are rare.

Progressive Stage: Generally starting from Week 2, some patients experience a significant worsening of fever, cough and fatigue. Severe patients develop difficulty in breathing and/or hypoxemia, and in severe cases, there can be rapid progression to acute respiratory distress syndrome, septic shock, refractory metabolic acidosis, coagulation dysfunction and multiple organ failure; some severe and critical patients may always have low to moderate fever during the course of the disease, or even no obvious fever at all.

Convalescent Stage: Moderate cases enter the convalescent stage from Week 2, with the body temperature returning to normal and cough and fatigue symptoms significantly relieved. Severe cases may enter the convalescent stage around Week 4–6, with a longer duration of symptoms such as fatigue and shortness of breath even after the body temperature returns to normal.

Mild cases may only exhibit low fever and slight fatigue through the course of the disease without any sign of pneumonia. Some patients even have no obvious symptom, but chest CT screening may reveal ground-glass shadows in the lungs.

The prognosis is generally good for most patients, while a small number of patients may experience severe conditions. The elderly and those with underlying chronic diseases tend to have a poor prognosis, while children and adolescents generally exhibit milder symptoms.

6.3.2 Clinical classification and clinical characteristics

(1) Mild: Clinical symptoms are mild, and no signs of pneumonia are observed on imaging.

(2) Moderate: It presents with symptoms such as fever, respiratory issues, and imaging shows signs of pneumonia.

(3) Severe: A case is classified as severe if any of the following is met.

① Shortness of breathing occurs with a respiratory rate (RR) of ≥ 30 breaths per minutes;

② At rest, the oxygen saturation is $\leq 93\%$;

③ Partial Arterial Oxygen Pressure (PaO_2)/Fraction of Inspired Oxygen (FiO_2) ≤ 300 mmHg (1 mmHg = 0.133 kPa).

In high-altitude areas (above 1000 m), PaO_2/FiO_2 shall be corrected according to the formula: PaO_2/FiO_2* [Atmospheric Pressure (mmHg)/ 760].

Patients with significant lesion progression (>50%) within 24−48 h in lung imaging are managed as severe cases.

(4) Critical: A case is classified as critical if any of the following is met.

① Respiratory failure occurs and mechanical ventilation is required;

② Shock occurs;

③ ICU treatment is required with concomitant organ failure.

6.3.3 Clinical manifestations of COVID-19 in children

According to the pediatric case data available, there is a possibility of infection from newborns onwards, with most cases having a close contact history with infected individuals or being part of family clustered cases. Infected children may be asymptomatic or present with fever or cough, a small number of children may experience upper respiratory symptoms, including nasal congestion, running nose, sore throat, expectoration, fatigue, headache and others; besides, gastrointestinal symptoms such as abdominal discomfort, nausea, vomiting, abdominal pain and diarrhea may occur in some infected children. The body temperature is mostly low to moderate fever, the duration lasts 1–2 day(s), with some cases showing no obvious fever. Compared with adults, most infected children show a mild clinical manifestation and generally recover within 1–2 week(s) since the onset of the disease. Very few of them progress to lower respiratory tract infection. Currently, no child death has been reported, but it is not ruled out that, as the epidemic progresses and the number of infected patients increases, there may be reports of severe cases and deaths in children in the later stage. At present, no evidence of mother-to-child transmission has been reported yet.

6.3.4 Complications

(1) ARDS: It is a manifestation of the lung infection progressing to a severe stage and one of the main causes of death in critically patients. Investigations of early cases in some hospital in Wuhan find that, the incidence of Acute Respiratory Distress Syndrome (ARDS) in COVID-19 patients ranges from 17% to 29%, and among ICU patients, the incidence of ARDS is even higher, ranging from 61.1% to 85.0%.

(2) Arrhythmia: Arrhythmia is a common complication in severe and critical patients, especially in those with underlying heart diseases. It is reported that the incidence of arrhythmia in hospitalized patients is 16.7%, and it increases to 44.4% in ICU patients.

(3) Acute myocardial injury can occur in patients without preexisting heart disease, but the incidence may be higher in patients with preexisting heart disease. It is reported that the incidence of acute myocardial injury ranges from 7.2% to 12.0%, while in ICU patients, it is reported to be 22.2%–31.0%.

(4) Shock: Patients with severe COVID-19 may develop secondary septic shock or cardiogenic shock due to heart injury. It is reported that the incidence of shock in hospitalized patients ranges from 4.0% to 8.7%, and it increases to 23.0%–30.6% in ICU patients.

(5) Acute Kidney Injury (AKI): It can be manifested as increased blood urea nitrogen and serum creatinine, decreased glomerular filtration rate, and in severe cases, renal replacement therapy is required. The incidence of AKI reported in the literature is lower than that of ARDS and shock. It is reported that the incidence of AKI in hospitalized patients ranges from 3% to 7%, and it increases to 8.2%–23.0% in ICU patients.

(6) Thrombocytopenia: The incidence of thrombocytopenia and coagulation dysfunction is relatively high in severe and critical patients, and severe thrombocytopenia often indicates a poor prognosis.

(7) Secondary Infection: Hospital-acquired infections can occur in hospitalized patients. The incidence of secondary infection in hospitalized patients is reported to be around 10%, while in ICU patients, it can be as high as 31%. The reported incidence of ventilator-associated pneumonia in hospitalized patients is 1%.

Bibliography

[1] Calza L, Manfredi R, Verucchi G, et al. SARS: una nuova emergenza sanitaria mondiale (SARS: a new emergency in the world health). Recenti Prog Med 2003;94(7–8):284–94.

[2] Kuiken T, Fouchier RA, Schutten M, et al. Newly discovered coronavirus as the primary cause of severe acute respiratory syndrome. Lancet 2003;362(9380):263–70.

[3] Lee N, Hui D, Wu A, et al. A major outbreak of severe acute respiratory syndrome in Hong Kong. N Engl J Med 2003;348(20):1986–94.

[4] Hui DS, Chan MC, Wu AK, et al. Severe acute respiratory syndrome (SARS): epidemiology and clinical features. Postgrad Med J 2004;80(945):373–81.

[5] Bowen L, Puxuan LU, Zeng YJ, el at. Clinical imaging research of the first Middle East respiratory syndrome in China. Radiol Infect Dis 2015;2:173–6.

[6] Zhou B, Tang X, Lu P. Infectious atypical pneumonia. Beijing: People's Medical Publishing House; 2004. p. 128–88.

[7] Lu P, Zhou B. Diagnostic imaging of emerging infectious diseases. Beijing: People's Medical Publishing House; 2013. p. 37–48.

[8] Special Committee on Radiology of Infection and Inflammation of Chinese Research Hospital Association. Committee of the infectious diseases Radiology Group of Chinese Society of Radiology; Committee of the Infectious Diseases Radiology Section of Chinese Medical Doctor Association; Infectious Disease Imaging Group. Guideline for medical imaging in auxiliary diagnosis of coronavirus disease 2019. Chin J Med Imaging Technol 2020;36(3):1–11.

[9] Zheng Q, Lu Y, Lure F, et al. Clinical and radiological features of novel coronavirus pneumonia. J X Ray Sci Technol 2020;28(3):391–404.

[10] Yang Q, Liu Q, Xu HB, et al. Imaging of coronavirus disease 2019: a Chinese expert consensus statement. Europ J Radiol 2020;127:1–6.

[11] Huang CL, Wang YM, Li XW, et al. Clinical features of patients infected with 2019 novel coronavirus in Wuhan, China. Lancet 2020;395:497–506.

[12] Chen NS, Zhou M, Dong X, et al. Epidemiological and clinical characteristics of 99 cases of 2019 novel coronavirus pneumonia in Wuhan, China: a descriptive study. Lancet 2020;395:507–13.

[13] Chen HX, Ai L, Lu H, et al. Clinical and imaging features of COVID-19. Radiol Infect Dis 2020;7(2):43–50.

CHAPTER 7

Laboratory test and other tests

Dang Yan and Lou Jinli
Beijing YouAn Hospital, Capital Medical University, Beijing, China

7.1 Blood routine test

Anticoagulated blood specimen in the acute stage within 7 days after the onset of the disease is collected. The collection volume is 5 mL, and vacuum blood collection tubes containing EDTA anticoagulant are advised for blood collection.

In the early stage of the disease, the total white blood cell count in peripheral blood is normal or reduced, the lymphocyte count is decreased, and the monocyte count is increased or normal. The degree of lymphocyte reduction is related to the severity of the disease. If the absolute value of lymphocytes is less than $0.8 \times 109/L$, or if there is a significant decreased in CD+ and CD+ T cell counts, special attention is needed. It is generally recommended to rechecking the blood routine changes after 3 days. Research has shown that the Neutrophil-to-lymphocyte Ratio (NLR) is an independent risk factor affecting the occurrence of severe cases and has good predictive value. Patients with bacterial infections or critical conditions may experience elevated white blood cell count and neutrophil count. In severe cases, D-dimer and prothrombin time are elevated, the platelet count is reduced and coagulation disorder occurs, and the detection of thrombosis-related molecular markers is helpful for early judgment of DIC in critical patients. When critical patients undergo Extracorporeal Membrane Oxygenation (ECMO), it is necessary to monitor four blood coagulation indices, coagulation factors and anticoagulant enzyme activity to assess the patient's coagulation status.

7.2 Blood biochemistry test

Fasting blood samples are collected for blood biochemistry test. When specimens are centrifuged, the operator can not leave the centrifuge and

The Pathology and Imaging of Coronavirus Pneumonia
ISBN 978-0-443-32930-2
https://doi.org/10.1016/B978-0-443-32930-2.00012-X

shall pay attention to observe if the centrifugation process is normal (e.g., abnormal sounds); after the centrifugation, let the specimen stand at least 10 min; then open the centrifuge lid and disinfect it with 75% ethanol spray before proceeding with the subsequent testing process; if any specimen container is or suspected to be damaged, stop the centrifugation process at once, let it stand for 30 min, carefully open the lid, disinfect it with 75% ethanol spray and then proceed with the cleaning process; do not use a cap remover during the process of centrifugation, and remove the cap of the specimen in a biosafety cabinet.

C-reactive protein, erythrocyte sedimentation rate and ferritin are elevated. In most patients, and some patients may also experience elevated liver enzymes, lactate dehydrogenase, creatine kinase and myoglobin. In some critical patients, myocardial injury can be observed with elevated levels of cardiac troponin. Some patients have varying degrees of renal impairment, accompanied by elevated blood urea nitrogen or serum creatinine. If it progresses to sepsis, acidosis, hyperlactate or hyperlactacidemia may occur. The progressive elevation of lactate acids requires vigilance for worsening of the disease.

7.3 Etiological examination

The medical institution receiving the cases shall collect relevant clinical specimens and promptly send them to the designated local disease control agency, medical institution or third-party testing laboratory for relevant pathogen detection. The collected clinical specimens include samples from the patient's upper respiratory tract (URT), such as throat swabs, nasal swabs, etc., samples from the lower respiratory tract (LRT), such as respiratory aspirates, bronchoalveolar lavage fluid, deep sputum, etc., conjunctival swabs, fecal specimens, anticoagulated blood, and serum specimens, etc. The collection, transportation, storage, and testing of specimens shall be temporarily managed according to the regulations for Biosafety Class II high-pathogenic microorganisms and executed in accordance with the Regulations on Biosafety Management of Pathogenic Microorganism Laboratories, Provisions on the Transport of Strains or Samples of Highly Pathogenic Bacteria (Virus) that can Infect Humans [Order of the Ministry of Health of the People's Republic of China (No. 45)], and other relevant requirements.

Novel coronavirus nucleic acid can be detected in various specimens, including nasopharyngeal swabs, sputum, lower respiratory tract secretions, blood, and feces, etc.

If clinically indicated and testing conditions permit, blood and other specimens can be collected before using antibiotics to perform bacterial culture to determine if there is concurrent bacterial pneumonia or sepsis and identify the pathogen. However, this shall not delay empirical antimicrobial therapy. At the same time, for patients who test positive for other pathogens, screening for novel coronavirus shall also be conducted to avoid misdiagnosis.

Currently, in cases of SARS and MERS, dual infections coexisting with other respiratory viruses have been discovered. Studies have found that approximately 1%–5% of COVID-19 patients also experience concurrent infections with other pathogens. Therefore, detailed microbiological investigations are required for all suspected cases. Both URT and LRT specimens can be tested for other respiratory viruses, such as influenza A virus, influenza B virus (including zoonotic influenza A virus), respiratory syncytial virus, parainfluenza virus, rhinovirus, adenovirus, enterovirus (e.g., EV-D68), human metapneumovirus, and endemic human coronaviruses (such as HKU1, OC43, NL63, and 229E). LRT specimens can also be used to detect bacterial pathogens, including legionella pneumophila and others.

7.4 Serum immunological examination

Serum of acute stage and convalescent stage are collected. The first serum sample shall be collected as early as possible (preferably within 7 days after onset of illness), and the second serum sample shall be collected in the third to fourth week after onset of illness. The collection volume is 5 mL and a vacuum blood collection tube without anticoagulant is advised. Serum specimen can be stored at 4°C for 3 days or frozen at −20°C for a long time. Repeated freezing and thawing shall be avoided during the specimen transportation.

Most patients infected with the novel coronavirus have normal procalcitonin levels. However, when bacterial infections are present, procalcitonin levels may increase.

Viral particles trigger cytokine storms in the body, leading to a series of immune responses. Apart from causing changes in peripheral blood leukocytes and lymphocytes, abnormal clinical levels of cytokines may also occur. The levels of inflammatory cytokines are normal or slightly elevated, but in patients with organ failure, cytokine levels can significantly increase. Researches have shown that the initial levels of Interleukin 1B (IL-1B), interleukin 1RA (IL-1RA), interleukin 7 (IL-7), interleukin 8 (IL-8), interleukin 9 (IL-9), interleukin 10 (IL-10), basic fibroblast growth factor (bFGF), granulocyte colony-stimulating factor (GCSF), granulocyte-macrophage colony-stimulating factor (GMCSF), interferon- γ (IFN- γ), interferon-inducible protein-10 (IP10), monocyte chemoattractant protein (MCP1), macrophage inflammatory protein 1A (MIP1A), macrophage inflammatory protein 1B (MIP1B), platelet derived growth factor (PDGF), tumor necrosis factor α (TNF-α) and vascular endothelial growth factor (VEGF) in patients are higher than those in healthy adults. The levels of interleukin 5 (IL-5), interleukin 12p70 (IL-12p70), interleukin 15 (IL-15), chemotactic factor for eosinophil and reduced upon activation, normal T cell expressed and secreted factor (rantes) are similar to between healthy adults and patients. The plasma levels of interleukin 2 (IL-2), IL-7, IL-10, GCSF, IP10, MCP1, MIP1A and TNF-α in critical patients are higher than those in noncritical patients. Other studies have indicated that levels of interleukin-6 (IL-6) and IL-10 are elevated in critical patients, and cytokines can serve as one of the indicators for predicting the transition from mild to severe cases. Patients with progressive increase of peripheral blood inflammatory factors such as IL-6 and C-reactive protein shall be vigilant for the worsening of the disease. The first pathological autopsy of the deceased patient revealed a significant reduction in the number of peripheral blood $CD4^+$ and $CD8^+$ cells, while their status was over-activated, with a high proportion of double-positive HLA-DR (CD_4 3.47%) and CD_{38} (CD_8 39.4%). Furthermore, the excessive activation of T cells, characterized by an increase in $CCR4^+CCR6^+$ Th17 cells with highly proinflammatory effects in CD_4+ T cells and CD_8+ T cells with high concentrations of cytotoxic granules, can partially explain the severe immune damage in patients. This study also provides references and basis for more ideal prognostic markers in severe cases.

As the course of infection progresses, the human immune system produces antibodies, among which IgM is an early antibody produced after infection, indicating a current or recent infection. On the other hand, IgG suggests a past infection. The novel coronavirus-specific IgM antibodies typically start to appear positive around 3—5 days after the onset of the disease, and the IgG antibody titer increases by 4 times or more during the recovery period compared with the acute stage. By detecting the levels of novel coronavirus IgM antibodies, IgG antibodies, and total antibodies, it is possible to effectively assess whether the target population has been infected with SARS-CoV-2. Serum-specific IgM and IgG antibodies to the novel coronavirus are considered positive. The conversion of IgG antibodies from negative to positive or an increase of 4 times or more in the IgG antibody titer during the recovery period compared with the acute phase can be used as the diagnostic criteria. Novel coronavirus antibodies can serve as a supplement to nucleic acid testing and help reduce the issue of false negatives in nucleic acid testing. However, antibody testing cannot completely replace nucleic acid testing. In some patients with weaker immune function, the antibody levels may be very low, or there may be a delay in antibody production due to low antigen levels at the early stage of infection. Additionally, interference factors in antibody testing methods and the possibility of cross-reactivity with other coronaviruses need further research, all of which can result in missed or false testing. The WHO also specifically points out that serological diagnosis shall only be considered when RT-PCR is not available.

In addition, screening for antigen and antibodies of other respiratory pathogens, such as influenza A virus, influenza B virus, mycoplasma pneumoniae, chlamydia pneumoniae, respiratory syncytial virus, adenovirus, and coxsackievirus, shall be conducted to exclude the possibility of infection by other respiratory pathogens and detect any coinfections.

7.5 Molecular biological examination

Molecular biology testing mainly includes the following two points: ① Real-time fluorescent RT-PCR detection of the novel coronavirus nucleic acid in respiratory or blood specimens, yielding a positive result; ② Viral gene sequencing of respiratory or blood specimens, showing high homology with known novel coronaviruses. Although CT imaging and virus antibody testing can serve as the diagnostic evidence, the coronavirus nucleic acid testing remains the "gold standard" for diagnosis.

In respiratory virus infections, the viral load is generally highest in the nasopharyngeal region during the early stages. However, as the disease progresses, the virus in the upper respiratory tract may be cleared, and the viral load in the lower respiratory tract increases. Therefore, it is advisable to collect specimens from both the upper respiratory tract (URT) and lower respiratory tract (LRT) for testing to reduce the occurrence of false negatives. In situations where obtaining LRT samples is convenient (e.g., in mechanically ventilated patients), clinicians may choose to collect only

LRT samples. The general sampling priority is considered as follows: bronchoalveolar lavage fluid > tracheal aspirate or sputum > nasopharyngeal or oropharyngeal swab > nasal swab. At the same time, the low virus detection rate in upper respiratory tract specimens may also indicate a low viral load in the upper respiratory tract. Patients with low viral load in the upper respiratory tract may have lower infectivity, making it more difficult for the virus to spread.

Specimens used for nucleic acid detection shall be detected as soon as possible, and specimens that can be detected within 24 h can be stored at 4°C; Specimens that cannot be detected within 24 h shall be stored at −70°C or below (if there is no condition for storage at −70°C, they shall be stored in −20°C refrigerator temporarily). Any testing for novel coronaviruses shall be conducted in a laboratory with appropriate conditions by personnel who have received relevant technical safety training. The personnel involved in specimen preprocessing and nucleic acid extraction shall take Level 3 protective measures. Before nucleic acid amplification, the specimens can be predisinfected, including incubating at 56°C for 30 min and adding proteinase K, etc. When handling mucus sputum with pancreatic enzyme or 4% sodium hydroxide, avoid generating aerosols. Maintain a negative pressure environment (if feasible) and promptly dilute and remove aerosols.

7.5.1 Real-time fluorescence RT-PCR detection of novel coronavirus nucleic acid

The routine detection method for novel coronavirus infection is using the Real-time Fluorescence RT-PCR method. When using novel coronavirus nucleic acid testing (real-time fluorescence RT-PCR method), it is recommended to use primers and probes targeting the open reading frame 1ab (ORF1ab) and nucleocapsid protein (N) gene regions in the novel coronavirus genome.

Target 1 (ORF1ab):

Forward Primer (F): CCCTGTGGGTTTTACACTTAA.

Reverse Primer (R): ACGATTGTGCATCAGCTGA.

Fluorescence Probe (P): 5′-FAM-CCGTCTGCGGTATGTGGAAAGGTTATGG-BHQ1-3′

Target 2 (N):

Forward Primer (F): GGGGAACTTCTCCTGCTAGAAT.

Reverse Primer (R): CAGACATTTTGCTCTCAAGCTG.

Fluorescence Probe (P): 5′-FAM-TTGCTGCTGCTTGACAGATT-TAMRA-3′

For nucleic acid extraction and real-time fluorescence RT-PCR reaction system and reaction conditions, refer to the instructions of reagent kit provided by the relevant manufacturer. Negative: No Ct value or Ct $\geq$ 40.

Positive: Ct value $<$ 37, reported as positive.

Gray Zone: Ct value is between 37 and 40 (the specific value may vary depending on the reagent kit from different manufacturers, please refer to the instructions of relevant reagent kit).

Interpretation of Specimen Results:

1. Conditions interpreted as "detected"

Any of the follow is required to be met: ① When both ORF1ab and N genes are positive, it is considered positive; ② If only one of them is positive, the nucleic acid of original specimen shall be extracted for retesting. If either ORF1ab or N gene remains positive after retesting, it is considered as positive.

2. Handling of results interpreted as "suspicious":

Under any of the following circumstances, the result is interpreted as "suspicious": ① When the Ct value of two target sites is between the positive and negative Ct value, or one target site is positive and the Ct value of the other target site is between the positive and negative Ct value (for details, refer to the instructions of relevant reagent kit); ② When one target site is negative, and the Ct value of the other target site is between the positive and negative Ct values. For "suspicious" positive results, it is recommended to perform a re-extraction of the specimen's nucleic acid and simultaneously amplify and test it with the nucleic acid extracted in the previous procedure. Based on the results of the two tests, if both target sites can be confirmed as "positive," then it can be reported as "detected." Otherwise, it shall be reported as "suspicious." When the report is "suspicious," the laboratory shall take the following measures: ① Repeat the experiment using reagents from a different manufacturer or employ a more sensitive method for further confirmation; ② It is recommended that the clinical staff re-collect specimens or obtain samples from a different site for retesting.

3. Handling of results interpreted as "not detected"

When there is no amplification reaction in both target sites, it can be reported as "not detected," and the result shall be explained. This may be a case where the viral load is below the detection limit, and shall be combined with clinical analysis. When there is a high suspicion of SARS-CoV-2 infection based on clinical symptoms and other examinations, it is recommended to recollect specimens or obtain samples from a different site for retesting.

A negative result cannot rule out a novel coronavirus infection, and it is necessary to exclude possible factors that may lead to false negatives, including: ① Poor sample quality, such as respiratory samples from the oropharynx, or improper sampling leading to the viral load in the sample falling outside the detection range of the methodology; ② Sample collection too early or too late in the course of infection; ③ Improper storage, transportation, and handling of samples; ④ Efficiency issues in the process of viral nucleic acid extraction, which can be influenced by individual differences and the expertise level of the testing operator; ⑤ Technical factors such as viral mutations, PCR inhibition, etc.; ⑥ Efficiency issues related to the reagent kit.

Therefore, for patients with a strong suspicion of novel coronavirus infection but initially tested negative, it is advisable to collect samples from multiple respiratory sites several times. Additionally, other types of samples (such as blood, urine, and feces) can be collected to test the presence of the virus in different parts of the body.

Currently, in addition to the real-time fluorescence RT-PCR method, the virus nucleic acid can also be tested through digital PCR methods. Compared with RT-PCR, the digital PCR method features a lower detection limit, which can reduce the issue of false negatives to some extent. It can also enable quantitative detection of viral nucleic acids, making it more meaningful for disease monitoring.

7.5.2 Virus gene sequencing

The first case of the novel coronavirus in China is discovered through pathogen metagenomics. Pathogen metagenomics has certain advantages in detecting virus with high mutation or low load viruses in samples, which is an important complement to RT-PCR in practical work. Metagenomics not only allows for the analysis of viral load and variations in samples from critical patients, but also provides information on bacterial and fungal coinfections, and further analyze drug resistance genes, virulence factors, and host inflammatory responses. However, considering that most hospitals currently lack sequencing equipment, bioinformatics personnel to assist with interpreting sequencing results, established workflows during the development stage, optimized reagent ratios, the establishment of positive and negative reference materials, and initial support for preclinical trials, along with the high cost and relatively long testing period of sequencing, metagenomics is not yet suitable for routine clinical testing.

In addition, screening for other respiratory pathogens (such as adenovirus, parainfluenza virus, respiratory syncytial virus, mycoplasma, chlamydia, influenza A virus, influenza B virus, etc.) is required to exclude the possibility of other viral infections and detect any coinfections. Using multiple testing platforms such as gene chips, multiplex PCR, and gene sequencing can improve the efficiency of testing potential pathogens. Especially for suspected cases, it is advisable to use multiple methods such as multiplex PCR to detect common respiratory pathogens whenever possible.

7.6 Blood gas analysis

Blood gas analysis helps in assessing the patient's oxygenation status, and when combined with elevated lactate levels, it can screen for high-risk patients with oxygenation disorders.

Patients with a resting state peripheral oxygen saturation (SpO2) $\leq$93% or a PaO_2/FiO_2 ratio $\leq$300 mmHg are diagnosed as severe cases and require hospitalization for treatment. If it develops into acute respiratory distress syndrome (ARDS), the PaO_2/FiO_2 ratio can be used for diagnosis and determining its severity.

The oxygenation status (in adults) for ARDS is classified as follows: Mild ARDS: 200 mmHg $<$ PaO_2/FiO_2 $\leq$ 300 mmHg (with PEEP or CPAP $\geq$5 cmH_2O or in nonmechanical ventilation state). Moderate ARDS: 100 mmHg $<$ PaO_2/FiO_2 $\leq$ 200 mmHg (with PEEP $\geq$5 cmH_2O or in nonmechanical ventilation state). Severe ARDS: PaO_2/FiO_2 $\leq$ 100 mmHg (with PEEP $\geq$5 cmH_2O or in nonmechanical ventilation state). When PaO_2 is not available, a PaO_2/FiO_2 $\leq$315 mmHg suggests ARDS, including nonmechanically ventilated patients.

Oxygenation status (in children): BiPAP or CPAP $\geq$5 cmH_2O with PaO_2/FiO_2 $\leq$ 300 mmHg or PaO_2/FiO_2 $\leq$ 264 mmHg. Mild ARDS (invasive ventilation): 4 $\leq$ Oxygenation Index (OI) $<$ 8 or 5 $\leq$ Oxygenation Saturation Index (OSI) calculated from SpO_2 $<$ 7.5; Moderate ARDS (invasive ventilation): 8 $\leq$ OI $<$ 16 or 7.5 $\leq$ OSI $<$12.3; Severe ARDS (invasive ventilation): OI $\geq$ 16 or OSI $\geq$ 12.3.

7.7 Pulmonary function examination

Pulmonary function testing is an important tool for the diagnosis and assessment of respiratory system diseases and is widely used in clinical diagnosis and treatment. During the outbreak of the epidemic, it is essential to strictly adhere to the indications for pulmonary function testing. It is strongly recommended to defer testing for confirmed or suspected COVID-19 cases during their infectious period. For other patients, nonurgent cases shall also be postponed. Pulmonary function laboratory staff shall strictly follow standard graded protective measures. Participants shall be isolated for testing in separate areas. Disposable respiratory filters must be used during the examination, and attention shall be given to the cleanliness and disinfection of the pulmonary function testing environment and equipment.

For suspected or confirmed COVID-19 cases, during the acute phase of isolation and within 14 days after the release from isolation during the recovery period, pulmonary function testing must be conducted in specific negative pressure wards or negative pressure areas. The pulmonary function testing personnel must strictly follow level three protective measures, which include wearing disposable caps, protective face shields (or full-face respirators or positive-pressure headgear), medical protective masks, work clothes, protective clothing, disposable latex gloves, and disposable shoe covers. When masks and gloves are contaminated, they shall be replaced promptly. After each subject is examined, the gloves shall be replaced, and strict hand hygiene disinfection shall be performed. The examination environment and equipment need to be thoroughly cleaned and disinfected. After the 14-day isolation period is lifted during the recovery phase, the examination shall be conducted following standard protective measures for regular patients.

A study involving 110 COVID-19 patients found that 51 cases (47.2%) had abnormal lung diffusion function, 27 cases (25.0%) had abnormal total lung capacity (TLC), 15 cases (13.6%) had abnormal forced expiratory volume in one second (FEV1), 10 cases (9.1%) had abnormal forced vital capacity (FVC), 5 cases (4.5%) had abnormal FEV1/FVC ratio, and 8 cases (7.3%) had abnormal small airway function. There were significant differences in the proportion of abnormal lung diffusion function among different severity subgroups. In the mild subgroup, the proportion was 30.4%, in the moderate subgroup, it was 42.4%, and in the severe subgroup, it was 84.2%. As the disease severity increased, the impairment of lung diffusion function became more pronounced. The TLC (total lung capacity) in severe patients was significantly lower than in mild or moderate patients, indicating that severe pneumonia patients have more severe restrictive lung ventilation dysfunction. The study showed that some patients had varying degrees of lung function impairment, mainly manifested as pulmonary diffusion dysfunction, followed by restrictive ventilatory dysfunction. The extent of impairment was correlated with the severity of the disease during hospitalization.

Bibliography

[1] Zhongnan Hospital of Wuhan University Novel Coronavirus Management and Research Team, Evidence-Based Medicine Chapter. China international exchange and promotive association for medical and Health care. A rapid advice guideline for the diagnosis and treatment of 2019 novel coronavirus (2019-nCoV) infected pneumonia (standard version). Med J Chin Peoples Lib Army 2020;45(1):1–20.

[2] Huang C, Wang Y, Li X, et al. Clinical features of patients infected with 2019 novel coronavirus in Wuhan, China. Lancet 2020;395(10223):497–506.

[3] Wang D, Hu B, Hu C, et al. Clinical characteristics of 138 hospitalized patients with 2019 novel coronavirus–infected pneumonia in Wuhan, China. JAMA 2020;323(11):1061–9.

[4] Chen N, Zhou M, Dong X, et al. Epidemiological and clinical characteristics of 99 cases of 2019 novel coronavirus pneumonia in Wuhan, China: a descriptive study. Lancet 2020;395(10223):507–13.

[5] Guan W, Ni Z, Hu Y, et al. Clinical characteristics of 2019 novel coronavirus infection in China. N Engl J Med 2020;382:1708–20.

[6] Liu J, Liu Y, Pan X, et al. Neutrophil-to-lymphocyte ratio predicts severe illness patients with 2019 novel coronavirus in the early stage. J Transl Med 2020;18(1):206.

[7] Chinese Society of Laboratory Medicine. Biosafety protection guidelines for 2019 novel coronavirus pneumonia in clinical laboratory testing (trial first edition). 2020.

[8] National Health Committee, State Administration of Traditional Chinese Medicine. The COVID-19 diagnosis and treatment program. Trial 8th ed. 2020.

[9] National Health Commission. Diagnosis and treatment protocol for severe and critical cases of novel coronavirus pneumonia. Trial 2nd ed. 2020.

[10] National Health Committee. The prevention and control of 2019 novel coronavirus pneumonia. 4th ed. 2020.

[11] World Health Organization. Laboratory testing for 2019 novel coronavirus (2019-nCoV) in suspected human cases: interim guidance. 2020.

[12] Wang M, Wu Q, Xu W, et al. Clinical diagnosis of 8274 samples with 2019-novel coronavirus in Wuhan. medRxiv 2020. https://doi.org/10.1101/2020.02.12.20022327.

[13] Wan SX, Yi QJ, Fan SB, et al. Characteristics of lymphocyte subsets and cytokines in peripheral blood of 123 hospitalized patients with 2019 novel coronavirus pneumonia (NCP). medRxiv 2020. https://doi.org/10.1101/2020.02.10.20021832.

[14] Xu Z, Shi L, Wang YJ, et al. Pathological findings of COVID-19 associated with acute respiratory distress syndrome. Lancet Respir Med 2020;8(4):420–2.

[15] World Health Organization. Clinical management of severe acute respiratory infection when Novel coronavirus (nCoV) infection is suspected: interim Guidance. 2020.

[16] Chinese Society of Laboratory Medicine. Expert consensus on nucleic acid testing of novel coronavirus pneumonia virus. Chinese Med J 2020;100:E003.
[17] Xi Mo, Qin W, Fu Q, et al. Understanding the influence factors in viral nucleic acid test of 2019 novel coronavirus. Chin J Lab Med 2020;43:E002.
[18] Wang X, Shi J, Ding W, et al. Research status and application discussion on nucleic acid test of 2019 novel coronavirus. Chin J Clin Lab Sci 2020;38(2):81−4.
[19] Yu F, Yan L, Wang N, et al. Quantitative detection and viral load analysis of SARS-CoV-2 in infected patients. Clin Infect Dis 2020;71(15):793−8.
[20] Mo X, Jian W, Su Z, et al. Abnormal pulmonary function in COVID-19 patients at time of hospital discharge. Eur Respir J 2020;55(6). 2001217.

CHAPTER 8

Imaging examination technologies

Liu Jiaping[1], Huang Kuanlong[2], Tang Xiaoli[2], Hou Dailun[1], Fang Wenchun[3], Xu Bingren[2], Zeng Lingzi[2], Zhang Yaming[2] and Zheng Guangping[4]
[1]Beijing Chest Hospital, Capital Medical University, Beijing, China; [2]Shenzhen Qianhai Shekou Free Trade Zone Hospital, Shenzhen, China; [3]Comprehensive Clinic of Shenzhen Healthcare Committee Office, Shenzhen, China; [4]The Third People's Hospital of Shenzhen, Shenzhen, China

8.1 Coronavirus pneumonia imaging scanning scheme as well as CT and DR quality control in operating procedures

8.1.1 Fever clinic or independent DR examination scheme

8.1.1.1 Examination precautions

Install a dedicated Digital Radiography (DR) machine for COVID-19 pneumonia patients. Strictly follow the guidelines for equipment and room disinfection provided by the Radiology Department.

8.1.1.2 Imaging examination scheme

1. Preparation before Examination

After completing the examination for a diagnosed patient, all single-use items used for the previous patient shall be replaced before conducting the examination for the next one.

Patient Preparation: Carefully check the examination consultation form and precautions, clarify the examination purpose and requirements; the patient shall wear a medical mask throughout the examination process; remove necklaces, chest ornaments, and other high-density items (such as bras, clothing with zippers, buttons, and painted surfaces).

2. Examination Method

(1) Photographic Distance: 180 cm.

(2) Grid: Use a grid with a minimum grid ratio of 10:1.

The Pathology and Imaging of Coronavirus Pneumonia
ISBN 978-0-443-32930-2
https://doi.org/10.1016/B978-0-443-32930-2.00014-3

(3) Exposure Conditions: Typically, high-kilovoltage radiography (120 kV) is used with automatic exposure technique.

(4) Photographic Posture: The patient is in a posterior-anterior (PA) position, standing in front of the image receptor with the bulb tube facing back, the anterior chest is firmly pressed against the image receptor, with the hands placing on the iliac crests or hugging the detector; the shoulders are drooped, upper arms internally rotated (scapulae pulled apart), palms facing backward (externally rotated), and the head slightly extended backward with the chin resting on the upper edge of the image receptor; the central ray is aligned with the level of the fifth thoracic vertebra.

(5) Protective Requirements: Use lead aprons or other protective measures to shield other body parts that do not require radiation exposure as much as possible.

(6) Breathing Requirement during Collection: Take a deep breath and hold it at the end of expiration for sample collection.

8.1.2 Bedside photographic examination scheme

8.1.2.1 Examination precautions

A mobile DR (Digital Radiography) unit is designated as a dedicated equipment placed in the isolation ward for emergency use. When conducting bedside X-ray examinations for critical patients, it is essential to ensure that this mobile device is exclusively used for this purpose. The mobile unit shall not be moved outside the isolation ward until the end of the epidemic. After the epidemic is over, the mobile DR unit can only be returned to its original workplace after comprehensive cleaning and disinfection.

8.1.2.2 Imaging examination scheme

1. Preparation before the examination is the same as for DR examination
2. Examination Method

(1) Photographic Distance: 100 cm.

(2) Exposure Conditions: 75–85 kV is recommended if the grid is used; and 65–70 kV is suggested when no grid is used.

(3) Photographic Posture: Place the detector tightly against the patient's back, and align the central line to the center of the vertical line connecting both nipples for the X-ray beam projection. If the patient cannot lie flat, the positioning and angle of the X-ray beam can be adjusted according to the specific situation.

(4) Protective Requirements: Use lead aprons or other protective measures to shield other body parts that do not require radiation exposure as much as possible.
(5) Breathing Requirement during Collection: Take a deep breath and hold it at the end of expiration for sample collection.

8.1.3 CT examination scheme

8.1.3.1 Examination precautions

(1) In the isolation ward equipped with a dedicated CT scanner: The radiology technologist and accompanying nurse shall take Level 2 protection measures; and patients are required to wear disposable medical masks throughout the whole examination.
(2) In the isolation ward without a dedicated CT scanner: A diagnosed patient is examined with a designated CT scanner by a designated personnel at a designated time; Under the coordination of the Infection Control Department, Medical Affairs Department, Radiology Department, and patient wards, a closed-loop process and pathway for transporting, examining, and returning diagnosed patients are established; prior to the examination, in collaboration with hospital security and logistics staff, the area around the designated CT scanner and in the pathway shall be cleared of all unnecessary personnel and other patients; after each examination session, the CT technologist responsible for the setup of the equipment shall spray, wipe, and disinfect the equipment, as well as activate the ultraviolet (UV) light for further disinfection; if the same CT scanner is used for diagnosed and suspected cases in fever clinics, the examination time of both shall be differentiated strictly, with a minimum interval of 1 h, so as to ensure thorough cleaning and disinfection of the CT room and its surrounding environment.

8.1.3.2 Common examination scheme

8.1.3.2.1 Preparation before examination

(1) Staff Preparation: It is recommended to assign two technologists with one for operating the CT equipment and the other for positioning the patients in the CT room and instructing them on breathing and breath-holding techniques. If conditions do not permit, family members or clinical staff accompanying the patient may be asked to provide assistance, but it is essential to ensure that they wear personal protective equipment. During patient reception, it is recommended to use voice communication as much as possible when explaining the examination

instructions. If objective circumstances require direct contact between the technologist and the patient, a distance of at least 1 m shall be maintained; when positioning the patient, the head of the technologist shall be as far away from the patient's respiratory tract as possible. Before and after contact with the patient, proper hand hygiene shall be performed promptly.

(2) Preparation of Auxiliary Equipment: Designate a specific CT machine for suspected or confirmed cases, and preferentially select CT models that can raise and lower the bed on the console. If there is no such CT machine that can raise and lower the bed on the console, an auxiliary equipment with a step-like pedal can be placed in front of the examination bed to assist patients in getting on and off. Select a separate operating room and do not be share with other machines. If it is not possible to meet the above conditions, after the examination, disinfection shall be carried out, including air disinfection in other rooms connected to the operating room. If possible, renovate the CT room to have a negative pressure ventilation system, which means that under special equipment, the air pressure inside the room is lower than that outside the room, allowing only fresh air outside to flow into the room. The air contaminated by the patients inside the room will not leak out, thereby avoiding the risk of infection for healthcare personnel.

Remove the headrest from the examination bed to reduce cross-contamination (avoid having the subject's head fixed in one position and in contact with the inner side of the headrest, which may lead to infection).

Cover the examination bed with disposable sheets. The protective equipment shall include lead caps and rectangular lead aprons, and when used, they shall be separated from the patient's body and clothing by disposable sheets.

(3) Patient Preparation: Carefully check the CT examination consultation form, understand the conditions, and clarify the examination purpose and requirements. The patient shall wear a disposable medical mask throughout the examination process, perform hand hygiene and remove necklaces, chest ornaments, and other high-density items (such as bras, clothing with zippers, buttons, and painted surfaces) before entering the CT room.

8.1.3.2.2 Scanning process

The patient is in a supine position, with the head end entering first, arms raised to hold the head, and the body placed in the center of the bed. The scanning range starts from the apex pulmonis and extends to the base pulmonis. In cases of suspected rib fractures, the scanning range extends to the end of the ribs. Before scanning, the patients are asked to remove any metallic objects within the scanning range, and informed to maintain the required position during the examination until it is completed; at the same time, the training of inspiration-breath holding-exhalation is carried out to ensure the image quality and avoid motion artifacts caused by inadequate breath-holding. Scanning Position Baseline: The X-axis horizontal line of the external positioning line is parallel to the acromion connecting line; the Z-axis line of the external positioning line coincides with the median sagittal plane; the coronal line of the external positioning line is positioned at the level of the mid-axillary line. For obese patients, the midpoint of the maximum anterior-posterior diameter within the scanning field shall be used as a reference point to prevent the loss of surface images of the skin and the potential missed diagnosis of skin and chest wall lesions. For elderly patients and those who cannot hold their breath, the scanning protocol shall be modified to use a larger pitch and shorten the scanning time. When necessary, the technologist may need to capture the breath-holding interval to perform the scan in the caudocranial direction.

8.1.3.2.3 CT scanning parameters (Table 8.1)

Body Mass Index (BMI): Small is <19 kg/m^2, Medium is 19–24 kg/m^2, Large is [1jeemaa1]24 kg/m^2; for weight recommendations, Small is [0jeemaa0]70 kg, Medium is 70–90 kg, Large is [1jeemaa1]90 kg.

Optimized chest scanning parameters include: tube voltage, tube current, pitch, field of view, rotation time, and current modulation technology of automatic tube etc.

(1) Tube Voltage: Generally set at 120 kV, for small-sized patients, it can be set at 100 kV, and for large-sized patients, it can be set at 140 kV. The thoracic anatomical structure is rich, requiring sufficient penetration. Low voltage scanning can increase image noise, making it difficult to identify exudative lesions in the lung.

Table 8.1 Technical parameter setting of chest CT.

Item	Technical parameter
Tube voltage	Body size (BMI or body weight) Small 100 kV Medium 120 kV Large 140 kV
Tube current[a]	Body size (BMI or body weight) Small 100–133 mAs Medium 133–144 mAs Large 144–216 mAs
Pitch	1 (not greater than 1 if there is no specific requirement)
Rotation time	0.4 ~ 0.6 s
Field of view	33 ~ 35 cm
Thickness of reconstructed layer	1 mm

[a]For models with automatic exposure, refer to milliampere-seconds (mAs) or average mAs values. For models without automatic exposure, select a fixed mAs within the clinically appropriate range based on clinical experience.

(2) Tube Current: The normal range for automatic tube current is 133–144 mAs for patients with a regular body size. For small-sized patients, it can be set as 100–133 mAs, and for large-sized patients, it can be set as 144–216 mAs. (Body Mass Index (BMI): Small is <19 kg/m^2, Medium is 19–24 kg/m^2, Large is >24 kg/m^2; for weight recommendations, Small is <70 kg, Medium is 70–90 kg, Large is >90 kg). It is recommended to use an iterative algorithm to improve the image quality.

(3) Pitch: It is recommended to set it to 1. If there are no specific requirements, the value is recommended to not be greater than one to avoid losing small lesions or image details.

(4) Field of View (FOV): It is recommended to select a fixed value (33–35 cm) to improve the accuracy of follow-up diagnosis for COVID-19 pneumonia patients. For obese patients, a unified reconstruction setting of 40–45 cm can be used to ensure that image information is not lost.

(5) Rotation Time: 0.4–0.6 s.

(6) Current Modulation Technology of Automatic Tube: Reduce the patient's radiation dose.

For the low-dose setting of children, refer to the automatic tube current technology for adults. If the hospital's hardware conditions permit, low-dose wide-detector volume scanning can be used on high-end CT machines as an alternative to conventional X-ray examinations.

8.1.3.2.4 CT image reconstruction

Conventional Image Reconstruction: Typically, images are reconstructed with a thickness of 5 mm to create lung window images (lung window: window width 1000–1500 HU, window level −650 to −500 HU, using the lung algorithm) and mediastinal window images (mediastinal window: window width 250–300 HU, window level 30–50 HU, using the standard algorithm). Thin-layer image reconstruction: Typically, images are reconstructed with a slice thickness of 1 mm to create thin-section (high-resolution) lung window images (using the lung algorithm, with a window width of 1000–1500 HU and a window level of −650 to −500 HU).

8.1.3.2.5 Disinfection after examination

According to the above protection requirements, the environment, equipment, and personnel shall be disinfected regularly on a daily basis.

8.1.3.2.6 CT image postprocessing technology

Maximum Intensity Projection (MIP): MIP images can better display the longitudinal axis of blood vessels compared with thin-layer scanning, thus enhancing the stereotactic effect, facilitating the differentiation between blood vessels and nodular lesions and improving the detection rate of nodular lesions. MIP can not only distinguish nodules from adjacent vascular branches, but also accurately display nodules around the lobular center and bronchovascular structures, which is of great significance for precise differential diagnosis in the later stages.

Minimum Intensity Projection (MinIP): Its principle is similar to MIP, except that the air-containing organs (e.g., bronchi, etc.) with significantly low density are highlighted, which has distinct advantages in displaying airway structures with densities close to that of the air. MinIP is beneficial in displaying the pulmonary emphysema and mosaic sign resulting from the alveoli hyperinflation and the destruction of alveolar walls.

CT Virtual Endoscopy (CTVE): A method that reconstructs the structure relationship between the tracheal lumen and the wall based on the volume data obtained from the scan and observe the tracheal wall structure inside using a simulated endoscope method. CTVE can assist in accurately locating the bronchoscope before the procedure, which has greatly improved the diagnostic rate of Transbronchial Needle Aspiration (TBNA).

Volume Rendering (VR) is a visualization technique that displays anatomical structures realistically, providing clear visualization of the shape and spatial relationships of organs. It is primarily used for bronchial imaging, vascular imaging, and displaying complex three-dimensional anatomical relationships.

Multiplanar Reformation (MPR): MPR is an image postprocessing technology that recombines the precalculated pixel data into the section at any angle (coronal, sagittal, etc.) to display cross-sectional images. MPR can display the relationship between lesions and blood vessels, pleura, and chest wall more accurately, aiding in precise lesion localization. MPR can also better demonstrate the positional relationship between lesions and bronchi.

8.1.4 Imaging examination scheme for special population

Infants, young children, adolescents, pregnant women, and women of childbearing age are considered high-risk populations for radiation injury, and radiation exposure shall be used with caution or avoided whenever possible.

8.1.4.1 Imaging examination for children

Considering radiation protection, chest X-ray radiography is recommended as the preferred imaging modality. When detecting lung changes specific to pediatric interstitial pneumonia, this diagnostic indicator shall be considered positive. Chest X-ray radiography has low sensitivity for diagnosing coronavirus pneumonia. For definitive diagnosis, low-dose CT examination may be necessary.

(1) DR Examination: Photographic Distance: 100—110 cm. Girds are not recommended, tube voltage: 55—60 kV, automatic exposure technique (using a middle ion chamber); take a supine anterior-posterior position, with the detector placed tightly against the patient's back, and the central line aligned vertically to the center of the nipple line.

(2) CT Examination: For positioning images, axial scanning parameters are generally selected, and it is recommended to use 80 kV and 25 mA. Transverse scanning is generally performed using helical scanning with low radiation dose and a tube voltage of 100 kV, employing intelligent radiation dose tracking technology. The thickness of the acquisition layer is set 2—5 mm, and the thickness/spacing of the reconstruction layer is set 0.5—1.0 mm. The rotation speed of the tube ranges from 0.27 to 0.8 r/s, and the pitch is set 0.5 to 1.0. Iterative reconstruction technology is used. Radiation protection measures shall be strictly followed according to the radiation protection regulations, shielding other parts of the body, especially those sensitive to radiation such as children's gonads. During the examination of infants and young children, active protection measures shall be taken for accompanying personnel. For patients who cannot cooperate with breath-holding or require sedation, the scan shall be conducted during calm respiration.

8.1.4.2 Chest CT examination for pregnant women

(1) Background: People are generally susceptible to SARS-CoV-2, and as the epidemic spreads, cases of infection in pregnant women have been reported in China. Because pregnant women are in a special state of immune tolerance, especially in the late stages of pregnancy with increased tidal volume and gas filtration in the lungs, the disease can progress rapidly once infected with the virus, and they may evolve into severe or critical cases, endangering both maternal and fetal health. As a special health group, pregnant women are different from others, thus requiring greater attention to infection control and radiation protection during CT examinations. During CT examinations, it is advisable to choose low-dose scanning protocols whenever possible. At work, it is recommended to perform low-dose chest CT examinations on pregnant women with their informed consent, and to take abdominal radiation protection measures for pregnant women.

(2) Scanning Protocol: Take transverse scanning with helical scanning mode enabled, and utilize intelligent radiation dose tracking technology (10–300 mA) and iterative reconstruction technology; 120 kV; adjust the noise index to 15, with the pitch of 1.375:1, detector 0.625 × 64 with collimation, reconstruction layer thickness of 1.25–10.00 mm, iterative reconstruction parameters for lung window of 40%, iterative reconstruction parameters for mediastinal window of 50%, and iterative reconstruction parameters for thin-layer images of 60%. The total dose length product (DLP) controlled by the adjusted CT scanning protocol is between 50 and 150 (mGy cm), equivalent to approximately one-third of the dose used in conventional chest scanning.

Note: Whether X-rays can cause fetal developmental abnormalities depends on the gestational age at the time of exposure and the radiation dose received by the fetus. During early embryonic development, high-dose exposure to radiation (>1 Gy) can be lethal to the embryo. However, the radiation dose from medical diagnostic imaging examinations is much lower than 1 Gy. Currently, there are no reports of causing fetal malformation, growth restriction, or miscarriage with radiation exposure doses below 50 mGy. Radiation exposure at Week 8–15 of pregnancy has the greatest impact on the fetal central nervous system. Some scholars suggest that the minimum radiation threshold for causing intellectual disabilities is between 60 and 310 mGy. However, clinically documented cases of such patients with recorded exposure doses have all been above 610 mGy.

8.1.5 Quality control in CT and DR operating procedures

8.1.5.1 Strictly follow correct operating procedures

1. Registration and Waiting
   - **(1)** Accurately input the patient information at the registration reception and avoid duplicate image numbering; after completing the registration, perform hand disinfection immediately.
   - **(2)** Perform imaging examinations by appointment based on the number of patients (excluding emergencies), and limit the number of patients undergoing examinations to prevent crowded waiting.
   - **(3)** Provide the appropriate level of protection for patients from the fever clinic, and designate the specified route to the dedicated examination facility for the examination.
2. Quality Control in CT Operating Procedures
   - **(1)** Both the radiology technologists and patients shall strictly follow the infection control requirements for COVID-19 before entering the imaging room.
   - **(2)** Close the central air conditioning and switch to independent air conditioner; keep the scanning room and control room clean, and prohibit storing unrelated items inside.
   - **(3)** Explain the precautions to the patient before the examination and provide necessary breathing instructions, aiming for good cooperation.
   - **(4)** Follow the instructions on the examination request to position the patient and perform the scan accurately, adjust the window level and width correctly according to the requirements for high-quality images, and carefully design and complete the scanning plan.
   - **(5)** The radiology technologists shall learn about the imaging manifestations of coronavirus pneumonia, promptly communicate with the diagnostician, and report any issues or findings in a timely manner.
   - **(6)** Closely monitor the patient's conditions during the scan, and in case of any special circumstances or emergencies, immediately

stop the procedure and provide necessary assistance or medical intervention.

(7) Ensure proper ventilation, regular disinfection, and air purification indoors; after scanning suspected or confirmed patients, promptly disinfect the equipment and the scanning room.

3. Quality Control in DR Operating Procedures
   (1) Appropriate Image Density: Includes tissue background density and tissue image density. The standard for human tissue image density varies depending on different body positions and locations.
   (2) Tissue images show clear levels and have good contrast: In the chest, it is possible to distinguish the pulmonary fields from the mediastinum, the pulmonary fields from the chest wall, and the hierarchical structure of the peripheral lung patterns.
   (3) Proper Photographic Posture: The correct photographic position is indicated when the area to be examined is isolated or minimally overlapped with other structures in the image, and even if there is such overlap, its contours can still be clearly distinguished. The structural images observed show no significant distortion.
   (4) No technical operation defects: Clear left and right labeling; numbers do not overlap with the subject being photographed. There is no edge shadow of the shutter or in vitro artifact.

8.1.5.2 Environment management

(1) Environment Management of CT Room: The temperature and humidity of the scanning room and control room shall meet the requirements of the CT machine, with the temperature generally controlled at (22 ± 4)°C and the relative humidity below 65%. Stabilized power supply and uninterrupted power supply (UPS) shall be equipped and proper grounding shall be ensured.

(2) Environment Management of DR Room: Keep the indoor temperature 18–25°C and humidity around 45% during operation; the movement of the tube suspension system shall be slow to prevent damage to mechanical parts; no heavy objects shall be pressed or hit on the chest panel to avoid damaging the detector.

8.1.5.3 Equipment maintenance management

1. CT Machine Maintenance Management.
   (1) CT machine operators shall possess the necessary skills and qualifications for their position.
   (2) Maintenance and servicing of the CT machine shall be carried out by qualified technicians or personnel who have received proper training.

(3) The accuracy of CT values shall be calibrated every week, and regular stability checks shall be performed. If necessary, status inspection shall also be conducted.
(4) If any equipment abnormality is found, do not attempt to handle it blindly. Immediately report it to the department head and logistics support department, and contact the manufacturer for timely repairs.

2. DR Machine Maintenance Management
   (1) Maintain the machine room dry and clean, ensure normal operation during startup and shutdown, and reboot the DR system once every 24 h to ensure that the entire computer is in the specified initial state.
   (2) Regularly calibrate the detector, and perform detector calibration once a week to improve image resolution.
   (3) Strictly follow the operating procedures, and promptly report for maintenance in case of equipment malfunction.

8.1.6 Other imaging examinations

Currently, the screening and diagnosis of coronavirus pneumonia primarily rely on conventional X-ray and CT examinations. Other imaging examinations such as magnetic resonance imaging (MRI) are mainly used as auxiliary diagnostic tools when complications arise during clinical treatment. During the process of corresponding examination, imaging department staff shall strictly adhere to the hospital's infection control regulations to prevent cross-infection between patients and between patients and medical staff. After the examination, they shall rigorously perform the necessary equipment cleaning, disinfection, and air purification. The relevant examination technicians, if necessary, can isolate themselves for 2 weeks according to the incubation period of coronavirus pneumonia before resuming normal work.

8.2 Infection control and management measures for imaging equipment of primary hospitals and mobile cabin hospitals

8.2.1 Primary task of imaging department in primary hospital

The imaging department of primary hospitals mainly faces the screening of patients with coronavirus pneumonia, and hospitals with the necessary conditions shall use dedicated DR/CT examination equipment. For hospitals without dedicated DR/CT equipment, it is necessary to scientifically manage the examination order of patients. High-risk patients, such as those with epidemiological history, from fever clinics, or suspected or confirmed patients in isolation wards, shall be examined in batches at different time intervals. Strict disinfection measures shall be implemented, and emergency management plans shall be established. For hospitals with fewer patients, the principle of "one person, one disinfection" shall prevail.

8.2.2 Mobile cabin hospitals

1. The Site Selection and Environment of Mobile Cabin Hospital

   The mobile cabin hospital primarily serves confirmed cases of mild and moderate patients, ARE often converted from temporary structures without standard equipment rooms or suitable rooms for modification. The mobile advantages of the cabin CT allow it to be deployed as needed. The site selection of CT needs to consider the patient examination process, radiation protection, and infectious disease control. The principles for site selection include: ensuring convenience for patients to reach the examination room while minimizing contamination; maintaining an appropriate distance for radiation protection to reduce harm to the public; facilitating medical staff to don and doff protective equipment, ensuring safety in infectious disease control; and allowing easy arrangement of internal power and network facilities within the equipment room.
2. CT Design and Quality Control in Mobile Cabin Hospitals

   According to the diagnosis and treatment protocol for coronavirus pneumonia and the management regulations for mobile cabin hospitals, the mobile cabin CT is used to perform high-resolution CT scanning of the lungs for mild and moderate patients. The scanning tasks are relatively straightforward and specific, allowing for targeted equipment configuration and the design of scanning protocols. It is recommended to equip with a CT machine with 16 or more detector rows, the detector width shall be $\geq$20 mm, and the actual heat capacity of the tube shall be $\geq$2 MHu. The scanning time for routine whole-lung shall be $\leq$10 s. Low-dose scanning techniques can be used to protect certain patients during short-term re-examination, and iterative reconstruction technology can improve image signal-to-noise ratio while reducing radiation dose. After turning on the CT machine, the first-time quality control test shall be performed, including CT positioning light accuracy, diagnostic bed positioning accuracy, layer thickness deviation, noise, uniformity, CT values, CT value linearity, high contrast resolution, and low contrast detectability, with the high contrast resolution of $\geq$11 LP/cm, noise of <0.45% and the uniformity of being within $\pm$5 HU.

3. Personnel Protection and Disinfection

Radiographer shall wear standard protective equipment at level II or above when entering the cabin. In accordance with the Technical Guidelines for the Prevention and Control of 2019 Novel Coronavirus Infections in Medical Institutions (First Edition) issued by the National Health Commission, entering and exiting the cabin shall follow the prescribed procedures for donning and doffing protective clothing under the supervision of infection control personnel in the hospital. Relevant infection zoning and passage management regulations shall be strictly adhered to, and all CT rooms need to be managed as contaminated areas. Technicians shall try to minimize close contact with patients. They can use voice intercom systems to communicate with patients in separate rooms. If it is necessary to have direct contact with patients, they shall immediately use quick-drying hand disinfectant after the contact before operating the equipment. Technicians on duty shall enhance self-protection awareness, conduct mutual supervision, and strictly adhere to standard protective requirements. In case of accidental exposure, they shall immediately disinfect and replace protective equipment as soon as possible. The air and surface in the CT room shall be disinfected using a 30–40 W medical ultraviolet lamp during equipment suspension period. Ultraviolet lamp can cause the breakage and cross-linking of nucleic acid chains, thereby altering the biological activity of nucleic acids and killing viruses. No personnel shall be present within the range of ultraviolet light exposure. Equipment surfaces can be wiped with 75% ethanol disinfectant or 1000 mg/L chlorine-containing disinfectant.

For floor disinfection, use a mop soaked with 1000 mg/L chlorine-containing disinfectant. Avoid using spray disinfectants or high-concentration Sodium Hypochlorite Disinfectant (84 Disinfectant) for indoor disinfection to prevent damage to electronic components.

8.2.3 Management requirements for imaging diagnosis inspection equipment

1. Installation Phase

During the installation phase of medical imaging equipment, it is essential to complete the installation according to the specified requirements and ensure successful debugging. Regardless of the type of medical imaging equipment, it shall first satisfy the safe and stable operating environment, ensuring that the surrounding temperature, humidity, and power conditions meet the installation requirements. Additionally, it is essential to follow the technical requirements of the installed equipment during the operation to avoid damaging the machine. In addition, it is important to note that after ensuring that the installation conditions and requirements are met, it is necessary to conduct testing and debugging to ensure that all instruments and equipment remain in good condition, thereby guaranteeing their work quality.

2. Use Stage

(1) Establish Procedures and Operating Regulations: The establishment of standardized procedures is an essential prerequisite for ensuring the diagnostic quality of medical imaging equipment. Firstly, before the patient undergoes the examination, it is necessary to repeatedly verify their basic information, examination site, etc., to avoid misdiagnosis due to name repetition, mispronunciation, or incorrect examination site. Secondly, the imaging technologist shall also instruct the patient to prepare for the examination, including removing metallic jewelry, high-density nonmetallic accessories, rubber bands, stamps, patches, etc., and removing excessive clothing, leaving only a single layer of cotton underwear (female patients need to remove bras for chest examinations); for vascular imaging examinations, it is essential to inquire about the patient's history of iodine allergy, liver and kidney functions, fever, etc., and determine whether the patient can safely undergo the examination based on the indications and contraindications. For example, patients with severe liver or kidney diseases, iodine allergy (for contrast-enhanced scans), high fever, etc., do not meet the examination requirements. In addition, before the examination, the staff shall inquire whether the reproductive-age patient is pregnant or planning to conceive. Comprehensive understanding of precision medical imaging equipment by operators is essential for establishing and improving operating procedure, including knowledge of the equipment's working principles, environmental requirements, structure performance, adjustment methods, and technical parameter requirements, etc. First and foremost, operators shall fully grasp the standard requirements for conducting examinations on different body parts using different equipment to ensure imaging quality and provide reliable information for clinical diagnosis and treatment.

(2) Emphasize on Maintenance and Upkeep: Any medical imaging equipment may experience malfunctions, and in many hospitals, measures are often taken to remedy the damage promptly upon detection. However, this situation can potentially increase the risk of medical equipment and even lead to more serious malfunctions. In order to ensure the efficient operation of medical imaging equipment, after evaluating the quality of the manufacturer's after-sales service, the hospital also needs to indicate the requirements for after-sales services, to ensure that during the clinical use phase, apart from ensuring that the examination environment meets the standard, maintenance and upkeep of the equipment shall also be carried out, and adequate preparations for handling any malfunctions shall be made promptly. This requirement is beneficial for the repair work in case of malfunctions, and also helps identifying potential issues in a timely mode and taking preventive measures, thereby ensuring the normal operation of medical imaging equipment. To effectively maintain medical imaging equipment, the staff shall first enhance their awareness of prevention, especially for components that have a high risk of damage and conduct regular replacements so as to avoid severe consequences.

(3) Specific Action Plan: Firstly, develop a detailed schedule for the regular inspection and quality control of medical imaging equipment, emphasizing the importance of daily maintenance and upkeep by the imaging technologists; Secondly, the technologists shall seek technical training and guidance from the manufacturers, tailored to the specific models of different equipment, to gain a deeper understanding of the working principles and familiarize themselves with the quality control manuals to ensure the efficient implementation of quality control work; Lastly, the technologists are strictly required to report the execution of the plan on time, ensuring compliance with relevant regulations and completion of all assigned tasks before proceeding with further services.

(4) Implement Department Collaboration System: The department collaboration system can enhance communications among different departments, thereby effectively improving the management and quality control of medical imaging equipment. The specific operations shall include: Firstly, establish a comprehensive medical imaging equipment archive in the hospital, including details such as equipment name, manufacturer, model, production date, purchase date, parameter configuration, performance evaluation, common malfunctions, and corresponding solutions, etc, which shall be shared among all relevant departments and serve as a reference for future maintenance and upkeep; Secondly, implement a three-level maintenance system, including routine maintenance, first-level maintenance, and second-level maintenance. Routine maintenance focuses on surface cleaning; first-level maintenance involves in cleaning and antioxidation treatment of equipment circuit boards; second-level maintenance emphasizes on the replacement of equipment components and the resolution of major systemic or functional issues. The implementation of the three-level maintenance system is a crucial objective of the department collaboration, and it also serves as the foundation for ensuring the efficiency of medical imaging equipment management and quality control, which shall be given sufficient attention. Through the above analysis, it is evident that the management and quality control of medical imaging equipment are crucial in ensuring the proper functioning of the equipment and providing reliable diagnostic information for clinical diagnosis and treatment. Furthermore, effective management and quality control can also safeguard the safety of both patients and medical staff.

Bibliography

[1] Chinese Society of Imaging Technology. Expert consensus on radiological examination scheme and infection prevention and control of COVID-19. Electron J Emerg Infect Dis 2020;1(5):1–9. 1st ed.
[2] Mingguo S, Wang M, Jianming Y. A clinical guide for radiological technologiest [M]. Beijing: People's Medical Publishing House; 2013. p. 43–5.
[3] Tuberculosis Group, Professional Committee on Infectious Disease Imaging Technology of Chinese Society of Radiology, Professional Imaging Committee of Chinese Medical Association Tuberculosis Branch. Expert consensus on chest CT scan normalization. China Med Equip 2020;35(2):185–9.
[4] Jianming Y, Zeng Y. Medical imaging examination technology. Beijing: People's Medical Publishing House; 2016. p. 139–42.
[5] Hongjun L. Radiology of infectious diseases. Beijing: People's Medical Publishing House; 2014. p. 1–3.
[6] Lu P, Shi Y, Lu Z, et al. The development and achievements of radiology of infectious diseases in China. Electron J Emerg Infect Dis 2020;5(4):223–8.
[7] Lv Q, Mei S, Zhang R, et al. Construction and effect evaluation on precise prevention and control system of COVID-19. Electron JEmerg Infect Dis 2020;5(4):229–34.
[8] Wu F, Xingxing M, Caiyun W, et al. Prevention and control of COVID-19 epidemic and construction on long-term mechanism of occupational health in infectious disease hospital. Electron J Emerg Infect Dis 2020;5(3):198–202.

CHAPTER 9

Imaging manifestations of SARS

Zhao Dawei[1], Li Li[1], Yuan Hong[2] and Guo Jiachen[1]
[1]Beijing YouAn Hospital, Capital Medical University, Beijing, China; [2]The Third People's Hospital of Shenzhen, Shenzhen, China

9.1 Imaging manifestations in early stage

9.1.1 Imaging manifestations

9.1.1.1 X-ray

Single or multiple lesions featuring small patchy shadows and low density, mostly seen as single lesion, and sometimes accompanied by local increased and thickened lung markings. The X-ray has a lower detection rate for smaller or low-density lesions, and it can sometimes be challenging to visualize lesions that overlap with the mediastinum and cardiac silhouette.

9.1.1.2 CT

Small patchy ground-glass opacities, mostly seen as single lesion, accounting for 82.1%. Mostly round, sometimes resembling the anatomy of pulmonary lobules, with a conical shape. Traversing shadows of blood vessels with lightly higher density can be observed, along with a reticular pattern of thickened interlobular septa. Localized small patchy, multiple, or larger consolidations can also be observed. Pulmonary lobular lesions can merge to involve a segment of the lung, and they may present as ground-glass opacities with consolidations, but this is less common. Some lesions show increased and thickened vascular shadows around them, and the lesions are mostly located in the lower lobes of both lungs and the subpleural regions.

9.1.2 Case description

Case 1: Patient, female, 31 years old. Diagnosed with SARS, the imaging manifestations on the first day of the disease course are shown in Fig. 9.1.

Case 2: Patient, female, 38 years old. Diagnosed with SARS. See Fig. 9.2 for the imaging manifestations on the second day of the disease course.

Case 3: Patient, male, 49 years old. Diagnosed with SARS. See Fig. 9.3 for the imaging manifestations on the third day of the disease course.

The Pathology and Imaging of Coronavirus Pneumonia
ISBN 978-0-443-32930-2
https://doi.org/10.1016/B978-0-443-32930-2.00008-8

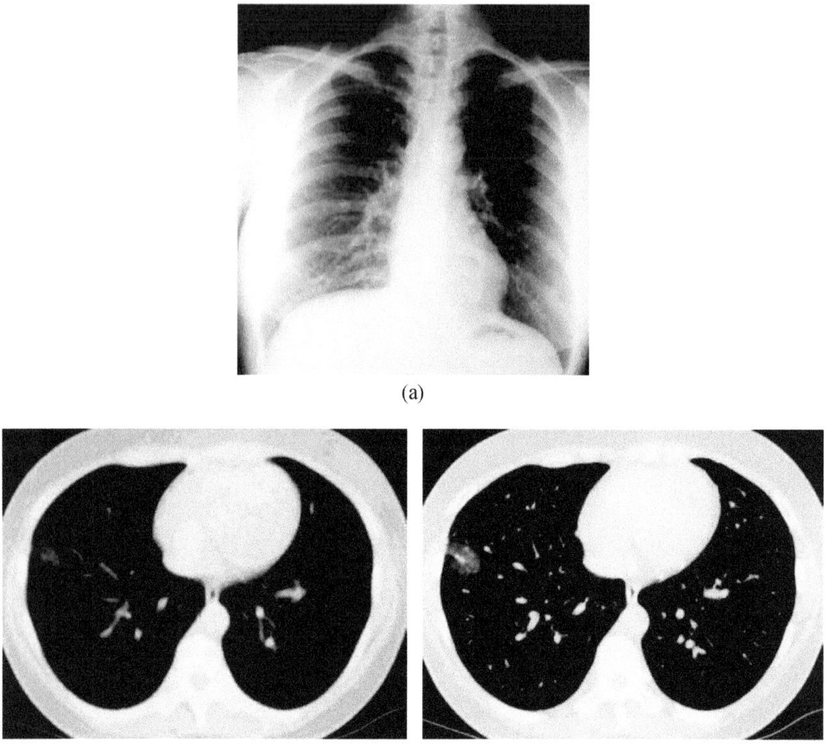

(a)

(b) (c)

Figure 9.1 Early imaging manifestations of SARS (I). (a) No abnormalities observed on chest X-ray; (b) CT shows ground-glass opacities beneath the pleura in the right lower lobe; (c) HRCT shows ground-glass opacities beneath the pleura in the right lower lobe, with slightly higher density traversing shadows of blood vessels visible within the lesion.

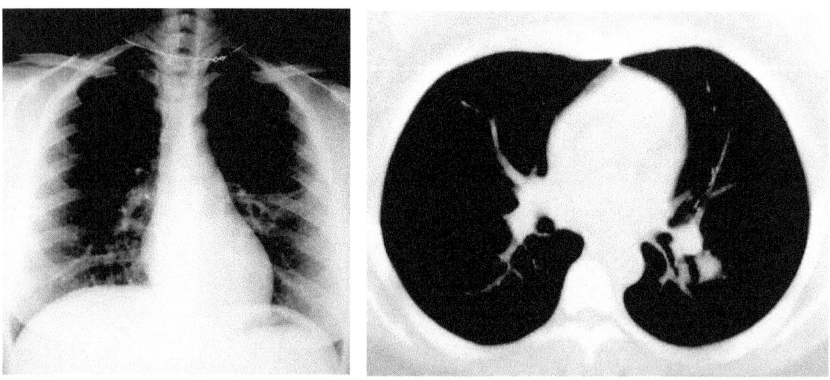

(a) (b)

Figure 9.2 Early imaging manifestations of SARS (II). (a) Chest X-ray shows small patchy shadows at the left hilum; (b) CT showns small patchy ground-glass opacities in the left lower lobe.

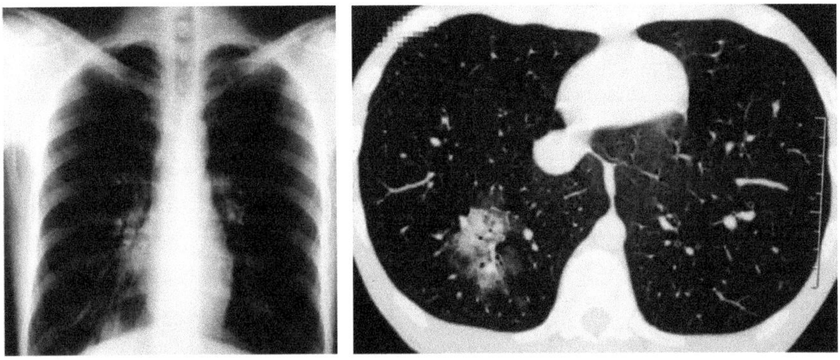

(a) (b)

Figure 9.3 Early imaging manifestations of SARS (III). (a) Chest X-ray shows subpleural ground-glass opacities in the right lower lobe; (b) HRCT shows the merging of small lung lobules into large patchy ground-glass opacities with consolidation in the right lower lobe.

9.1.3 Diagnostic key points

(1) In the early stage, the lesions present as subpleural single or multiple small patchy shadows with low density, mostly seen as single lesion, and sometimes accompanied by local increased and thickened lung markings.

(2) In the early stage, small patchy ground-glass opacities may show thickened vascular shadows with slightly higher density. Larger lesions can involve a segment of the lung, but they are less common. Ground-glass opacities with consolidations may be observed at the level of pulmonary lobules. Some lesions show increased vascular shadows around them. The lesions are mostly located in the subpleural regions of the lower lobes of both lungs and the subpleural regions of the lung periphery.

9.2 Imaging manifestations in progressive stage

9.2.1 Imaging manifestations at progressive stage

The majority of patients experience a rapid progression and worsening of the lesions within 14 days after the onset of the disease. In the early stages, small patchy shadows can evolve and merge into larger, multiple, or diffuse lesions within 3–7 days. The lesions progress from one lung to involve both lungs and extend from a single lung segment to involve multiple segments merging together. In severe cases, significant progression can occur within 1–2 days after the onset of the disease.

9.2.1.1 X-ray

The main manifestations are large patchy and widespread ground-glass opacities and consolidations, and can also show clumpy shadows. Ground-glass opacity refers to an area of decreased lung translucency with indistinct boundary. Lung consolidation is an area of increased density with either blurred or relatively clear borders. The lesions are mostly distributed diffusely in both lungs, but they can also be located in one side of the lungs, with a higher prevalence in the lower lung fields. The course of the disease is relatively long and may lead to pneumothorax, mediastinal emphysema, and subcutaneous emphysema. The amount of pneumothorax is generally small. In some cases, pneumothorax occurs after using a ventilator, but it can also happen in patients who have not been on a ventilator. Lesions adjacent to the pleura may be accompanied by localized pleural thickening or mild filmy adhesions. Pleural changes may regress with the absorption of the pulmonary lesions.

9.2.1.2 CT

The CT examination mainly shows ground-glass opacities along with areas of lung consolidation.

(1) Ground-glass Opacity: The lesion morphology can be patchy, resembling the shape of a pulmonary lobe or segment, or rounded opacities in different sizes. Lesions with a diameter of 3 cm or larger account for more than 50%. The lesions are often multiple and have a diffuse distribution. Various forms of lesions can coexist, with rounded ground-glass opacities being more common. In some cases, from the onset of the disease until the absorption of the lesions, the manifestations are solely ground-glass opacities. The density of ground-glass opacities can be uneven. Within the areas of lower density ground-glass opacities, fine pulmonary vascular branches may be visible. Some ground-glass opacities may show thickened interlobular septa and intralobular interstitium, with fine linear shadows and reticular structures beneath the pleura.

(2) Ground-glass Opacity with Lung Consolidation: Within large patchy, small patchy, or rounded ground-glass opacities, there may be areas of higher density representing lung consolidation.

(3) Lesion with a Predominance of Lung Consolidation: Lung consolidation appears as patchy high-density shadows or consolidations involving entire lobes or segments of the lung. Generally, they are multiple, but occasionally, there can be a single lobe consolidation. Lung consolidation is usually accompanied by varying degrees of ground-glass opacity, and cases of lung consolidation without ground-glass opacities are rare. According to the literature, 69.8% of patients show significant involvement of the lower lobes of both lungs. In 60.1% of patients, the lesions are distributed in a mixed pattern involving both the inner and outer regions of the lungs. Only 4.3% of patients exhibit a central distribution pattern.

(4) After discharge, chest CT follow-up may show varying degrees of pulmonary interstitial fibrotic changes in both lungs. Severe cases may suffer from residual lung fibrosis.

9.2.2 Case description

Case 1: The patient is diagnosed with SARS. See Fig. 9.4 for the imaging manifestations on the seventh day of the disease course.

Case 2: The patient is diagnosed with SARS. The imaging manifestations are shown in Fig. 9.5.

Case 3: The patient is diagnosed with SARS. The imaging manifestations are shown in Fig. 9.6.

9.2.3 Diagnostic key points

(1) Large patchy consolidations or widespread ground-glass opacities with consolidations are the main imaging manifestations during the

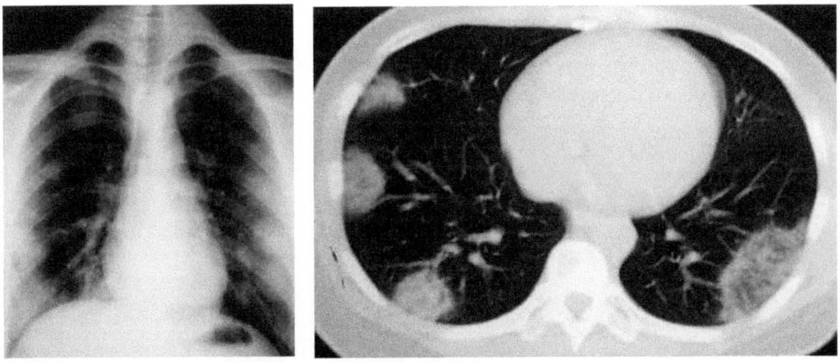

(a) (b)

Figure 9.4 Imaging manifestations of SARS at the progressive stage (I). (a) Chest X-ray shows bilateral subpleural large patchy ground-glass opacities; (b) CT shows bilateral subpleural large patchy ground-glass opacities, with slightly higher density lung consolidation.

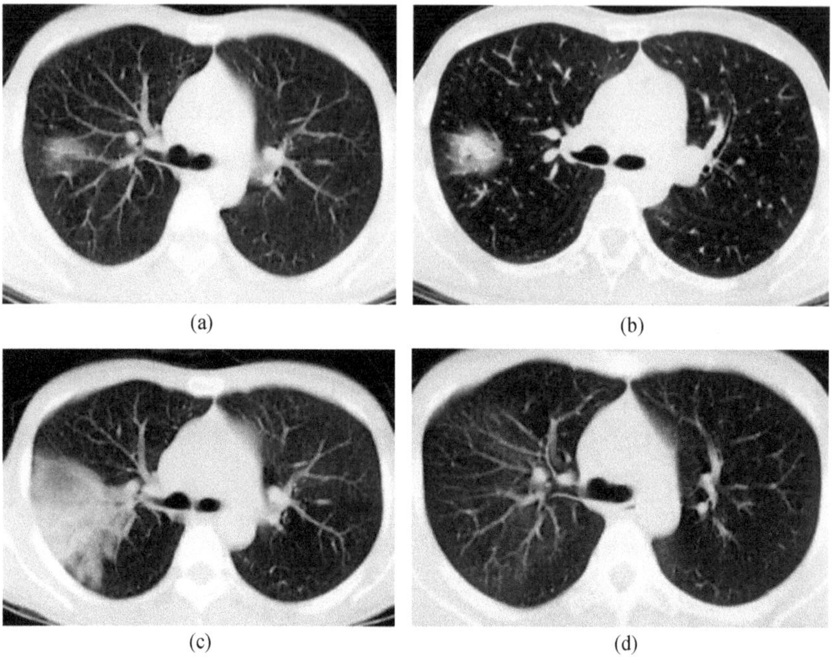

Figure 9.5 Imaging manifestations of SARS at the progressive stage (II). (a) On the second day of the disease course, CT shows initial faint small patchy ground-glass opacities in the right lung; (b) on the fifth day of the disease course, CT shows consolidations; (c) on the 11th day of the disease course, the area of consolidation expands significantly, and air bronchograms are visible in it; (d) on the 16th day of the disease course, there are residual few fibrous stripes in the lungs, and the lesions are mostly absorbed.

progressive stage of SARS, both of which can occur separately or coexist simultaneously.

(2) Air bronchogram sign and "paving stone" sign are common features during the progressive stage of SARS.

9.3 Imaging manifestations in convalescence stage

9.3.1 Imaging manifestations

Through continuous observation using HRCT scanning, the pulmonary lesions of SARS reach their peak at the fourth week. After the fourth week, 55% of patients show signs of early fibrosis, including reticular shadows with or without associated ground-glass opacities. Occasionally, traction bronchiectasis may be observed, with a higher prevalence in older and more

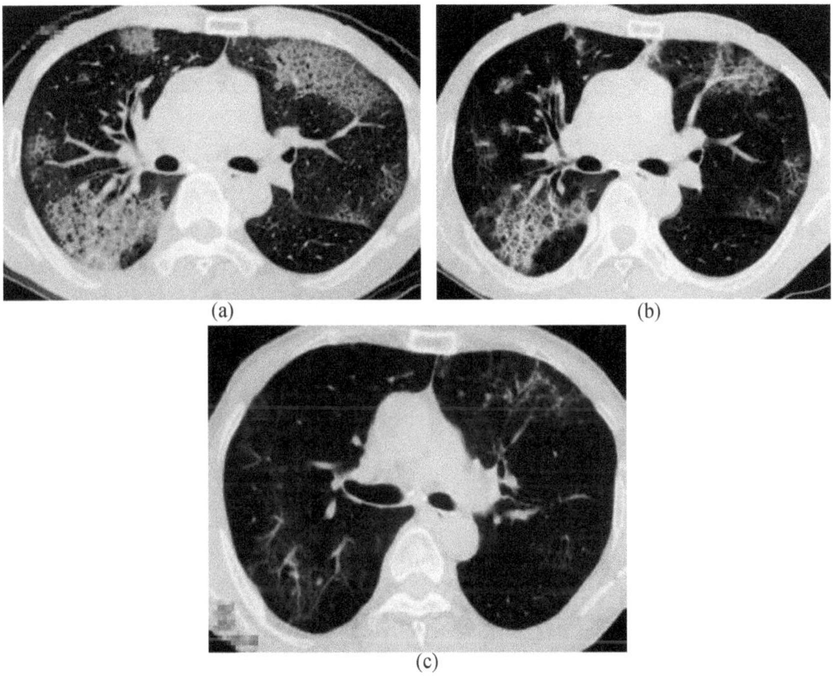

Figure 9.6 Imaging manifestations of SARS at the progressive stage (III). (a) on the seventh day of the disease course, CT shows multiple ground-glass opacities in both lungs, presenting a change in "paving stone" pattern; (b) on the 16th day of the disease course, CT shows that the ground-glass opacities in both lungs are partially absorbed than before; (c) on the 25th day of the disease course, CT shows significant absorption of ground-glass opacities in both lungs, with local residual fibrous stripes visible.

severe patients (62%). During the follow-up HRCT of patients discharged for 2 months, 49.0% of SARS patients at the convalescence stage show abnormal CT manifestations, including 95.9% with ground-glass opacities and 59.2% with reticular shadows. The follow-up of discharged patients shall involve a comprehensive assessment of overall functions and health status, as there may be varying degrees of impairment, which is important to avoid the occurrence of severe sequelae.

9.3.1.1 Dynamic evolution of SARS imaging

The dynamic changes in imaging are related to various factors, such as age, preexisting diseases, and treatment effectiveness. The absorption of lesions generally occurs in 7–14 days after the onset of the disease, characterized by a reduction in the size of the lesion shadows and a gradual decrease in density. Especially in cases with significant response to steroid treatment, there can be a noticeable improvement in lung imaging as early as 1 day after the treatment. During the absorption process of the lesions, some patients may develop new lesion shadows in other areas. Due to the influence of various individual factors, some patients may have a prolonged duration of lesion imaging, and their course of disease may be twice as long as that of ordinary patients, or even longer. Patients with worsening conditions show an increase in the range of the aforementioned imaging manifestations and the appearance of new lesion shadows.

Acute Respiratory Distress Syndrome (ARDS) is the leading cause of death in adult patients. It manifests as diffuse lung alveolar consolidation and high-density opacities within the lungs. Preliminary observations indicate that deceased cases often show widespread lung consolidations during the progression of the lesions.

9.3.1.2 Pathological basis of imaging manifestations

The autopsy pathology reveals that the lesions are mainly characterized by severe lung tissue damage and ARDS, with histological changes in both pulmonary interstitium and alveoli. The pulmonary interstitial lesions include thickened interstitium, edema, and infiltration of mononuclear and multinucleated cells. The air cavity changes include extensive consolidation of lung tissues, with evidence of type II alveolar epithelial cell degeneration and necrosis, alveolar edema, focal hemorrhage, and formation of pulmonary hyaline membranes. In the early stage of the pathological process, the alveoli are filled with mucus-like fibrinous exudate.

The pulmonary interstitial lesions form a reticular ground-glass opacity on X-ray and CT imaging. The pulmonary air cavities are not completely filled, and there are still air-containing spaces, or a mixture of filled and unfilled alveoli, leading to the formation of ground-glass opacities. On HRCT, the degree of residual pulmonary alveoli varies, and the density of ground-glass opacities also differs. When the alveolar spaces are completely filled, it results in lung consolidation on imaging.

Patients can also exhibit rapid and transient changes resembling edema, possibly due to the body's immune response, with more prominent pulmonary edema changes.

9.3.1.3 Imaging manifestations of complications

(1) Pulmonary Interstitial Proliferation: More significant pulmonary interstitial proliferation appears around 30–40 days after the onset of the disease, initially showing thickened interlobular septa, intralobular interstitium, and subpleural curvilinear opacities on imaging. The patchy shadows in the lungs decrease in volume and increase in density. Gradually, high-density stripe opacities and honeycomb-like images appear within the lungs. In severe cases, the lung volume decreases, and there may be mediastinal shift. Pulmonary interstitial proliferation can be widespread within the lungs, with thickened interlobular septa, intralobular interstitium, and subpleural curvilinear opacities as the main manifestations. It can also present as irregular high-density patches and stripe opacities in some localized areas. The presence of honeycomb-like opacities indicates pulmonary interstitial fibrosis.

(2) Infection: Complicated infections can lead to the formation of cavities and pleural effusion. Typically, this occurs after 3 weeks from the onset of the disease.

(3) Pneumothorax, mediastinal emphysema, and subcutaneous emphysema.

(4) Bone ischemic necrosis.

9.3.2 Case description

Case 1: On the 62nd day of the disease course, the imaging manifestations of SARS with pulmonary interstitial complications, as shown in Fig. 9.7.

Case 2: Imaging Manifestations of SARS Complicated with Staphylococcus Aureus Infection, as shown in Fig. 9.8.

Case 3: Imaging Manifestations of SARS Complicated with Mediastinal Emphysema, as shown in Fig. 9.9.

Case 4: Imaging Manifestations of SARS Complicated with Bone Ischemic Necrosis, as shown in Fig. 9.10.

9.3.3 Diagnostic key points

(1) When clinical respiratory symptoms worsen continuously, and imaging shows nontypical SARS manifestations, one shall be vigilant about the occurrence of complications.

(2) SARS-related skeletal complications and sequelae are not uncommon and shall be taken seriously.

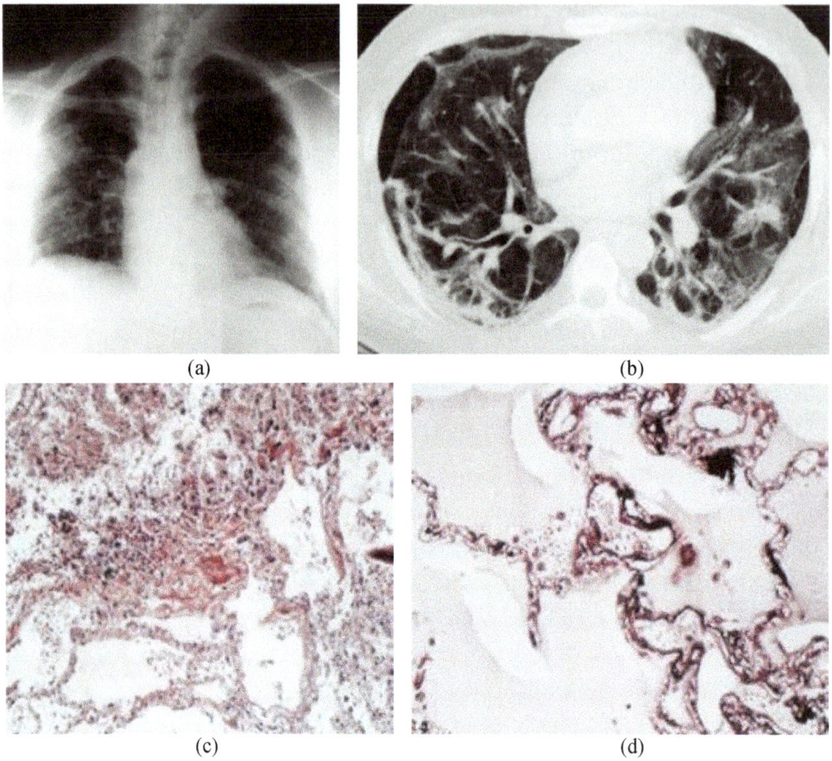

Figure 9.7 Imaging manifestations of SARS with pulmonary interstitial complications. (a) Chest X-ray shows increased lung markings and fibrous stripe opacities in both lungs; (b) CT shows residual subpleural curvilinear opacities, fibrous stripes, and ground-glass opacities in both lungs; (c) histopathology demonstrates reticular fibrous proliferation in the alveolar capillary walls, shows to irregular thickening of the capillary walls (reticular fibers ×200); (d) histopathology shows collagen fiber proliferation in the alveolar septa, appearing as irregular rose-red structures (PTAH ×200).

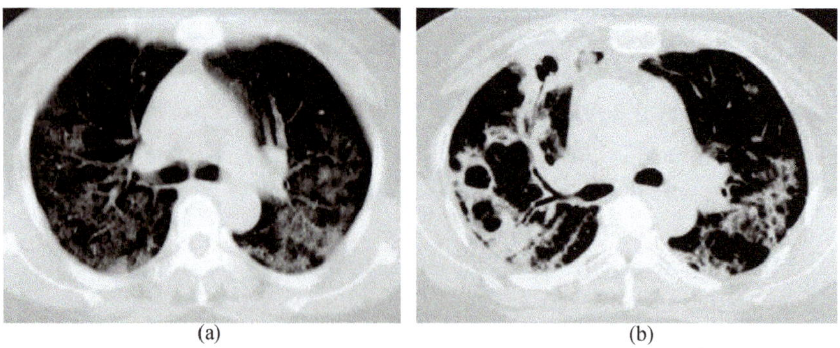

Figure 9.8 Imaging manifestations of SARS complicated with Staphylococcus aureus infection. (a) On the 16th day of the disease course, CT scan shows diffuse ground-glass opacities in both lungs; (b) on the 25th day of the disease course, CT shows consolidation in the right lung and the formation of multiple cavities.

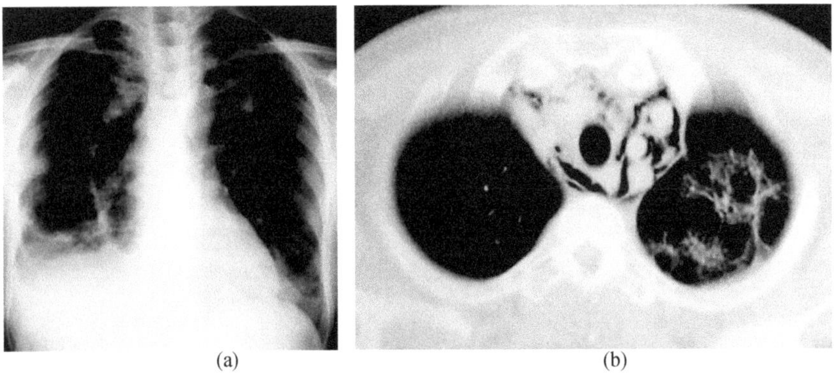

Figure 9.9 Imaging manifestations of SARS complicated with mediastinal emphysema. (a) Chest X-ray shows irregular lucency in the mediastinum; (b) CT shows multiple gas-density shadows in the mediastinum, multiple patchy shadows in the left upper lobe, and fibrous stripes.

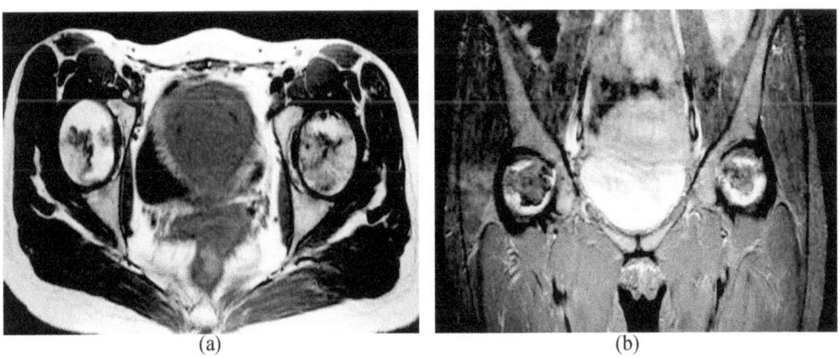

Figure 9.10 Imaging manifestations of SARS complicated with bone ischemic necrosis. (a) MRI T2WI sequence shows uneven signals of bilateral femoral heads, with irregular patchy and linear low signals within the high signal ring. The femoral heads do not show deformation; (b) hip joint coronal T1Wi fat-suppressed sequence shows a linear low signal band in the femoral head and an uneven low signal area below the femoral articular cartilage.

Bibliography

[1] Dawei Z, Ma D, Wang W, et al. Early X-ray and CT manifestations of SARS. Chin J Radiol 2003;37(7):597–9.

[2] Wang W, Daqing M, Dawei Z, et al. CT appearances and dynamic changes in severe acute respiratory syndrome. Chin J Radiol 2003;37(8):686–9.

[3] Wang Z, Lu P, Zhou B, et al. Chest X-ray manifestations of severe acute respiratory syndrome. J Clin Radiol 2003;22(6):449–51.

[4] Liu J, Jiang S, Chen B, et al. Image appearances of severe acute respiratory syndrome (A preliminary study of 260 cases). Chin J Med Imaging Technol 2003;19(7):790–2.

[5] Xie LX, Liu YN, Xiao YY, et al. Follow-up study on pulmonary function and lung radiographic changes in rehabilitating severe acute respiratory syndrome patients after discharge. Chest 2005;127(6):2119–24.
[6] Lu P, Yang G, Weiye Y, et al. Imaging follow-up of SARS patients complicated with pulmonary fibrosis. Chin J Med Imaging Technol 2004;20(12):1901–3.
[7] Weiye Y, Lu P, Zhu W, et al. Imaging follow-up observations of 34 SARS patients. Chinese Imag J Integrat Tradit West Med 2005;3(3):175–7.
[8] Zhenwei L, Zhang L, Zhang S, et al. An autopsy pathological analysis on 3 cases of severe acute respiratory syndrome. Chin J Pathol 2003;32(3):201–4.

CHAPTER 10

Imaging manifestations of MERS

He Yulin[1], Lu Puxuan[2], Liu Jun[3] and Zheng Qiuting[2]
[1]The First Affiliated Hospital of Nanchang University, Nanchang, China; [2]Shenzhen Center for Chronic Disease Control, Shenzhen, China; [3]The Second Xiangya Hospital of Central South University, Changsha, China

10.1 Overview of MERS

As of the end of April 2020, a total of 2519 laboratory-confirmed cases of MERS (Middle East Respiratory Syndrome) have been reported globally, with 866 deaths, resulting in a fatality rate of 34.3%. Among the confirmed cases, Saudi Arabia reported 2121 cases, accounting for approximately 84.2% of all confirmed ones, with 788 deaths and a fatality rate of 37.1%. Outbreaks occurring in countries outside the Middle East are relatively rare, and patients usually have a travel history to the Middle East region. There were no significant changes in the demographic and epidemiological characteristics of the reported cases when comparing the same period from 2013 to 2018. The age group of 50—59 years old remains the highest-risk population for primary infection of the disease and has the highest number of fatalities. The age group of 30—39 years old is most susceptible to becoming secondary cases, while the age group of 70—79 years old has the highest number of fatalities among secondary cases. Since 2015, the number of hospital-acquired MERS (Middle East Respiratory Syndrome) cases has significantly decreased due to the improvement of infection prevention and control measures in hospitals. In 2015, there was one imported case of MERS (Middle East Respiratory Syndrome) in China.

Bats may be the origin of MERS-CoV (Middle East Respiratory Syndrome Coronavirus), but as research has progressed, it has been discovered that the main host of the pathogen is primarily dromedary camels, and the transmission is often limited to sporadic human-to-human cases. Human infection with MERS-CoV typically occurs through inhalation of respiratory droplets from infected dromedary camels or through frequent and close contact with infected dromedary camels and patients. In

The Pathology and Imaging of Coronavirus Pneumonia
ISBN 978-0-443-32930-2
https://doi.org/10.1016/B978-0-443-32930-2.00016-7

addition, MERS infection is not limited to adults and can also affect children. Most cases in children are mild or asymptomatic, but severe cases and fatalities can occur in children with underlying health conditions. There have been reports of pregnant women infected with MERS-CoV, but it is currently not clear whether the virus can be transmitted through the placenta.

10.2 Imaging manifestations and diagnosis of MERS

10.2.1 Imaging manifestations of MERS

Imaging examination is a crucial component of the comprehensive clinical diagnosis of MERS (Middle East Respiratory Syndrome) and also serve as important guidelines for treatment. This includes early detection of the disease, differential diagnosis, monitoring dynamic changes, and detecting complications. Taking full advantage of imaging diagnosis, X-ray chest radiography and CT examinations are the main methods of for detecting MERS. Conventional X-ray examination typically involves posteroanterior (PA) chest radiographs in the standing position. Bedside chest radiography, when feasible in the patient's conditions, shall be performed in the seated position with posteroanterior chest radiographs. Digital imaging technologies such as computed radiography (CR) and digital radiography (DR) can enhance the diagnostic quality of chest X-ray examinations. CT can detect lesions that are difficult to identify on conventional chest X-rays. Generally, high-resolution CT (HRCT) examinations shall be used for better resolution and detailed evaluation. The establishment of imaging workflow based on Picture Archiving and Communication System (PACS) can improve work efficiency and reduce cross-infection. Imaging Examination Procedure: ① Initial examination: For patients clinically suspected of having MERS, X-ray chest imaging shall be the first choice. If no abnormalities are found on the X-ray, prompt follow-up shall be conducted, and if possible, CT examination can be performed. ② Treatment Follow-up: During the treatment of MERS, follow-up chest X-ray examinations are needed to monitor the changes in the disease and the effectiveness of the treatment. Generally, a follow-up chest X-ray examination is conducted once every 1−2 days, or the timing of follow-up may be adjusted based on the patient's disease progression and treatment response. If the chest X-ray suggests lung lesions, patients with access to resources can undergo CT examination. ③ Discharge Examination: A chest X-ray is required before discharge. Regular follow-up examinations shall be conducted after discharge until the inflammatory imaging completely disappears. For cases in which the chest X-ray has returned to normal, CT can reveal lesions that cannot be detected by X-ray. The imaging changes in MERS progress rapidly and lack significant specificity, but they play a clear role in diagnosing MERS, monitoring disease progression, and predicting prognosis.

In MERS patients, X-ray chest films consistently show abnormal findings, which can range from small to extensive lesions, involving one or both lung fields. Common radiographic manifestations include increased lung markings, pulmonary ground-glass opacities, patchy infiltrative changes, interstitial changes, patchy or confluent consolidations, nodular opacities, reticular opacities, and pleural effusion, etc. Some patients may have no abnormal chest X-ray findings, which could be due to the early stage of infection when chest lesions are not prominent, and the limited sensitivity of chest X-rays in detecting small ground-glass opacities. See Fig. 10.1.

Bilateral subpleural ground-glass density changes are characteristic CT manifestations in MERS. However, this feature can also be observed in pneumonia caused by chlamydiae, mycoplasma, severe acute respiratory syndrome, influenza A H1N1, and other viral pneumonias. Relying solely on CT scans makes it challenging to distinguish MERS from the mentioned diseases, and the final diagnosis requires combining Laboratory examinations. For individuals with relevant exposure history and clinical symptoms, the early appearance of ground-glass opacities on chest CT shall raise a high suspicion of MERS-CoV infection.

10.2.2 Key points for imaging differential of SARS, MERS and COVID-19

MERS, SARS, and COVID-19 are all caused by coronaviruses and exhibit pulmonary inflammation with characteristic imaging manifestations of ground-glass opacities and lung consolidation. However, there are slight differences in the staging and progression of imaging, and their radiological manifestations vary slightly. Specific differential points are summarized in Table 10.1.

10.2.3 Case description

Patient, male, born in 1971, is a close contact of a MERS case. He started experiencing back pain on May 21, 2015, with no fever, cough, or sputum. No specific treatment was given. On May 25, 2015, he developed a fever with a temperature as high as 39.7°C. There was no chills, cough, sputum

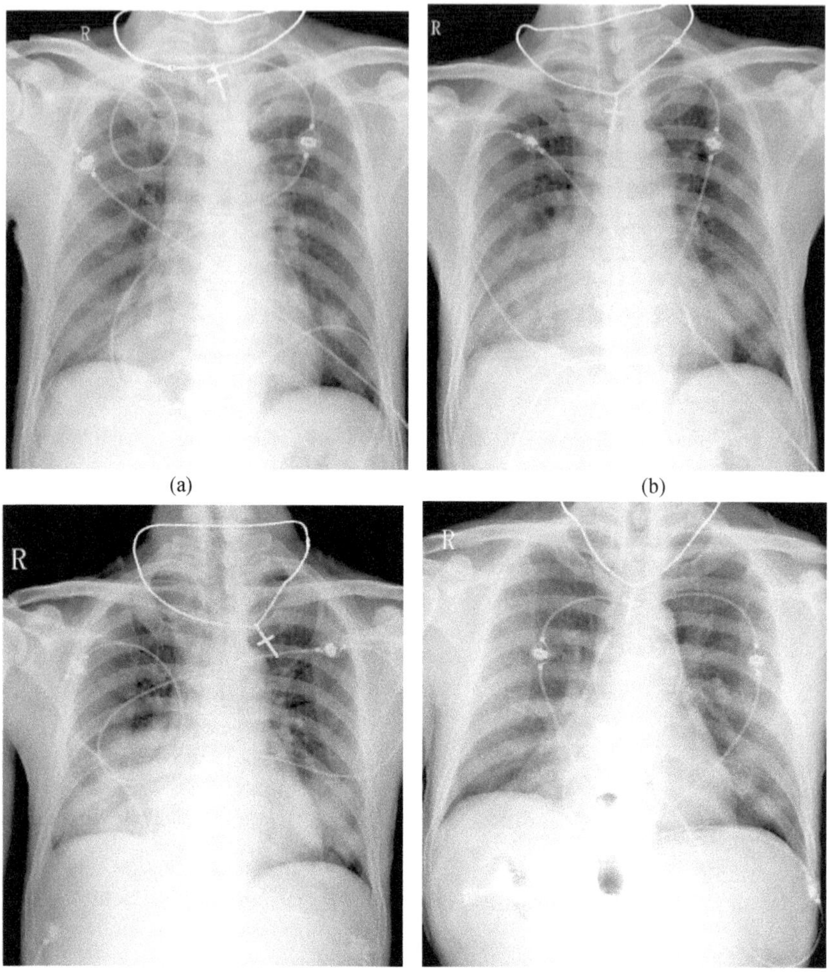

Figure 10.1 Chest X-ray. (a) On the seventh day of the disease, the chest X-ray upon admission shows small patchy increased density shadows in the lower lung fields near the cardiac border; (b) on the 10th day of the disease, the chest X-ray shows significant enlargement of the patchy shadows in both lower lung fields, with more pronounced progression in the right lower lung; (c) on the 13th day of the disease, the lung lesions continued to progress, showing large patchy consolidations; (d) on the 32nd day of the disease, the chest X-ray shows significant absorption of the lesions in the middle and lower lung fields, with only small patchy reticular shadows remaining.

production, shortness of breath, abdominal pain, diarrhea, or sore throat. At that time, self-administered cold medicine was ineffective. On June 1, 2015, the patient's sputum virus test was positive for MERS-CoV. He was diagnosed with MERS. See Fig. 10.1 for Chest X-ray.

Table 10.1 Imaging characteristics of three coronaviruses.

	SARS	MERS	COVID-19
Imaging staging (typing)	Initial stage, progressive stage, convalescence stage; type 1 (70.29%), type 2 (17.39%), type 3 (7.25%), type 4 (5.07%)	Type 1–4	Early stage, progressive stage, severe stage (critical stage), and convalescence stage (recovery stage)
Common manifestation	Ground-glass opacity (common and in early stage), consolidation, "paving stone sign" (in early stage)	Ground-glass opacity (more extensive), consolidation	**1** early stage: lesions are localized, presenting as patchy, subsegmental, or segmental ground-glass opacities, often accompanied by thickening of the interlobular septa **2** progressive stage: The lesions increase in number and expand in scope, involving multiple lung lobes. Some lesions become consolidations, and ground-glass opacities coexist with consolidations or linear opacities. Occasionally, the "paving stone sign" may appear **3** severe stage: Diffuse lesions within both lungs, with a minority exhibiting "white lung" appearance. Consolidations are predominant, accompanied by ground-glass opacities and frequently associated with fibrous stripe opacities and bronchial inflation sign

Continued

Table 10.1 Imaging characteristics of three coronaviruses.—cont'd

	SARS	MERS	COVID-19
Other manifestations	Mediastinal emphysema, pneumothorax	Pleural effusion, pneumothorax	"Halo sign", "reversed halo sign"
Rare manifestations	Pleural effusion, lymphadenectasis, cavity, calcification, reticulation, nodules	Cavity	Lymphadenectasis and pleural effusion are only seen in severe and critical cases
Distribution	Common in lung periphery and lower lung		
Progression evolution	Severe cases progress rapidly, and can be significantly worse within about 2 days; ground-glass opacity → mixed density → consolidation; focal → multifocal → diffuse; lower lobes → diffuse involvement; unilateral → bilateral: Peripheral → peripheral and central → diffuse (SARS); peripheral → peripheral and central → diffuse (COVID-19); eventually may present as "white lung"; in the convalescence stage, the extent of lesions decreases, and their density becomes lighter; in a few cases, pulmonary fibrosis may form		

10.2.4 Diagnostic key points

(1) The majority of MERS-CoV infections occur in Saudi Arabia, hence it is also known as Middle East Respiratory Syndrome (MERS).

(2) The source of MERS is primarily dromedary camels, with the main transmission channel of aerosol and close contact. Susceptible Population: People of all age groups can be affected. Patients with underlying diseases are more likely to progress to severe conditions.

(3) The main clinical manifestations include acute respiratory infection, with a sudden onset of high fever ranging from 39 to 40°C, possibly accompanied by chills, shivering, cough, sputum production, chest pain, headache, generalized muscle and joint pain, fatigue, and loss of appetite. Incubation Period: 2–14 days.

(4) The pathogen-related examinations mainly include virus isolation, viral nucleic acid detection, and serum MERS-CoV antibody testing. Various specimens can be collected for testing, including throat swabs, nasal swabs, nasopharyngeal or tracheal aspirates, sputum or lung tissue, blood, and feces. Among them, respiratory tract specimens have the highest positive detection rate.

(5) Confirmed cases shall meet at least one of the following criteria: ① Positive results on at least two-target PCR testing. ② Positive result on a single-target PCR product, confirmed by genetic sequencing. ③ MERS-CoV isolated from respiratory specimens. ④ Four-fold or greater rise in MERS-CoV antibody titer in convalescent-stage serum compared with acute-stage serum.

(6) The imaging characteristics of viral pneumonia mainly include ground-glass opacities and/or lung consolidations, predominantly involving both lungs or subpleural areas. On the basis of pneumonia, some cases may rapidly progress clinically, leading to respiratory failure, acute respiratory distress syndrome (ARDS), or multiple organ dysfunction syndrome (MODS), especially renal failure, and even life-threatening situations.

Bibliography

[1] Lu P, Zhou B. Diagnostic imaging of emerging infectious diseases. Beijing: People's Medical Publishing House; 2013. p. 37–48.

[2] Lan BW, Lu PX, Zeng YJ, et al. Clinical imaging research of the first Middle East respiratory syndrome in China. Radiol Infect Dis 2015;2(4):173–6.

[3] Lu PX, Zhou BP. Diagnostic imaging of emerging infectious diseases. Berlin: Springer; 2015. p. 5–28.
[4] Lu P, Yang G, Weiye Y, et al. Imaging follow-up of SARS patients complicated with pulmonary fibrosis. Chin J Med Imaging Technol 2004;20(12):1901–3.
[5] Lu PX, Zhou BP, Chen XC, et al. Chest X-ray imaging of patients with SARS. Chinese Med J 2003;116(7):972–5.
[6] Lu P. The characteristic of epidemiology and medical imaging in human avian influenza outbreak in China. Elect J Emerg Infect Dis 2017;2(2):124–6.
[7] Song L, Zeng B, Xiaoming G, et al. Imaging manifestations and differential diagnosis of novel Coronavirus pneumonia. Elect J Emerg Infect Dis 2020;5(2):82–6.
[8] Zheng Q, Lu Y, Tan L, et al. Advances in clinical and imaging studies of novel Coronavirus pneumonia. Elect J Emerg Infect Dis 2020;5(2):140–4.

CHAPTER 11

Imaging manifestations of COVID-19

Li Hongjun[1], Liu Bailu[2], Lv Zhehao[3], Song Lan[4], Li Di[5], Wang Haibo[2], Chen Tingting[2], Guan Ying[2], Li Ping[2], Chen Tianwu[6], Lu Hong[7], Liu Heng[8], Yin Xiaoping[9], Lu Zhiyan[10], Lu Yibo[11], Xing Lihong[9], He Yulin[12], Gong Xiaoming[13], Zhang Yu[9], Li Li[1], Ren Meiji[1], Kong Lili[14], Li Hongyan[15], Shi Yanbin[16], Ma Xi[9], Zhao Lingling[16], Wu Lianming[17], Shi Yuxin[18] and Xu Jianrong[17]

[1]Beijing YouAn Hospital, Capital Medical University, Beijing, China; [2]The Second Affiliated Hospital of Harbin Medical University, Harbin, China; [3]The First Affiliated Hospital of Harbin Medical University, Harbin, China; [4]Chinese Academy of Medical Sciences Peking Union Medical College Hospital, Beijing, China; [5]Beijing Children's Hospital, Capital Medical University, Beijing, China; [6]The Second Affiliated Hospital of Chongqing Medical University, Chongqing, China; [7]The Central Hospital Affiliated to Chongqing University of Technology, Chongqing, China; [8]Affiliated Hospital of Zunyi Medical University, Zunyi, China; [9]The Affiliated Hospital of Hebei University, Hebei, China; [10]Zhongnan Hospital of Wuhan University, Wuhan, China; [11]Nanning Fourth People's Hospital, Nanning, China; [12]The First Affiliated Hospital of Nanchang University, Nanchang, China; [13]Hubei Xianning Central Hospital, Xianyang, China; [14]Yantai Yeda Hospital, Yantai, China; [15]The First Affiliated Hospital of Nanyang Medical College, Nanyang, China; [16]The Sixth People Hospital of Zhengzhou, Zhengzhou, China; [17]Renji Hospital, Shanghai Jiaotong University School of Medicine, Shanghai, China; [18]Shanghai Public Health Clinical Center, Shanghai, China

11.1 Imaging manifestations in early stage

11.1.1 Clinical characteristics

COVID-19 patients may have no clinical symptoms in the early stage or may experience mild to moderate fever, cough, sputum production, dyspnea, and chest tightness as the main clinical symptoms. Some patients may also experience fatigue, nasal congestion, runny nose, headache, and diarrhea.

11.1.2 Imaging manifestations

It is more common within the first week of COVID-19 onset, often presenting as multiple lesions in both lungs, with single lesion being rare. The lesions are often located in the peripheral lung or subpleura, with the back and lateral segments of the middle and lower lungs being commonly affected. The density of the lesions is uneven, with early-stage lesions often

The Pathology and Imaging of Coronavirus Pneumonia
ISBN 978-0-443-32930-2
https://doi.org/10.1016/B978-0-443-32930-2.00009-X

presenting as faint and thin ground-glass opacities. Within these opacities, thickened blood vessels and bronchi with thick walls can be observed, sometimes accompanied or not accompanied by local interlobular septal thickening in a reticular pattern; the areas of consolidation are small and localized, with air bronchograms visible within them; ground-glass opacities can exist independently or coexist with areas of consolidation. The lesions are mostly irregular and wedge or fan-shaped, with their tips pointing toward the hilum of the lungs. Patchy or round-shaped lesions can also be observed, but the lesions typically do not involve the entire lung segment. In patients without other pulmonary diseases, no enlargement of mediastinal or hilar lymph nodes, pleural thickening, or pleural effusion is observed. If early CT images only show localized subpleural ground-glass opacities, they are often difficult to visualize on chest X-ray, which can easily lead to missed diagnosis.

11.1.3 Case description

Case 1: Patient, male, 62 years old. The patient had fever, fatigue and dry cough for 3 days. The patient had a history of traveling in an epidemic area. Physical Examination: Redness of pharynx. Laboratory Examinations: C-reactive protein 3 mg/L, white blood cell count 4.2×10^9/L, lymphocyte percentage 29.5%, neutrophil percentage 58.6%, procalcitonin <0.072 ng/mL, creatine kinase <0.5 U/L, cardiac troponin I < 0.017 μg/L, N-terminal probrain natriuretic peptide 10 pg/mL, influenza A virus, influenza B virus, and respiratory syncytial virus tests (−), SARS-CoV-2 nucleic acid RT-PCR (+) twice. See Fig. 11.1 for imaging manifestations.

Case 2: Patient, female, 54 years old. The patient had the fever for 6 days, accompanied by fatigue and headache. The patient developed fever, mild cough with a small amount of sputum, along with fatigue and intermittent headaches on January 24, 2020. The chest CT showed multiple subpleural ground-glass opacities in both lungs. Testing positive for SARS-CoV-2 nucleic acid, a diagnosis of COVID-19 is considered. The patient had a history of living in an epidemic area. Physical Examination: No positive signs. Laboratory Examinations (January 31, 2020): Blood Routine Examinations: White blood cell count 2.87 × 10^9/L, red blood cell count 4.14 × 10^12/L, hemoglobin 112 g/L, phospholipid transfer protein 206 × 10^9/L, and neutrophil percentage 61.7%. Serum potassium concentration: 3.22 mmol/L, serum sodium

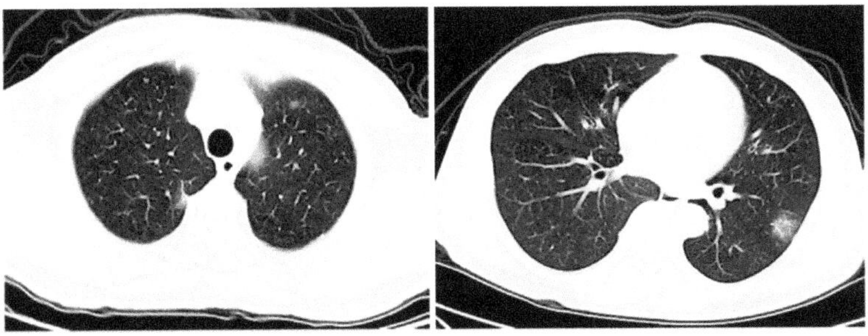

(a) (b)

Figure 11.1 Early imaging manifestations of COVID-19 (I). (a) CT shows small ground-glass opacities beneath the pleura in the posterior segment of the right upper lobe, with visible thickening and branching of blood vessels and slight traction of the adjacent pleura; (b) ground-glass density nodules with a round shape are observed beneath the pleura in the left lower lobe, with visible thickening and branching of blood vessels passing through them.

concentration: 137.5 mmol/L. Liver Function: Alanine aminotransferase (ALT) 43 U/L, aspartate aminotransferase (AST) 38 U/L, total bilirubin 9.6 μmol/L, albumin 38.5 g/L. See Fig. 11.2 for the patient's imaging manifestations.

Case 3: Patient, female, 39 years old. The patient had sore throat and runny nose for 8 days, without fever. She recently had contact with a confirmed COVID-19 patient. Physical Examination: Redness of pharynx. Laboratory Examinations: Highly sensitive C-reactive protein <0.5 mg/L, white blood cell count 6.07 × 109/L, lymphocyte percentage 22.1%, neutrophil percentage 74.1%, procalcitonin <0.072 ng/mL, creatine kinase 54 U/L, negative for

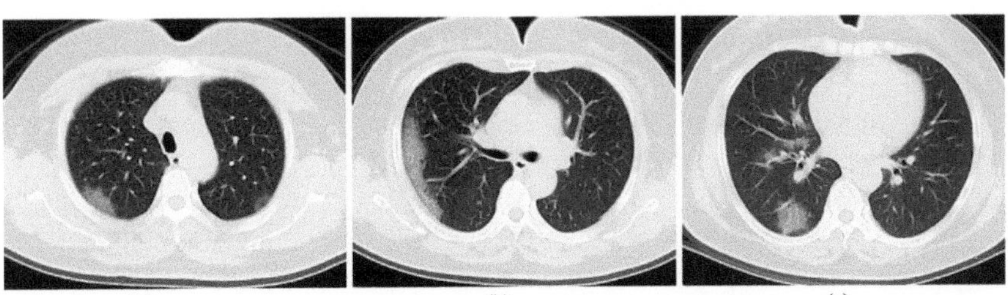

(a) (b) (c)

Figure 11.2 Early imaging manifestations of COVID-19 (II). (a–c) CT shows multiple round and patchy ground-glass opacities beneath the pleura in both lungs, with relatively faint density and blurred edges. Some lesions are oriented parallel to the pleura.

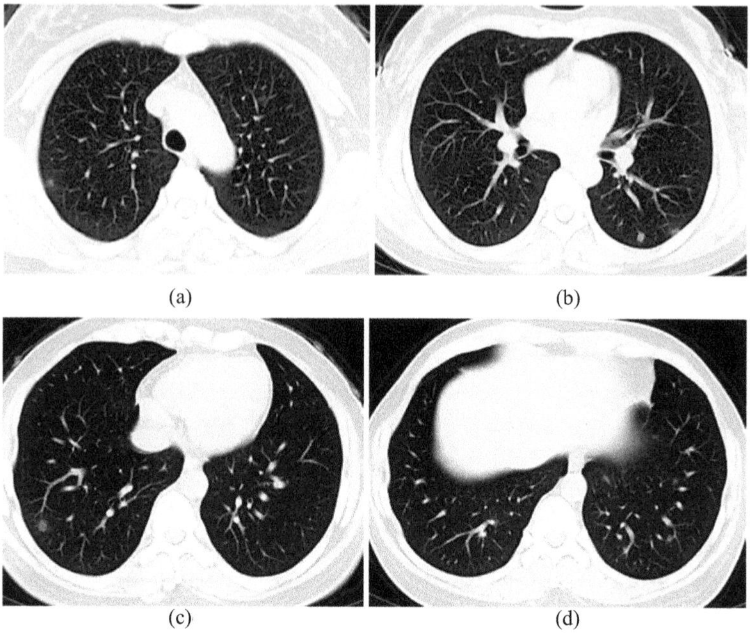

Figure 11.3 Early imaging manifestations of COVID-19 (III). (a and b) CT scan shows multiple nodules in the right upper lobe and both lower lobes of the lungs, most of which exhibit ground-glass opacity; (c) lesions adjacent to the pleura in the left lower lobe appear as ground-glass opacity, and the nodules beneath the pleura show partially solid density. The lesions are mainly distributed in the peripheral lung regions beneath the pleura; (d) small nodules in the posterior basal segment of the left lower lobe show a close relationship with bronchovascular bundles.

influenza A, influenza B, and respiratory syncytial virus detection, positive for SARS-CoV-2 nucleic acid RT-PCR (+) twice. See Fig. 11.3 for imaging manifestations.

11.1.4 Diagnostic key points

(1) In the early stage of COVID-19, CT imaging may only show scattered ground-glass opacity nodules or faint patchy opacities beneath the pleura.

(2) The internal part of the lesions often exhibits thickened and tortuous blood vessels, which are frequently accompanied by bronchial bundles.

11.2 Imaging manifestations in progressive stage

11.2.1 Clinical characteristics

In the progressive stage of COVID-19, the disease has worsened with obvious symptoms such as severe cough, fever, chest tightness, and dyspnea. Laboratory examinations show that the peripheral blood white cell count is either normal or decreased, and the lymphocyte count is either normal or reduced.

11.2.2 Imaging manifestations

The distribution of lesions increases, with a predominant involvement of the subpleural area, and multiple lung lobes can be affected. Partial ground-glass opacities can merge and expand, with increased density, irregular, wedge-shaped, or fan-shaped patterns and unclear boundaries. They may appear scattered and multifocal, even diffuse, often asymmetric on both sides. Thickened bronchovascular bundles or multifocal lung consolidation under the pleura can be observed. The lesions progress and change rapidly, with significant morphological changes within a short period. They may be accompanied by tissue necrosis leading to the formation of small cavities, and inflated bronchi can also be observed. In some cases, the lesions are all ground-glass opacities, usually without pleural effusion, and in very few cases, there may be enlargement of mediastinal and hilar lymph nodes. Patients without underlying diseases often have no pleural effusion, enlargement of mediastinal, or hilar lymph nodes.

11.2.3 Case description

Case 1: The patient is a 26-year-old female who was admitted to the hospital due to "chest and back pain for 3 days, and nasal congestion for 2 days". The fifth SARS-CoV-2 nucleic acid test was positive. White blood cell count: $4.45 \times 10^9/L$, Neutrophil count: $2.47 \times 10^9/L$, Lymphocyte count: $1.05 \times 10^9/L$, Neutrophil percentage: 55.5%, Lymphocyte percentage: 23.6%, Procalcitonin level: 0.15 ng/mL. See Fig. 11.4 for the imaging manifestations of dynamic changes in COVID-19 at the progressive stage.

Case 2: Patient, male, 65 years old. Chief Compliant: The patient had a fever for 2 days. Present Medical History: The patient developed a fever with a maximum temperature of 37.8°C, along with cough and sputum production, 2 days after returning to Beijing from a visit to an epidemic area by airplane. There was no chill, palpitation, or chest tightness. The patient

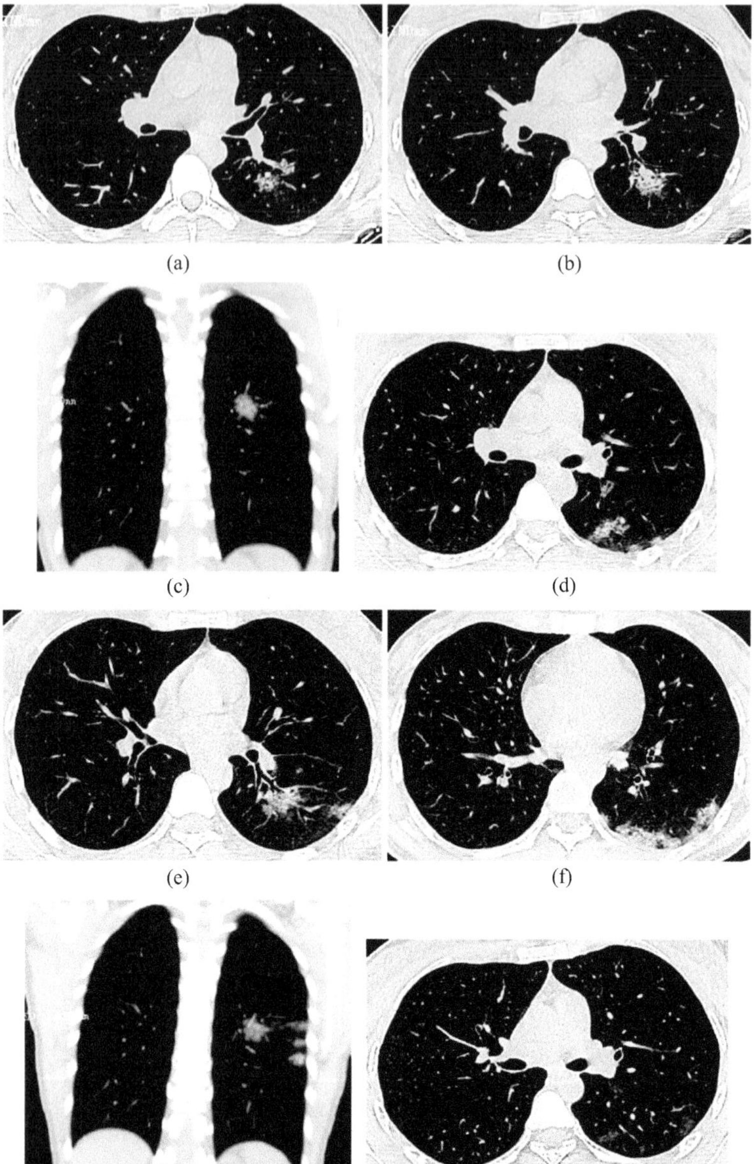

(a) (b) (c) (d) (e) (f) (g) (h)

Figure 11.4 Imaging manifestations of COVID-19 at the progressive stage (I). (a—c) On the third day of admission, the initial chest CT examination shows ground-glass opacities and patchy consolidation in the left lower lobe with "halo" sign around the lesions; (d—g) on the eighth day of admission, the consolidation in the left lower lobe increases in size, and new patchy consolidations appear beneath the pleura; (h—j) on the 13th day of admission, the density of the consolidation in the left lower lobe decreases, showing patchy ground-glass opacities with reduced extent; a new patchy ground-glass opacity with consolidation appears in the posterior basal segment of the right lower lobe; (k and l) on the 12th day after discharge, a follow-up CT scan shows that the original ground-glass opacities are mostly absorbed.

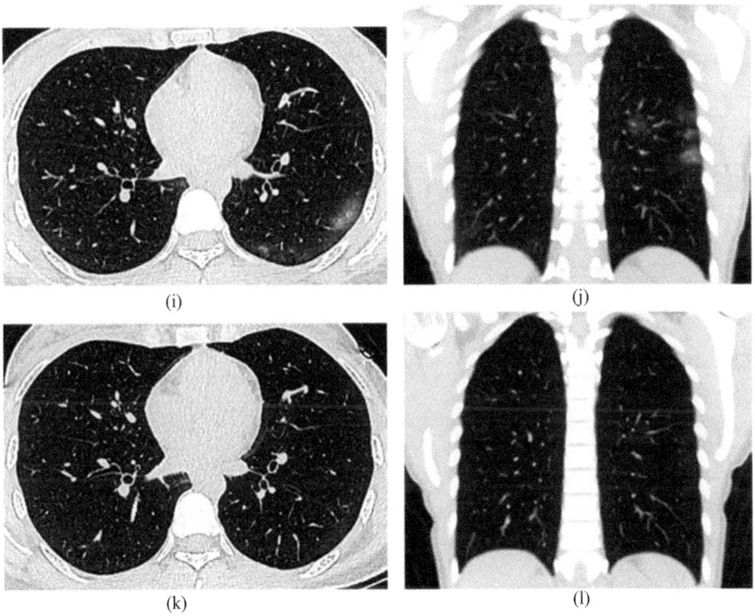

Figure 11.4 Cont'd

was admitted medical attention at an external hospital and diagnosed with COVID-19 after testing positive for SARS-CoV-2 nucleic acid. There was a history of travel to an epidemic area. Physical Examination: Coarse breath sounds. Laboratory Examinations: Partial pressure of carbon dioxide ($PaCO_2$): 33.3 mmHg↓, partial pressure of oxygen (PaO_2): 113 mmHg, oxygen saturation (SaO_2): 98.5%, base excess in extracellular fluid: −4.1 mmol/L↓. Blood routine examination (January 24, 2020): White blood cell count: 4.91 × 109/L↑, lymphocyte percentage: 19.1%↓, neutrophil percentage: 69.9%. Procalcitonin test (January 24, 2020): 0.11 ng/mL. C-reactive protein (January 24, 2020): 2.2 mg/L. Liver Function (January 24, 2020): Alanine aminotransferase (ALT): 25 U/L, Aspartate aminotransferase (AST): 14 U/L, Total bilirubin: 7 μmol/L, Direct bilirubin: 1.4 μmol/L. Blood Biochemistry (January 24, 2020) Examination: Creatinine (enzyme method): 86 μmol/L, Blood potassium: 3.61 mmol/L. Lactic acid (January 24, 2020): 1.84 mmol/L. See Fig. 11.5 for the patient's imaging manifestations.

Case 3: Patient, male, 35 years old. Chief Compliant: He had a fever for 6 days accompanied by fatigue. Present Medical History: He had a fever for 6 days before admission, with the highest body temperature reaching 38°C.

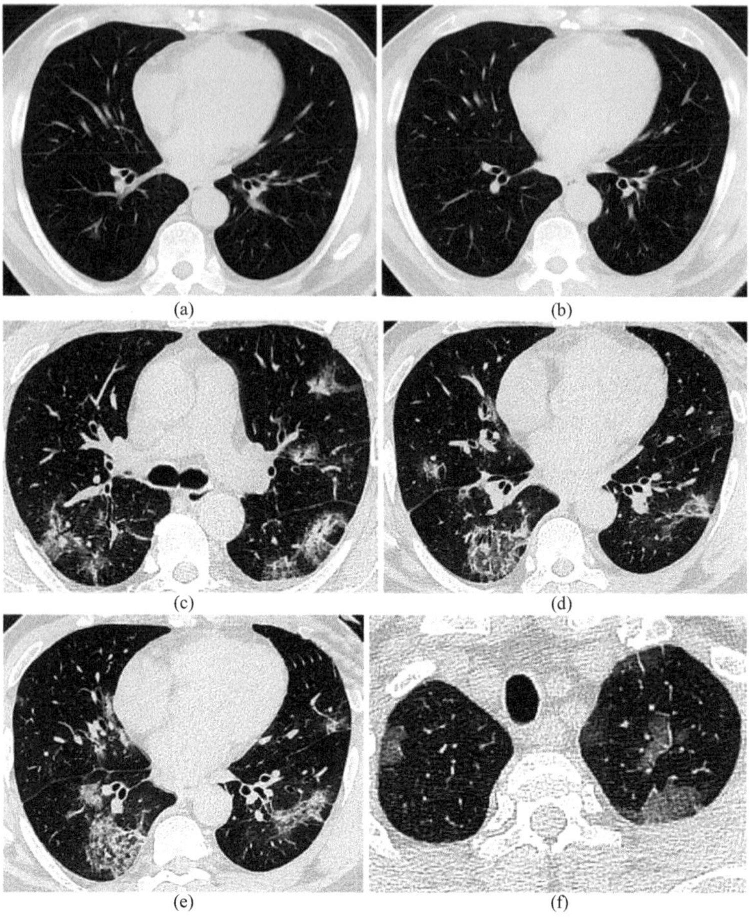

Figure 11.5 Imaging manifestations of COVID-19 at the progressive stage (II). Early stage: on January 24, 2020 (a and b) of the early stage of the disease, scattered pure ground-glass opacities were observed in both lower lobes of the lungs. The opacities showed relatively uniform density with blurred boundaries and were located between the pleura and the interlobular septa. The subpleural area had a higher number of opacities compared with the central part of the lobules. Progressive stage: on January 29, 2020 (c—e), (c) the lesions showed significant progression compared with previous images, extending to involve the upper lobes. Most of the lesions had blurred boundaries and showed the presence of a "halo" sign, and there were signs of air-bronchograms in the consolidation areas. (d) The extent of the lesions in both lower lobes had markedly increased, and there was a "reversed halo" sign in the right lower lobe. (e), the extent of the lesions in the lower lobes had further increased, and the lesions in the right lower lobe had fused. On February 2, 2020 (f—h), (f) the lesions showed further progression, and thickened interlobular septa with relatively smooth edges were visible. (g) The lesions showed diffuse progression, presenting a mixed distribution of ground-glass opacities and consolidations. The consolidation areas were mainly located centrally and showed a "halo" sign. (h) Large consolidation opacities were observed in both lower and middle lobes of the lungs. Convalescent

He had a history of traveling in an epidemic area before the onset of the disease. Physical Examination: Acute ill appearance. Laboratory Examinations: Alanine aminotransferase (ALT) 71 U/L, aspartate aminotransferase (AST) 52 U/L, total bilirubin 16.3 μmol/L, albumin 34.6 g/L. SARS-CoV-2 nucleic acid test is positive. See Fig. 11.6 for imaging manifestations.

Case 4: Patient, male, 21 years old. Next-generation sequencing (NGS) for NL-63 was performed. See Fig. 11.7 for imaging manifestations.

11.2.4 Diagnostic key points

(1) In the progressive stage of the disease, large areas of ground-glass opacities can be seen with fine internal reticular patterns and clear boundaries. The lesions tend to merge, and are distributed parallel to the pleura and located close to the hilar side in wedge-shape, and areas of consolidation may also appear, with a tendency for gravity-dependent distribution.

(2) In the early stage of COVID-19, it mostly presents as ground-glass opacities with a patchy appearance without consolidation. The lesions originate from secondary pulmonary lobules and involve the interlobular septa. The central part of the secondary lobule is

stage: on February 5, 2020 (i–k). (i) The extent of the lesions in the lung apex had increased slightly. (j) The consolidation area in the upper lobe showed significant absorption. (k) The lesions were absorbed, and there were signs of scar contraction and bronchial dilation. Some areas of absorption showed lung atelectasis and fibrosis, with clear boundaries and a stiff appearance in the affected regions. Convalescent stage: on February 9, 2020 (l–n). (l) The extent of the lesions in the lung apex had increased slightly. (m) The affected areas showed patchy ground-glass opacities. (n) The lesions in the early-onset regions had absorbed earlier, and the lesions in the central area and adjacent to the pleura were absorbed more rapidly, resulting in a "sandwich cookie" pattern with residual lesions. Convalescent stage: on February 12, 2020 (o–q). (o) The consolidation of the lesions in the lung apex had increased. (p) Fibrotic changes appeared in the affected areas. (q) Ground-glass opacities transformed into consolidations. Convalescent stage: on February 19, 2020 (r–t). During the recovery stage with recurrent symptoms, the patient experienced mild liver impairment and hypoalbuminemia. (r) The consolidations in some areas showed slight exacerbation compared with February 12, 2020. (s) Previously fibrotic regions reappeared as consolidation. (t) The extent of consolidations in the lung base had increased. Convalescent stage: on February 22, 2020 (u–v). Despite ongoing hypoalbuminemia, (u) the lesions showed slight absorption compared with February 19, 2020, with relatively clear edges in the fibrotic portions. (v) The consolidations within previously heavily fibrotic regions were absorbed again.

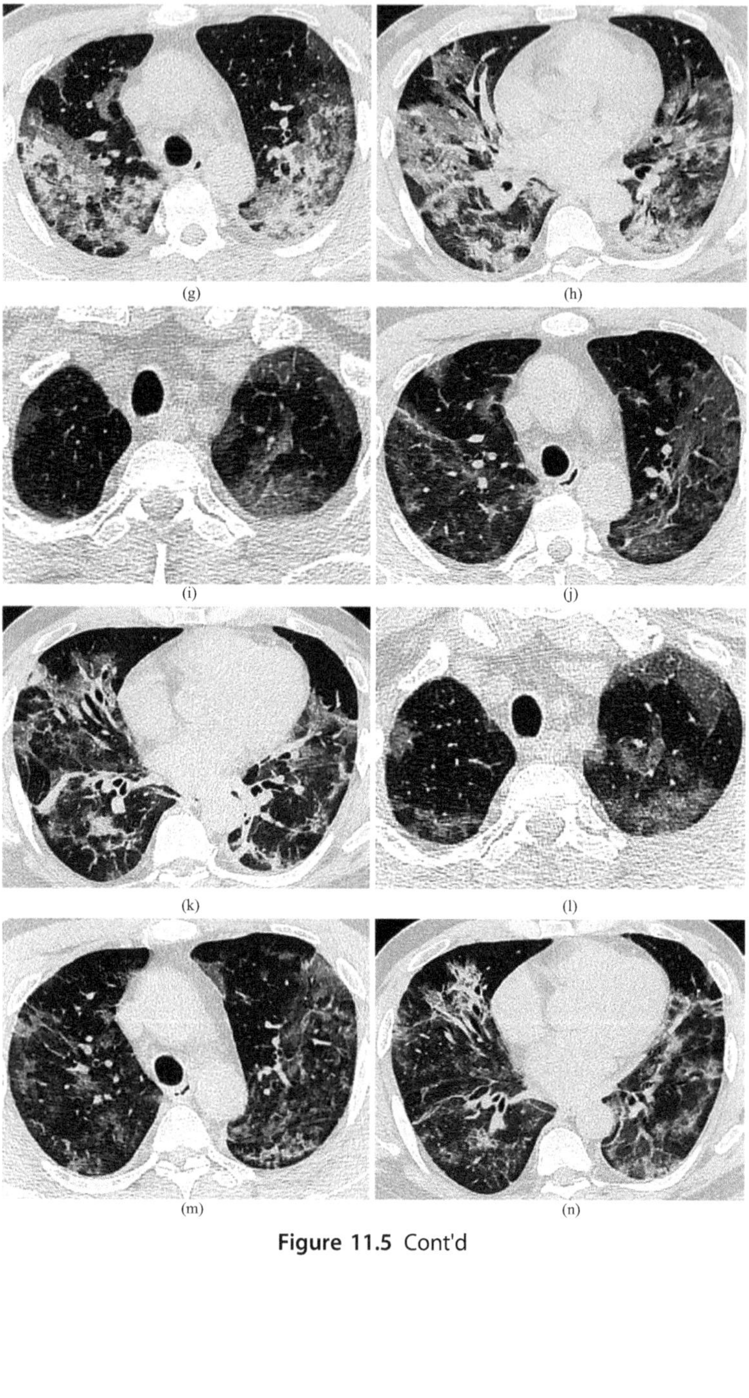

Figure 11.5 Cont'd

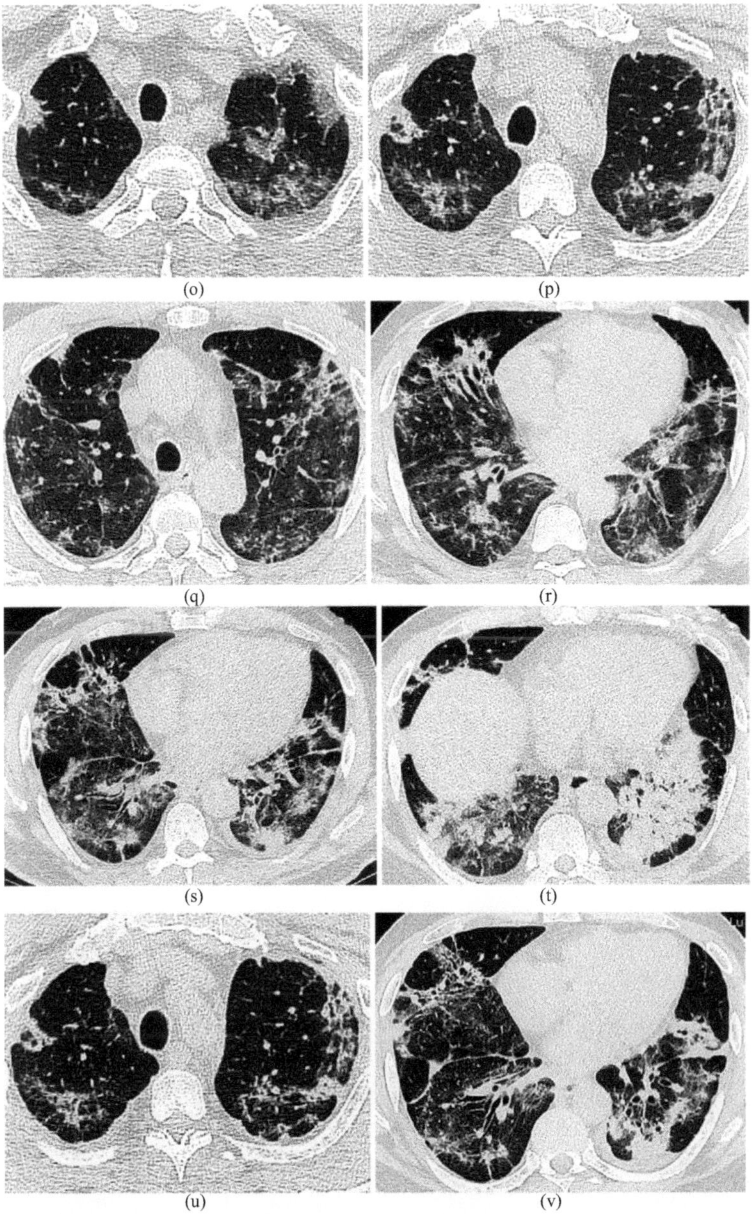

Figure 11.5 Cont'd

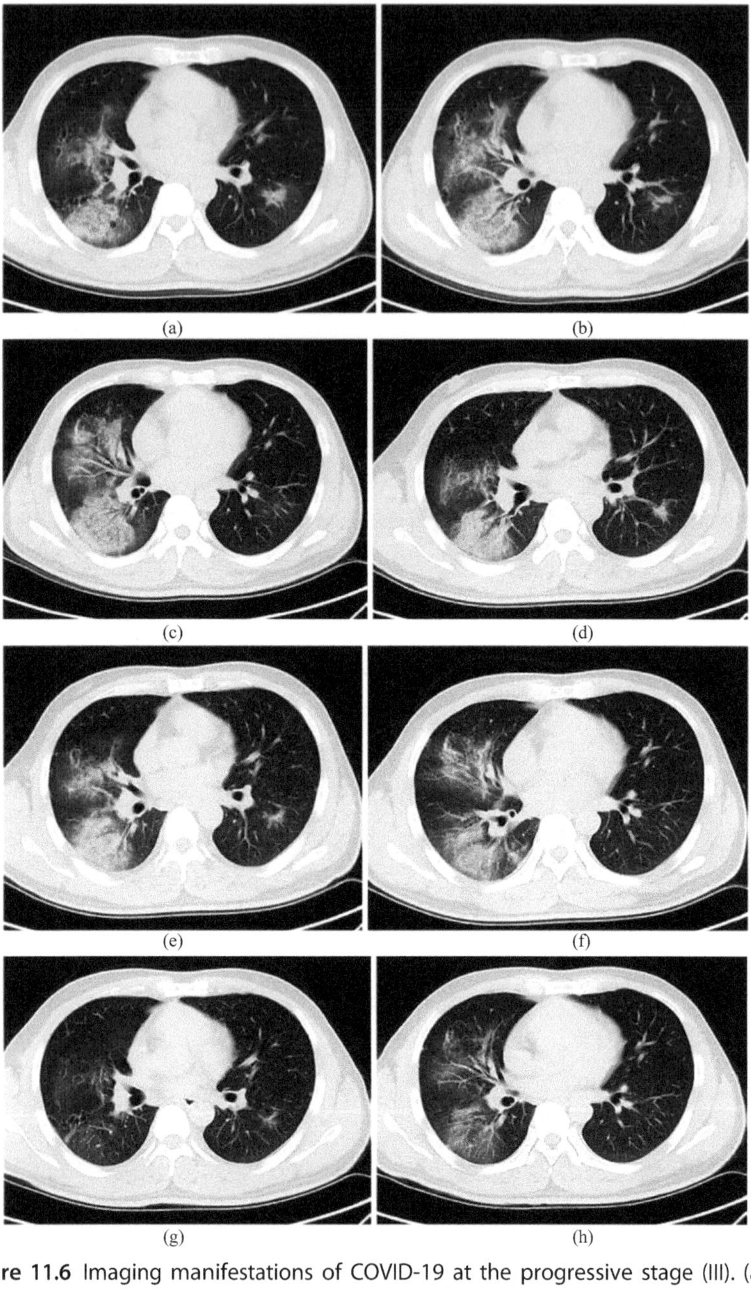

Figure 11.6 Imaging manifestations of COVID-19 at the progressive stage (III). (a–c) On January 26, 2020, the CT scan showed patchy ground-glass opacities in the right middle lobe and both lower lobes of the lungs. Within the lesions, blood vessels were visible, and some vessels appeared thicker. The interstitial and interlobular septa in the affected areas showed thickening; (d–f) on January 31, 2020, the CT scan showed ground-glass opacities in the right lower lobe, and compared with the previous images, the lesions had slightly become more solid. The extent of the lesions in the left lower lobe had slightly decreased; (g–i) on February 5, 2020, the CT scan showed a reduction in the extent of patchy ground-glass opacities in the right middle lobe and both lower lobes of the lungs. The density of the lesions decreased, and adjacent to the pleura in the right lower lobe, there was thickening observed.

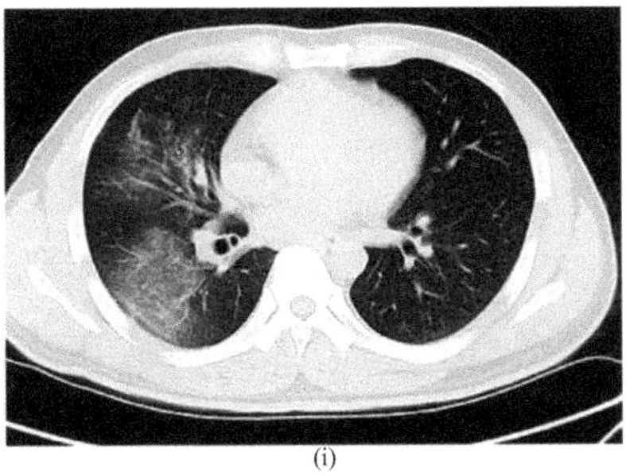

(i)

Figure 11.6 Cont'd

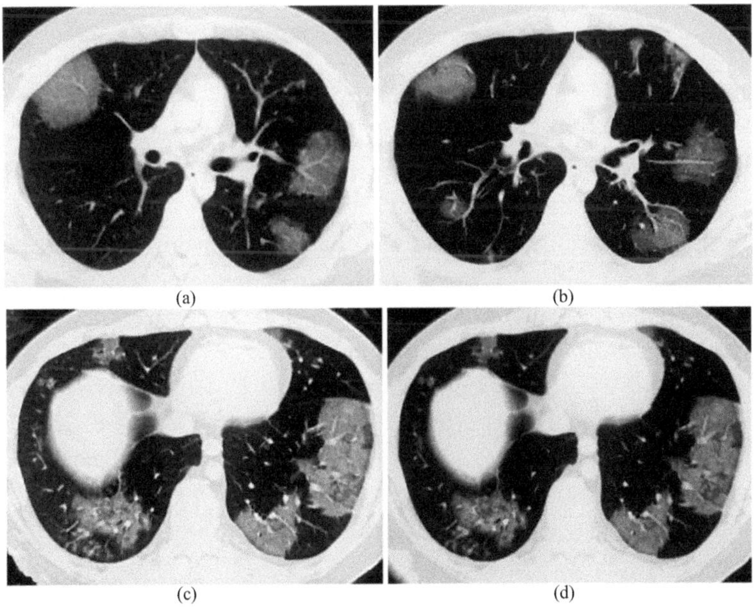

Figure 11.7 Imaging manifestations of COVID-19 at the progressive stage (IV). (a–d) The CT scan showed multiple round, patchy, and irregular ground-glass opacities in both lungs. The lesions were mostly located beneath the pleura and exhibited uneven density. Within the lesions, multiple thickened blood vessel shadows were visible.

more likely to form small round ground-glass opacities, which then spread diffusely outward, forming patchy opacities. The peripheral interlobular septa of the secondary pulmonary lobules can easily extend toward the pleural side and between the adjacent lobules, forming subpleural linear opacities.

(3) The lesions evolve rapidly, and within a few days, widespread consolidation can occur. Consolidation may be distributed in a segmental or lobar pattern. The lesions exhibited a characteristic of coexistence of multiple stages, where different lesions at different locations showed various phases of the disease process. Ground-glass opacities, consolidation, and fibrosis can be observed simultaneously.

(4) Within the lesions, small bronchi and blood vessels can be seen dilated. As the volume of the lesions increases, the "air bronchogram sign" can be observed within them. It may be accompanied by a small amount of pleural effusion.

11.3 Imaging manifestations in recovery stage

11.3.1 Clinical characteristics

COVID-19 infected patients mainly present with clinical symptoms such as fever, fatigue, and dry cough, and a few of them may be accompanied by nasal congestion, runny nose, sore throat, and diarrhea. Severe patients often develop respiratory distress and/or hypoxemia after 1 week of the onset of the disease. The recovery stage refers to 2–3 weeks after the onset of clinical symptoms. The above symptoms can be relieved after effective treatment, and chest CT examination in most patients indicates that the double lung disease has improved, and the white blood cell count, C-reactive protein and other indicators return to the normal range. Chest CT examination is highly sensitive to the diagnosis of COVID-19. Therefore, it is of great significance to judge the disease progression and outcome of patients in endemic areas by CT imaging manifestations.

11.3.2 Imaging manifestations in recovery stage

(1) The CT manifestations of recovery stage show a reduction in ground-glass opacities and consolidation, with decreased density, or the appearance of pulmonary fibrotic changes. In moderate cases of SARS-CoV-2 infection, after 1–2 weeks of active antiviral and antiinflammatory treatment, the extent of the lesions shrinks, density decreases, and consolidations reduce, and corresponding fibrotic strips may increase in the affected areas, and the boundaries become clearer. Additionally, new scattered small lesions may appear in other regions while the main lesions are resolving.

(2) In some patients, after the initial positive CT examination, the lesions show no progressive changes, indicating the transition to the recovery phase. During this stage, the lesions may appear as pure consolidations, resembling lung parenchyma density, or the density of the lesions may become similar to that of lung parenchyma with wrinkled edges.

(3) With the passage of time, the exudate is organized or absorbed by the body, and the lesions appear as grid-like thickening of the interlobular septa, twisting and thickening of the bronchial walls, showing stripe changes and scattered areas of increased density. The ground-glass opacities may be completely absorbed.

(4) Halo or reverse halo signs may appear during the recovery stage.

(5) Enlargement of mediastinal lymph nodes is rare.

(6) A small number of cases may have pleural effusion, which will completely resolve during the recovery stage. The thickening of the pleura will decrease or return to normal.

11.3.3 Case description

Case 1: A female patient, 46 years old, was admitted to the hospital 1 week ago after having close contact with individuals from an epidemic area. One day ago, the patient developed a fever without any apparent inducement. Self-measured body temperature reached a maximum of 38.5°C. There were no symptoms of chills, cough, phlegm, or chest discomfort. Upon admission, the patient's neutrophil count was elevated, and C-reactive protein (CRP) levels were increased. See Fig. 11.8 for imaging manifestations.

Case 2: The patient is a 77-year-old male who has experienced recurrent fever for 2 weeks. He reported body temperature fluctuations ranging from 37.0 to 38.5°C, without cough, sputum production, fatigue, or muscle pain. Two days before admission, chest CT examination showed bilateral pneumonia, suggesting viral pneumonia. See Fig. 11.9 for imaging manifestations.

11.3.4 Diagnostic key points

(1) There are clinical manifestations of fever. There is a history of travel to an epidemic area or contact with individuals from an epidemic area within 14 days before the onset of the disease.

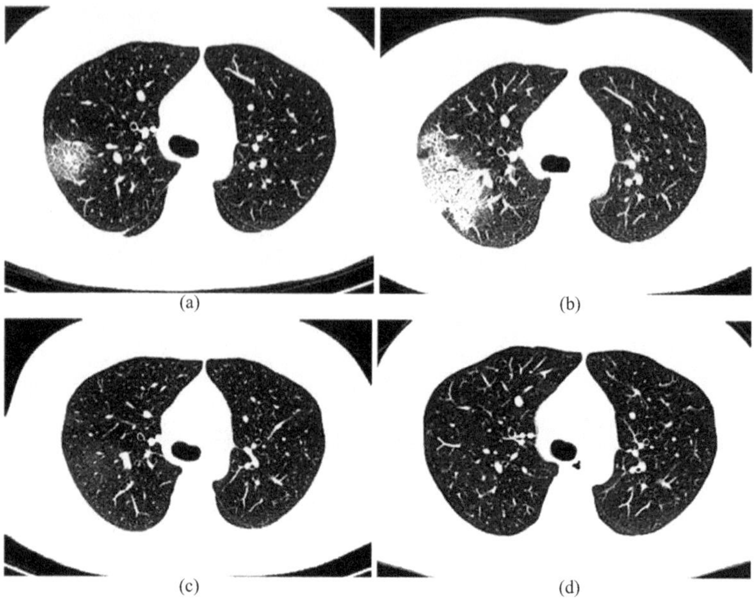

Figure 11.8 Imaging manifestations of COVID-19 at the progressive and recovery period (I). On day 1 after admission (a): Chest CT showed ground-glass opacities in the posterior segment of the right upper lobe with unclear boundaries; on day 4 after admission (b): Chest CT showed an enlargement of the lesion in the posterior segment of the right upper lobe, with increased density, indicating progression. 3 weeks after admission (c): Chest CT showed significant absorption of the lesion in the posterior segment of the right upper lobe, with the reduced size and decreased density; 1 month after discharge (d): Follow-up CT showed complete resolution of ground-glass opacities in the posterior segment of the right upper lobe.

(2) The patient's chest CT lesions show changes from early stage to progressive stage, and eventually to recovery stage. In the recovery stage, the density of the solid lesions decreases, and there are a few newly added ground-glass opacities in other areas of the lungs.

(3) As the density of consolidation lesions gradually decreases, their extent shrinks, and the original lesions are replaced by an increasing number of fibrous strands resulting from the absorption or organization by the body. Newly appearing ground-glass opacities also show a gradual decrease in density until they disappear entirely.

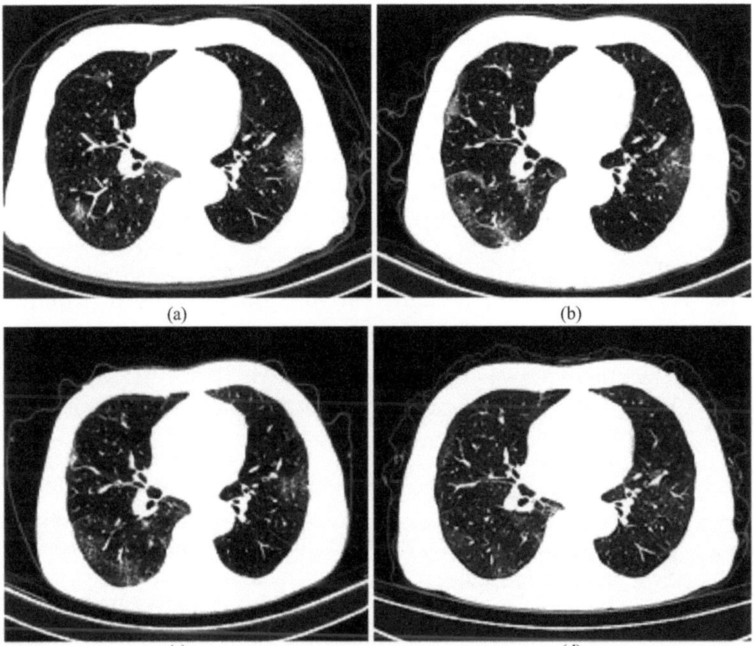

(a) (b) (c) (d)

Figure 11.9 Imaging manifestations of COVID-19 at the progressive and recovery period (II). On the day of admission (a): The patient's chest CT showed scattered patchy and ground-glass opacities in the subpleural regions of both lungs with unclear boundaries. On day 11 after admission, a follow-up chest CT (b) showed a slight decrease in density in some lesions, along with the appearance of ground-glass opacities in the right middle lobe. On day 14 after admission, another follow-up chest CT (c) showed a significant reduction in the extent of the lesions with decreased density compared with before. One month after discharge, a follow-up chest CT (d) showed a notable decrease in pneumonia-related opacities, and there were few residual fibrous strands at the site of the original lesions.

11.4 Imaging manifestations of severe and critical patients

11.4.1 Clinical characteristics

COVID-19 is clinically classified into four types: mild, moderate, severe, and critical. Severe and critical cases account for approximately 14% of the total. In adults, the diagnosis of severe COVID-19 can be made if any of the following criteria is met: 1. Presence of dyspnea with a respiratory rate (RR) of $\geq$30 breaths per minute. 2. Resting oxygen saturation

(SpO_2) ≤93%; ③ $PaO_2/FiO_2 \leq 300$ mmHg (1 mmHg = 0.133 kPa).④ Chest imaging showing significant progression of lesions (>50%) within 24–48 h. Adults with COVID-19 are considered critical if any of the following conditions is met: ① Respiratory failure requiring mechanical ventilation; ② Presence of shock; ③ Multiple organ dysfunction requiring intensive care unit (ICU) monitoring and treatment.

Severe and critical patients with COVID-19 are usually older and often accompanied by underlying diseases. The proportion of patients with multiple organ dysfunction is higher in this group. Critical patients may experience complications such as acute respiratory distress syndrome (ARDS), myocardial injury, liver and kidney dysfunction, gastrointestinal damage, and other conditions. Severe and critical patients are more likely to present with high fever, and their laboratory test results mainly show elevated C-reactive protein, normal or decreased white blood cell count, and reduced lymphocyte count.

11.4.2 Imaging manifestations

(1) The main CT manifestations of COVID-19 are patchy ground-glass opacities, primarily located beneath the pleura in the lungs. The lesions show thickening of blood vessel textures, with some areas accompanied by consolidation and "paving stone" sign.

(2) In severe and critical patients with COVID-19, the range of pulmonary lesions is broader, involving multiple lobes and segments in both lungs. These patients tend to have a greater number of lung lesions compared with those with mild or moderate symptoms, and the percentage of lung volume occupied by the lesions is higher.

(3) Severe and critical patients with COVID-19 have a higher incidence of lung consolidation, which can manifest as "white lung" appearance.

(4) Some severe and critical COVID-19 patients may present with pericardial effusion and pleural effusion, and a few patients may have enlarged mediastinal lymph nodes.

11.4.3 Case description

Case 1: A male patient, 64 years old, presented with fever after close contact with a confirmed COVID-19 patient. After admission, a throat swab was taken for real-time fluorescent PCR testing, which showed a positive result for SARS-CoV-2 nucleic acid. At rest, the oxygen saturation is between 92% and 93%. Laboratory Examinations: Lymphocyte percentage 14.2%, creatine kinase 3 U/L, C-reactive protein 7.87 mg/L, monocyte percentage 9.1%. He was diagnosed as COVID-19 (severe). See Fig. 11.10 for imaging manifestations.

Case 2: A male patient, 80 years old, had intermittent fever, cough, and sputum for 4 days. The patient had a recent history of contact with individuals returning from an epidemic area. The pharyngeal swab test was positive for SARS-CoV-2 nucleic acid using real-time fluorescent PCR. At rest, the oxygen saturation was between 90% and 95%, and noninvasive respiratory support is provided. Laboratory examinations showed a lymphocyte percentage of 14.8%, a monocyte percentage of 16.8%, and a C-reactive protein level of 16.3 mg/L. He was diagnosed as COVID-19 (critical). See Fig. 11.11 for imaging manifestations.

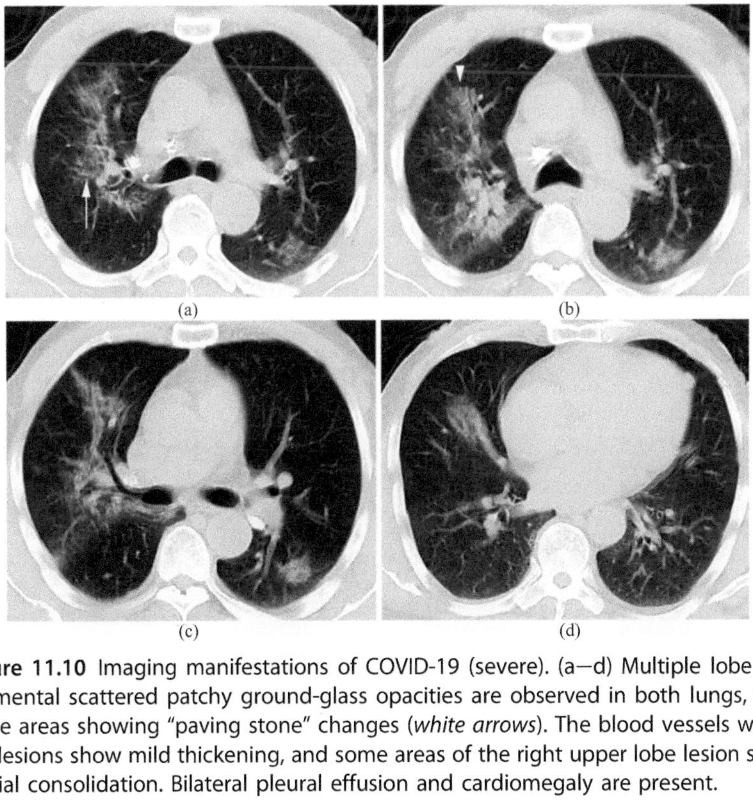

Figure 11.10 Imaging manifestations of COVID-19 (severe). (a–d) Multiple lobe and segmental scattered patchy ground-glass opacities are observed in both lungs, with some areas showing "paving stone" changes (*white arrows*). The blood vessels within the lesions show mild thickening, and some areas of the right upper lobe lesion show partial consolidation. Bilateral pleural effusion and cardiomegaly are present.

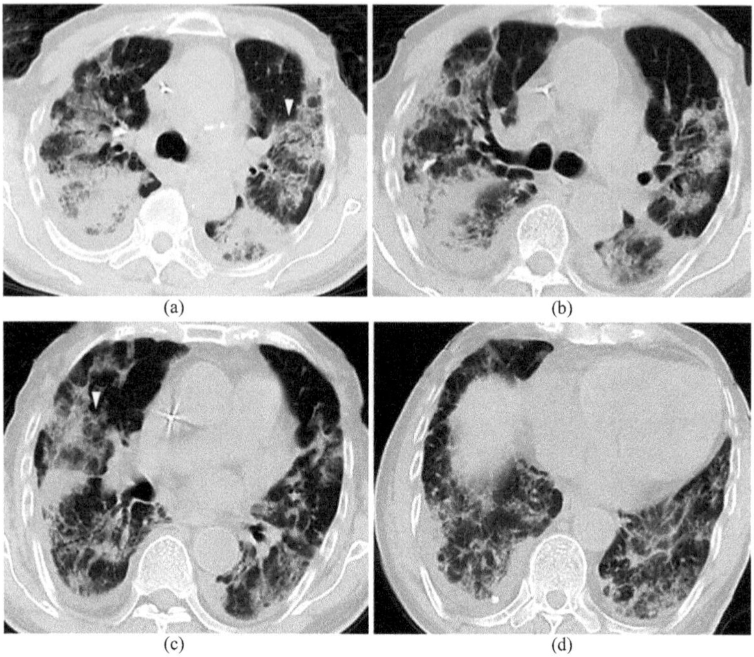

Figure 11.11 Imaging manifestations of COVID-19 (critical). (a–d) Diffuse multiple patchy ground-glass opacities are observed in both lungs involving several lobes and segments, with the presence of "paving stone" sign (*white arrows*). Multiple areas of consolidation are present within the lesions, with thickening of blood vessels in the affected areas and some showing a "white lung" appearance. Bilateral pleural thickening, slight pleural effusion in the chest cavity, and mild enlargement of the heart are present.

11.4.4 Diagnostic key points

(1) The patient had a clear epidemiological history.

(2) The patient had typical clinical symptoms (fever, cough, fatigue, etc.).

(3) Oxygen saturation is 90%–95%. Laboratory examinations show decreased percentages of lymphocytes and monocytes, and the SARS-CoV-2 nucleic acid test is positive.

(4) The chest CT scan shows multiple ground-glass opacities in various lung lobes and segments, with some areas exhibiting consolidation. The blood vessel patterns within the lesions appear thicker, and "paving stone" sign as well as "white lung" appearance are observed.

(5) Combined bilateral pleural effusion and cardiomegaly are present.

11.5 Imaging manifestations of special population

11.5.1 Infants and young children

11.5.1.1 Clinical characteristics

Epidemiological data shows that currently, infants and young children have a lower incidence of the disease, and the majority of cases are mild or moderate, with fewer severe or critical cases. The prognosis is generally favorable. Family clustering is the main mode of infection transmission.

11.5.1.2 Imaging manifestations

11.5.1.2.1 X-ray

Early Stage: Conventional X-rays may be negative, and in the initial stage of the disease, there are often no abnormal findings, leading to a high misdiagnosis rate. It can present as bronchitis or bronchiolitis. Progressive Stage: It can manifest as localized or nodular shadows in the lung field, predominantly in the periphery, without specificity.

Severe Stage: It presents with multiple diffuse consolidations in both lungs, and may even show "white lung" appearance, with or without pleural effusion on one side. Recovery Stage: The lesion range is decreased absorbed than before, with residual fibrous stripes.

11.5.1.2.2 HRCT

Early Stage: The lungs may show single or multiple subpleural patchy or ground-glass opacities with the center in the lobule. The central pulmonary small blood vessels may become thickened and increased. The lung texture may appear reticular (resembling "paving stone" sign). The lesions are mainly distributed along the bronchial bundles, posterior aspect, and sub-pleural region of the lung base, and the "air bronchogram" sign may be observed.

Progressive Stage: The area of ground-glass opacities increases and starts to consolidate. The opacities are often distributed along the bronchial bundles. Consolidation may be present with or without thickening of the interlobular septa. Pleural effusion is rare.

Severe Stage: Multiple bilateral ground-glass opacities, diffuse large consolidative opacities, accompanied by thickening of the interlobular and interlobar pleura, with or without pleural effusion on one side.

Recovery Stage: The lesions are reduced and absorbed than before, or only residual fibrous stripes are visible.

11.5.1.3 Case description

Case 1: A male patient, 7 months old, presented with fever 1 day ago, with a body temperature of 37.5°C, clear nasal discharge, and mild cough. The patient's elder brother had contact with a suspected case, and the suspected case's household has a confirmed case. The blood routine examination showed an increase in white blood cell count and lymphocyte count. See Fig. 11.12 for imaging manifestations.

Case 2: A female patient, 2 years old, presented with fever. Her mother had contact with individuals from the epidemic area and was diagnosed with COVID-19. The blood routine examinations showed a normal white blood cell count and lymphocyte count. See Fig. 11.13 for imaging manifestations.

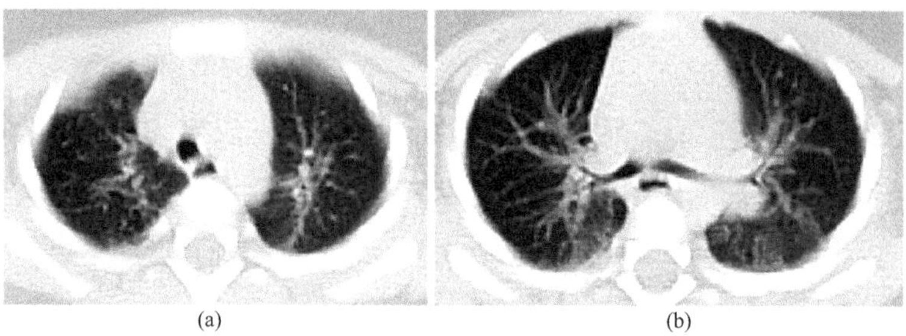

Figure 11.12 Imaging manifestations of infants with COVID-19 (moderate). (a and b) Both lungs show thickening of blood vessels and bronchial bundles, with patchy ground-glass opacities distributed along the vascular bronchial bundles. (This case is provided by the First Affiliated Hospital of Nanchang University).

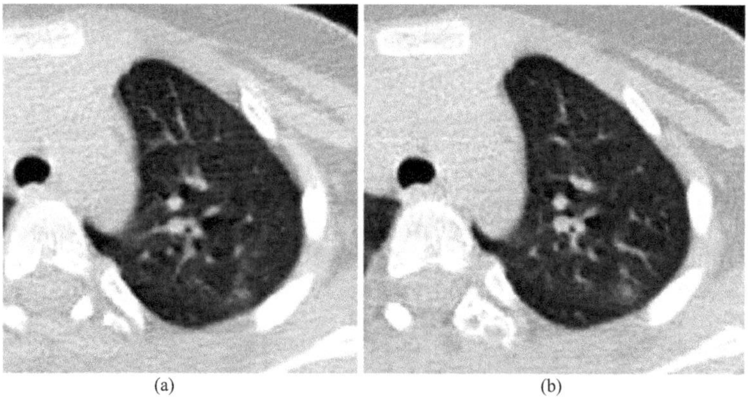

Figure 11.13 Imaging manifestations of young children with COVID-19 (moderate). (a and b) Patchy ground-glass opacities with blurred margins are observed in the subpleural region of the posterior segment of the left upper lobe. This case is classified as the moderate type with the imaging manifestations indicating as early stage. (This case is provided by Hejian People's Hospital).

11.5.1.4 Diagnostic key points

(1) Family clustering is the main route of infection. Family members are confirmed cases or have a history of contact with suspected cases.

(2) The clinical symptoms include fever, with possible accompanying runny nose and cough.

(3) Chest CT can present as single or multiple subpleural patchy or ground-glass opacities in the lungs. It can also show thickening of bilateral pulmonary vessels and bronchial bundles, with patchy ground-glass opacities distributed along the blood vessels and bronchial bundles.

11.5.2 Children and adolescents

11.5.2.1 Clinical characteristics

Due to the widespread susceptibility of COVID-19, children and adolescents with an epidemiological history and presenting symptoms such as fever, cough, mild sputum production, and muscle aches shall be considered for this disease. Family clustering is the main channel of transmission, which belongs to second or third-generation infections, indicating a weakened virulence of the virus. Children have lower immunity or less developed lung cell structures, leading to differences in ACE2 receptor distribution between children and adults. Most cases are clinically mild or moderate, while severe and critical cases are rare, and the prognosis is generally good. Based on a study conducted by domestic experts, including Academician Zhong Nanshan, using data from 1099 confirmed cases in 31 provinces (municipalities and autonomous regions) across the country, the incidence rate of children under 15 years old is approximately 0.9%.

11.5.2.2 Imaging manifestations

11.5.2.2.1 X-ray

Early Stage: Negative or both lungs show increased and blurred lung markings, and reduced lung transparency.

Progressive Stage: Patchy shadows appear in the lungs, mostly located subpleurally (under the pleura), often involving both lungs, but can also be unilateral, and may present as bronchopneumonia-like changes.

Severe Stage: The lungs show multiple patchy consolidation shadows, with air bronchogram sign visible inside, and even "white lung" appearance, with indistinct cardiac silhouette; pleural effusion is rare.

Recovery Stage: The lesion range becomes smaller, with clearer boundaries, reduced density, and visible fibrous stripes.

11.5.2.2.2 HRCT

Early Stage: The lesions are mostly located beneath the pleura in the lung periphery, relatively confined, and smaller in size. Ground-glass opacities and patchy shadows are common manifestations. Changes resembling bronchopneumonia show thickening and blurring of bronchovascular bundles, with patchy or nodular ground-glass opacities distributed around the bronchovascular bundles with fuzzy boundaries. Enlarged lymph nodes in the mediastinum and hilum, as well as pleural effusion are rare.

Progressive Stage: The ground-glass opacities in the lungs become denser, and consolidation occurs, with blurred and unclear boundaries. Multiple lesions may coalesce to form larger consolidated areas. Air bronchograms may be visible within the lesions. The volume of the lesions increases, and their number may also rise. The "reversed butterfly sign" is rare. Bronchopneumonia-like lesions may progress along the bronchovascular bundles. Enlarged lymph nodes in the mediastinum and hilum, as well as pleural effusion are rare.

Severe Stage: The lungs exhibit multiple patchy consolidations, with ground-glass opacities visible around the periphery. Air bronchograms may be observed within the consolidation areas. Both lungs may be extensively involved, but the "white lung" appearance is less common in children compared with adults. Pleural effusion is rare.

Recovery Stage: The consolidations within the lungs are absorbed, with decreased density, shrunk and concave edge and clear boundaries. Some residual fibrous stripes may be observed. The imaging presentation of mild cases in children can be negative.

The imaging presentation of adolescents is basically consistent with that of adults.

11.5.2.3 Case description

Case 1: A 9-year-old male patient presented with fever for 3 days apparent inducement, with the highest temperature reaching 38.9°C. He has been residing in an epidemic area for an extended period. Blood routine examination from an external hospital showed an elevated white blood cell count and a normal lymphocyte count. Upon admission, a reexamination of the blood routine showed normal white blood cell count and normal lymphocyte count. See Fig. 11.14 for imaging manifestations.

Case 2: A 4-year-old male patient had a fever and dry cough for 5 days, with body temperature fluctuating around 38.0°C. His father returned from an epidemic area 9 days ago and had close contact with the child, but

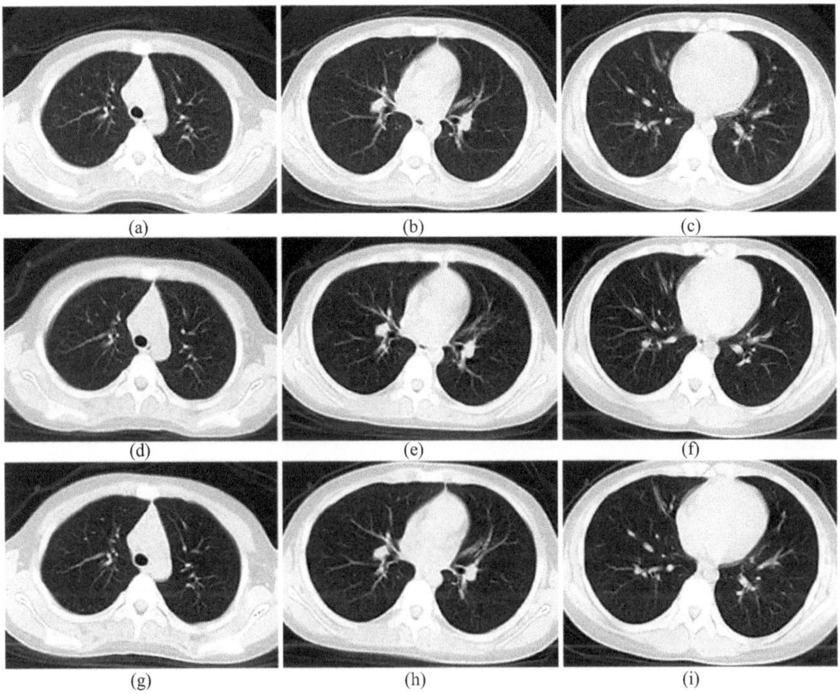

Figure 11.14 Imaging manifestations of children with COVID-19 (I). (a–c) On day 1 after admission, chest CT examinations show stripe opacities in the right middle lobe, while the rest of the lungs has no obvious abnormalities. This case was classified as mild. (d–f) CT images on day 3 after admission in the pediatric patient. (g–i) Chest CT images on day 5 after admission show no significant changes in lung imaging. (This case is provided by Baoding People's Hospital).

the father showed no obvious abnormalities. Laboratory Examinations: Both white blood cell count and lymphocyte count were normal. But C-reactive protein, procalcitonin, monocytes, creatinine, lactate dehydrogenase, alkaline phosphatase, and creatine kinase isoenzyme were elevated. Hemoglobin and total bilirubin were decreased. See Fig. 11.15 for imaging manifestations.

Case 3: The patient is a 17-year-old male and presented with fever and cough for 5 days, with the highest body temperature reaching 39°C. His father is a person who returned from an epidemic area and was diagnosed with SARS-CoV-2 infection. The Patient's Laboratory Examinations: Both the white blood cell count and lymphocyte count are normal. See Fig. 11.16 for imaging manifestations.

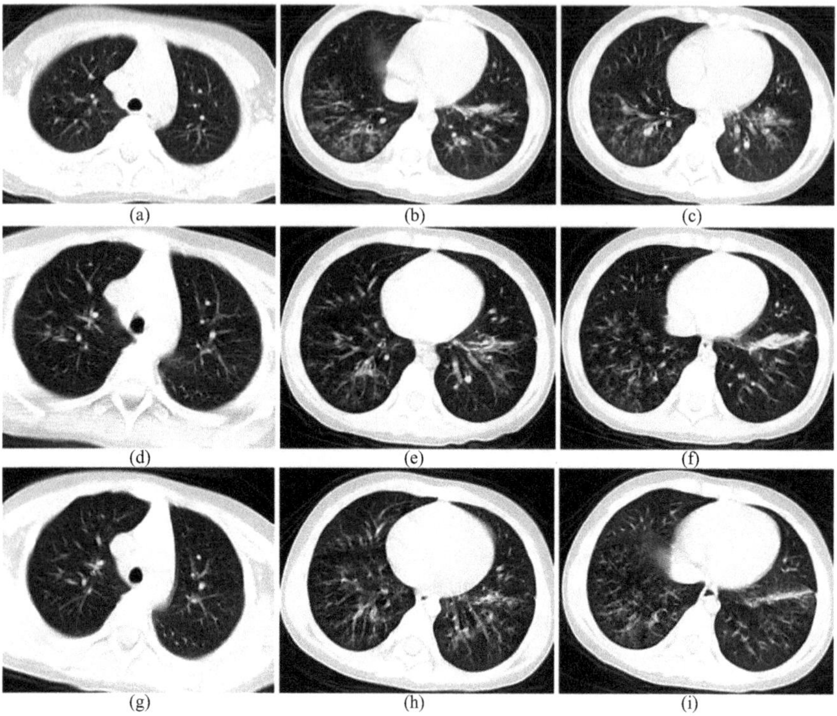

Figure 11.15 Imaging manifestations of child with COVID-19 (II). (a–c) On day 1 after admission, chest CT examinations showed multiple patchy ground-glass opacities distributed along the vascular and bronchial bundles in both lower lobes of the lungs, with partial consolidation in the left lower lobe, indicating a moderate type of the disease. (d–f) On day 3 after admission, chest CT examinations showed decreased density in some lesions in the right lower lobe, increased consolidation in the left lower lobe, and newly developed ground-glass opacities in the right upper lobe, indicating a progressive stage. (g–i) On day 6 after admission, chest CT examinations showed absorption and decreased density of lesions in both lower lobes of the lungs, indicating a recovery stage. (This case is provided by The Second Hospital of Xingtai).

11.5.2.4 Diagnostic key points

(1) Epidemiological History: He resided in an epidemic area and contacted history with individuals from the epidemic area within 14 days before the onset of the disease.

(2) Clinical Symptoms: Fever without obvious cause, with or without cough.

(3) Chest CT examinations mostly show mild and moderate types. It can present as bronchopneumonia-like changes, with thickening and blurring of bronchovascular bundles, and patchy or nodular ground-glass

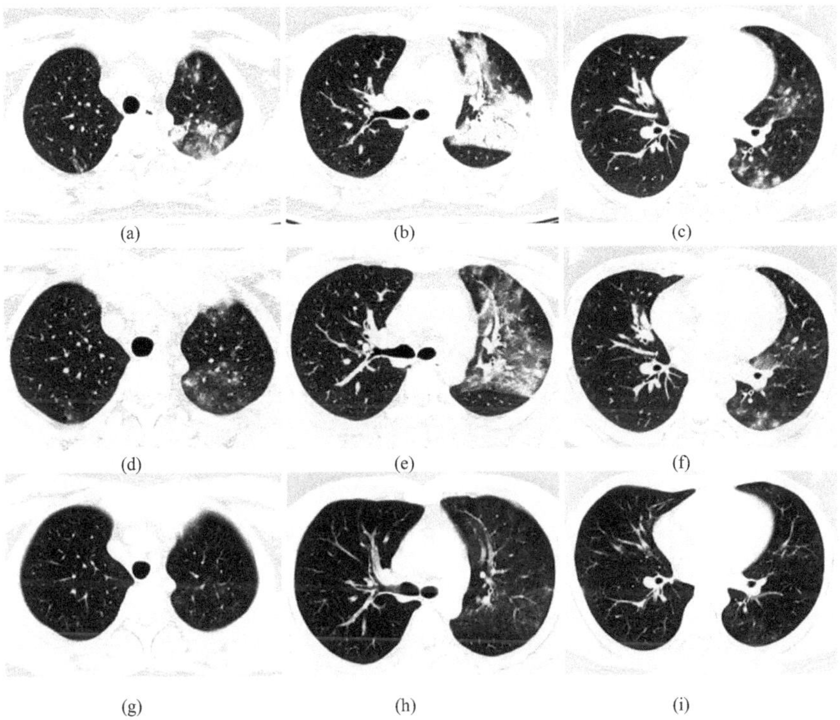

Figure 11.16 Imaging manifestations of adolescents with COVID-19. (a–c) On day 1 after admission, the chest CT showed multiple ground-glass opacities and patchy opacities in both lungs, with a predominance in the left lung. Partial consolidation was observed in the left lung, with air bronchograms visible within it. This case is considered as a moderate type, with imaging manifestations in the progressive stage; (d–f) on day 4 after admission, the chest CT showed that both lungs' ground-glass opacities and consolidations had been absorbed, with decreased density, indicating a transition to the recovery stage; (g–i) on day 7 after admission, the chest CT showed obvious absorption of lung lesions, with a small amount of residual ground-glass opacities remaining. (This case is provided by Shijiazhuang No. 5 Hospital).

opacities distributed along the bronchovascular bundles with indistinct borders. In the progressive stage, the consolidation area expands, but the "reverse butterfly" sign is less commonly seen. In the recovery stage, the lesions are absorbed, and the density decreases.

11.5.3 Patients in midtrimester pregnancy

11.5.3.1 Case description

Patient, female, 27 years old, with a history of fever and sore throat for 7 days. On January 25, 2020, the patient started to experience fever with a temperature of 37.6°C, accompanied by sore throat and headache. There were no symptoms of cough, sputum, chest tightness, or shortness of breath. After isolation and treatment, the patient's condition improved on its own. On January 29, 2020, the patient experienced fever again accompanied by chest tightness and shortness of breath, which worsened after physical activity. The treatment with intravenous fluids did not show significant improvement. The patient was admitted to the hospital after testing positive for SARS-CoV-2 nucleic acid. Epidemiological History: On January 22, 2020, the patient returned home from an epidemic area to celebrate the Chinese New Year, and his spouse worked at a pork market in the epidemic area. Patient's Physical Examinations: Slight dry rales heard in the left upper lung; abdominal distension consistent with the gestational age; pregnancy at 26 weeks of gestation. Laboratory Examination on January 28, 2020: Throat swab tested positive for SARS-CoV-2 nucleic acid. On February 2, 2020, blood routine + C-reactive protein showed: white blood cell count 13.04×10^9/L, absolute lymphocyte count 2.73×10^9/L, lymphocyte percentage 20.9%, absolute neutrophil count 9.51×10^9/L, neutrophil percentage 72.9%, and C-reactive protein 7.54 mg/L. On February 6, 2020, blood routine + C-reactive protein showed: white blood cell count 15.46×10^9/L, absolute lymphocyte count 2.44×10^9/L, lymphocyte percentage 15.8%, absolute neutrophil count 12.28×10^9/L, neutrophil percentage 79.4%, C-reactive protein 2.36 mg/L. On February 9, 2020, blood routine + C-reactive protein showed: white blood cell count 14.03×10^9/L, absolute lymphocyte count 3.07×10^9/L, lymphocyte percentage 21.9%, absolute neutrophil count 10.23×10^9/L, neutrophil percentage 72.9%, C-reactive protein 30.8 mg/L. Imaging Manifestations: Nodular consolidation is observed under the left upper lung pleura, with surrounding patchy ground-glass opacities, and visible air bronchograms. See Fig. 11.17 for imaging manifestations.

11.5.3.2 Diagnostic key points

(1) CT examination can clearly stage the lesions and evaluate treatment efficacy.

(2) Attention shall be paid to the radiation damage to the fetus, and CT examination shall be performed with the consent of the pregnant woman and appropriate abdominal protection measures.

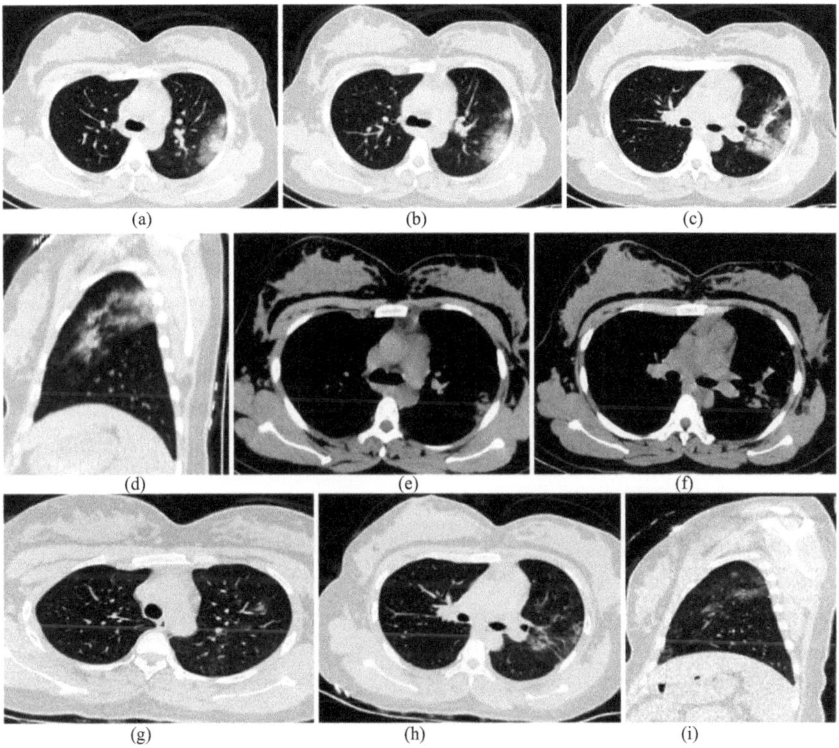

Figure 11.17 Imaging manifestations of patients midtrimester pregnancy with COVID-19. (a–d) On February 4, 2020, CT lung window showed large patchy ground-glass opacities in the left upper lung pleura, with partial consolidation. Within the ground-glass lesions, fine reticular opacities were observed, and air bronchograms were seen within the consolidation areas. (e–f) CT mediastinal window displayed the consolidated part of the lesion. (g–i) On February 12, 2020, CT lung window showed significant absorption of the original lesion after 8 days of treatment, with a slight increase in ground-glass opacities in the lung apex.

11.5.4 The elderly

11.5.4.1 Clinical characteristics

Due to the lack of immunity against COVID-19 in the human body, the general population is susceptible to the virus. The elderly individuals are more vulnerable to infection due to their weakened immune system and the presence of underlying diseases. Moreover, elderly patients have higher rates of severe illness and mortality. As of now, more than 80% of the reported death cases are in individuals aged 60 and above. Immune senescence is a critical factor contributing to the susceptibility of elderly individuals to severe conditions. The elderly individuals with underlying diseases may present with atypical clinical manifestations, laboratory findings, and pulmonary imaging manifestations. If there is an epidemiological history and the presence of symptoms such as fever, cough, mild sputum production, and muscle aches, along with abnormal laboratory and pulmonary imaging, manifestations, COVID-19 shall be considered. The clinical classification includes mild, moderate, severe, and critical types, with mild cases being less common.

11.5.4.2 Imaging manifestations

11.5.4.2.1 X-ray manifestations

Early Stage: Both lungs may show no abnormal changes or only exhibit increased and blurred lung markings, or localized patchy shadows in the peripheral areas of the lungs.

Progressive Stage: In the short term (1—3 days), it can rapidly progress, characterized by increased and disordered lung markings, and the presence of single or multiple patchy areas of increased density in the lung fields, most commonly found in the middle and outer regions of both lower lungs.

Severe Stage: The lung lesions significantly increase, showing multiple patchy areas and large consolidated opacities, or presenting as reduced lung transparency with diffuse and widespread consolidations in both lungs.

Recovery Stage: The extent of the lesions shrinks, with fibrous stripes remaining.

11.5.4.2.2 CT manifestations

Early Stage: In the early stage, patchy ground-glass opacities or small nodular opacities are commonly observed in both lungs, or faint reticular opacities can be seen in the background of ground-glass opacities, with lesions predominantly located subpleurally in the lungs.

Progressive Stage: In the progressive stage (1—3 days), the lesions rapidly expand and change. Ground-glass opacities and consolidation spread to involve multiple lung lobes, progressing from the periphery to the central areas of both lungs. The density of ground-glass opacities increases, and some may exhibit air bronchograms and thickened blood vessel shadows. Significant thickening of interlobular septa with accompanying linear opacities may be present, and a "halo sign" can be seen around the nodules. Newly developed lesions are mainly distributed subpleurally in the middle and lower lobes of both lungs, with some cases showing a small amount of pleural effusion.

Severe Stage: Within 48 h, the lesions increase by more than 50%. Extensive or diffuse consolidations dominate both lungs, with air bronchograms visible within the lesions. In a few cases, there is a significant increase in lung density, resulting in a "white lung" appearance. It may present with a small amount of pleural effusion on one or both sides.

Recovery Stage: The lung lesions become gradually absorbed, and the ground-glass opacities (GGO) decrease in size and density. Residual fibrous stripes and focal bronchiectasis may be visible, and in some cases, complete resolution can occur.

11.5.4.3 Case description

Case 1: The patient is a 70-year-old male who was admitted to the hospital due to "sore throat and cough for 7 days, fever and shortness of breath for 3 days." He reported that 7 days ago, he experienced sore throat without any apparent inducement, along with intermittent cough and yellow sputum, with a predominant dry cough. Two days later, he developed a fever, with the highest body temperature reaching 39.5°C. White blood cell count: 6.8×10^9/L, red blood cell count: 3.99×10^9/L, hemoglobin: 115 g/L, platelet count: 208×10^9/L, percentage of neutrophils: 79.3%, percentage of lymphocytes: 16.2%, percentage of eosinophils: 0.1%, high-sensitivity C-reactive protein: >5 mg/L, C-reactive protein quantification: 69.50 mg/L, quantification of procalcitonin: 1.58 ng/mL, creatine kinase MB isoenzyme: 27.00 U/L, creatine kinase: 387.70 U/L, lactate dehydrogenase: 607.00 U/L. The patient has a history of type 2 diabetes for 7 years, and his wife was diagnosed with COVID-19. See Fig. 11.18 for imaging manifestations.

Case 2: The patient is a 66-year-old female who was admitted to the hospital due to "fever, muscle aches, and fatigue for 5 days." Five days ago, she experienced chills, fever (body temperature of 37.9°C), muscle and joint aches in all four limbs, and fatigue. White blood cell count 3.1×10^9/L, lymphocyte count 0.9×10^9/L, and neutrophil percentage 62.1% , high-sensitivity C-reactive protein >5 mg/L, the quantitative C-reactive protein 55.60 mg/L and the quantitative procalcitonin <0.1 ng/mL.

Creatine kinase MB isoenzyme 10.00 U/L, Creatine kinase 55.90 U/L, Lactate dehydrogenase 312.00 U/L. He has a history of hypertension for more than 20 years, and 10 days before the onset, he attended a gathering of the elderly in an epidemic area. See Fig. 11.19 for imaging manifestations.

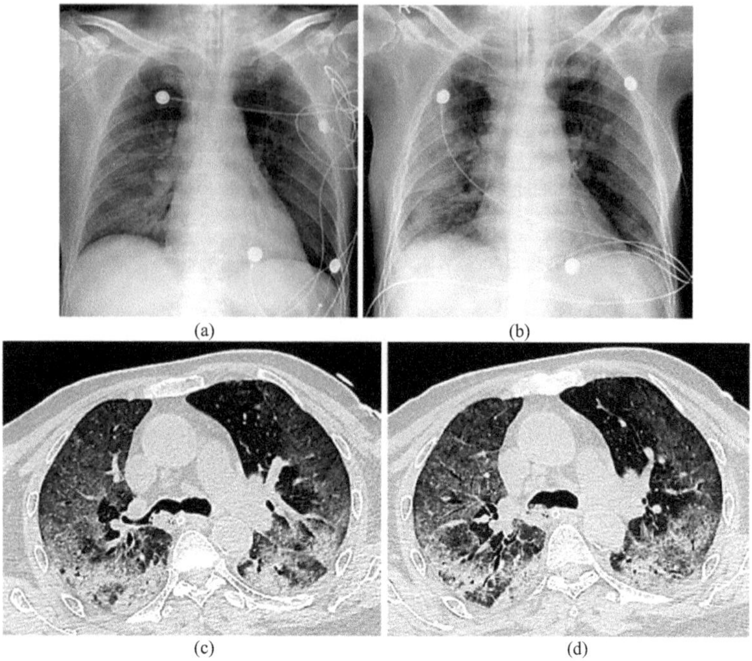

Figure 11.18 Imaging manifestations of the elderly with COVID-19 (I). (a) On the third day of onset, the posteroanterior chest X-ray showed increased transparency in both lung fields, with large areas of ground-glass opacities, and the lung markings appeared indistinct, mainly affecting the right middle and lower lung fields; (b) (follow-up X-ray on the ninth day of onset) the posteroanterior chest X-ray showed an increase in the number of lesions, with higher density of ground-glass opacities, primarily distributed in the middle and peripheral regions of the lung fields; (c) (follow-up CT on the 11th day of onset) the chest CT examinations showed widespread ground-glass opacities in both lungs, with partial consolidation. Inside the lesions, air bronchograms and thickened small blood vessels were visible.

11.5.4.4 Diagnostic key points

(1) There is a history of close contact with confirmed patients before the onset of the disease.

(2) The patient has an acute onset, presenting with sore throat, fever, dry cough, and muscle aches with no apparent inducement.

(3) Chest X-ray shows decreased lung transparency and extensive ground-glass opacities. CT scan reveals widespread ground-glass opacities with partial consolidation, and air bronchograms are visible inside it. In the recovery stage, the lesions are absorbed, initially showing residual fibrous stripes and subpleural lines, and eventually complete resolution of the lung.

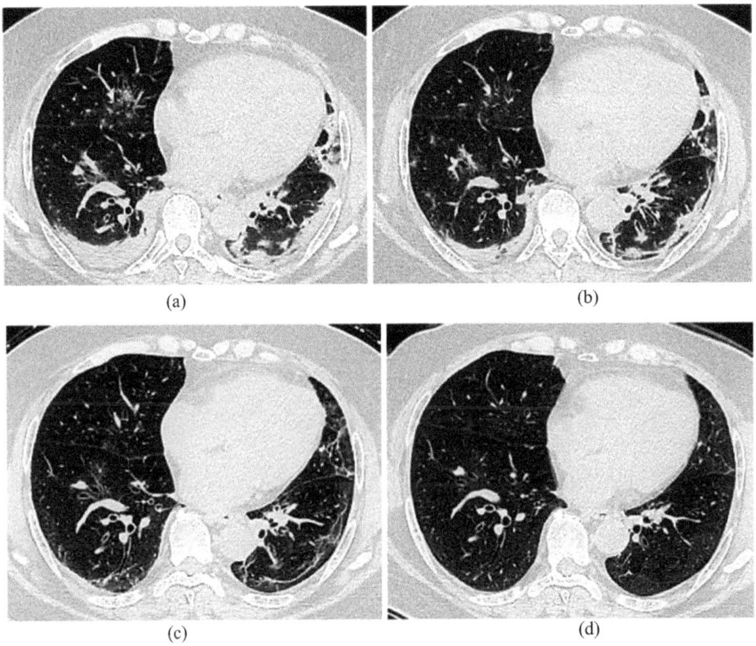

Figure 11.19 Imaging manifestations of the elderly with COVID-19 (II). (a) Scattered patchy ground-glass opacities and consolidation were observed in both lungs, with blurred margins, mainly in the left lower lobe, and a small amount of pleural effusion on the right side; (b) (follow-up after 17 days) the lesions had been absorbed, the consolidation in the left side had decreased, with a few residual fibrous stripes, and a small amount of pleural effusion on the left side; (c) (follow-up after 20 days) the lesions were absorbed further, with scattered residual fibrous stripes and subpleural lines in both lower lobes, the right pleural effusion is absorbed, and a thin band-like pleural thickening is visible; (d) (follow-up after 60 days) the lung lesions had been basically absorbed completely.

11.5.5 Patients with underlying diseases

11.5.5.1 Clinical characteristics

Epidemiological studies show that at least 25.2% of patients have one or more underlying diseases, including diabetes, hypertension, chronic liver disease, chronic obstructive pulmonary disease (COPD), heart disease, malignant tumors, and cardiovascular diseases, among others. Individuals with underlying diseases, as well as the elderly, particularly elderly males, are at higher risk due to their weakened immune systems. After infection, their conditions may worsen rapidly, leading to severe or critical respiratory symptoms, resulting in poor prognosis and higher mortality rates. Based on epidemiological research on 70,000 cases of COVID-19 patients by the

National Center for Disease Control and Prevention, the results show that the overall crude mortality rate of COVID-19 patients in China is approximately 2.3%, among of which individuals aged 60 and above account for 81%. The mortality rate is significantly higher for patients with underlying diseases. The clinical classification is common in moderate, severe and critical.

11.5.5.2 Imaging manifestations

11.5.5.2.1 X-ray

Early Stage: The lung images may be negative or show changes caused by underlying diseases.

Progressive Stage: In the short term (1—3 days), it can rapidly progress, manifesting as increased, thickened, and disordered lung textures, interwoven into a mesh-like or honeycomb pattern, particularly obvious on the lower lobes of both lungs.

Critical Stage: On the basis of diffuse interstitial lesions in both lungs, there are patchy and nodular areas with increased density interspersed. Recovery Stage: The extent of the lesions shrinks, with fibrous stripes remaining.

11.5.5.2.2 HRCT

Early Stage: Patients with underlying diseases, especially the elderly, show prominent interstitial lesions in the early stage, some of which appear as patchy ground-glass opacities, mainly located in the lower lobes of both lungs.

Progressive Stage: The extent of lung lesions increases, advancing from the periphery toward the center, and may be accompanied by a small amount of pleural effusion.

Critical stage: Diffuse thickening of the interlobular septa in both lungs, presenting a honeycomb pattern when associated with intralobular consolidation. There are often scattered patchy and nodular consolidations, and air bronchograms may be visible within the consolidations. Pleural effusion may be present unilaterally or bilaterally.

Recovery stage: After treatment or improvement in the patient's own immune response, the lesions in both lungs become gradually absorbed. There may be residual fibrous stripes, and some small bronchioles may experience traction dilation.

11.5.5.3 Case description

Case 1: The patient is an 80-year-old male who has experienced profuse sweating, palpitations, and slow response for 8 days, along with a fever for 4 days. He has a history of close contact with a confirmed COVID-19 patient and a medical history of chronic bronchitis for 10 years and hypertension for 5 years. Laboratory examinations revealed a white blood cell count of 4.48×10^9/L, lymphocyte count of 0.65×10^9/L, and C-reactive protein of 34.3 mg/L. See Fig. 11.20 for imaging manifestations.

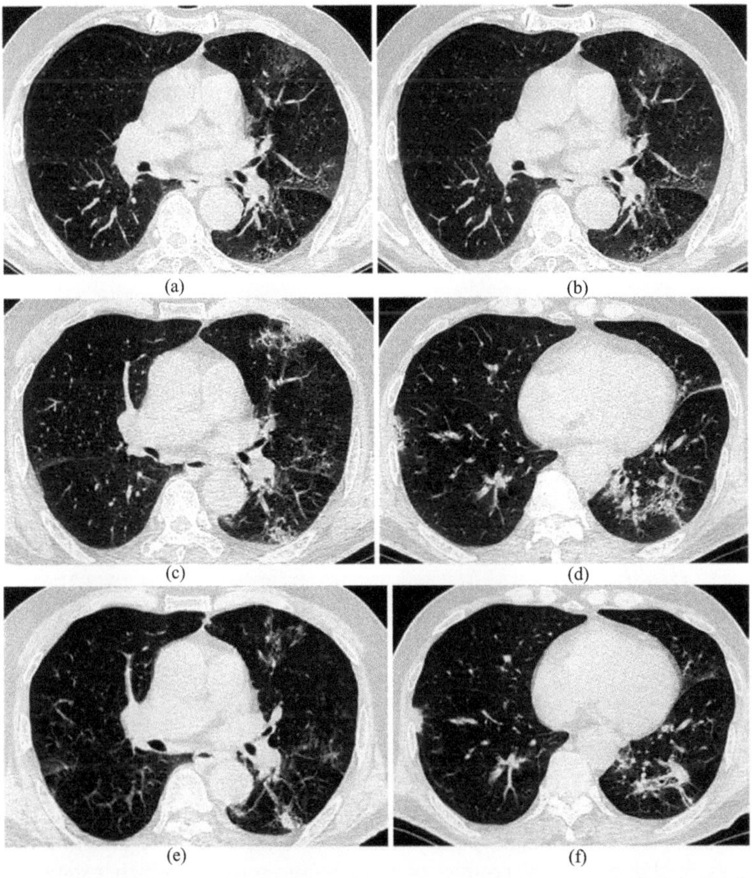

Figure 11.20 Imaging manifestations of chronic bronchitis. (a and b) On the fourth day after the onset, CT scan showed ground-glass opacities in the left upper lobe and both lower lobes of the lungs, accompanied by thickening of interlobular and intralobular septa. (c and d) On the 11th day after the onset, the lesions progressed, with an increase in the extent of involvement and the appearance of consolidations. (e and f) On the 22nd day after the onset, the lesion scope decreased, and fibrous stripes appeared, indicating improvement.

Case 2: The patient is a 73-year-old female who has been experiencing fever, cough, and sputum production for 15 days, along with shortness of breath for 6 days. She has a history of coronary heart disease for 10 years and no history of contact with COVID-19 patients or travel to epidemic areas. Laboratory examinations revealed a white blood cell count of 5.29×10^9/L, with a neutrophil percentage of 79.8%, a lymphocyte percentage of 13.8% and C-reactive protein level of 48.3 mg/L. See Fig. 11.21 for imaging manifestations.

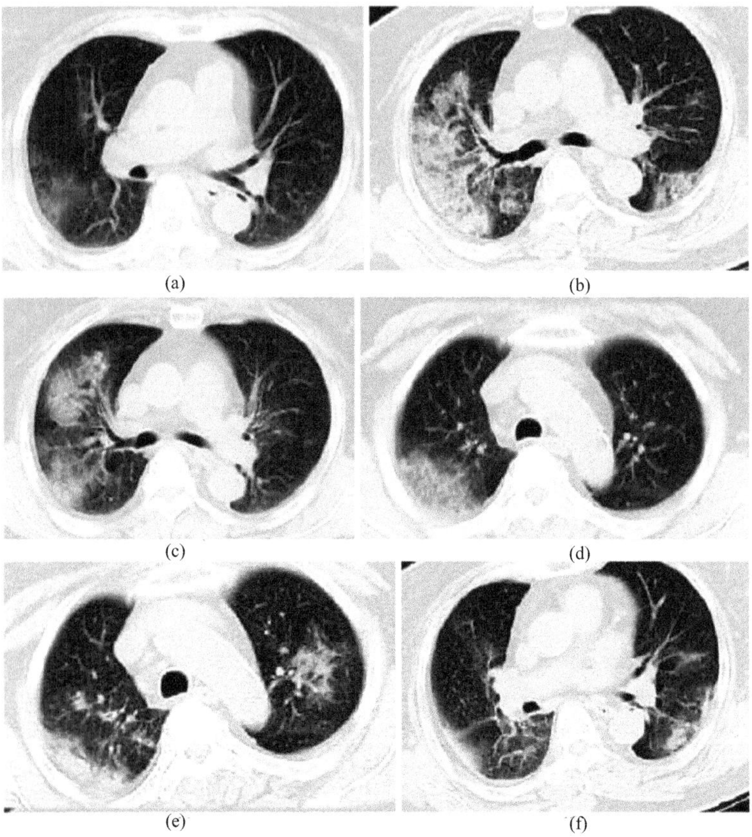

Figure 11.21 Imaging manifestations of coronary Heart disease with COVID-19. (a–c) The CT scan showed multiple patchy areas of increased density and ground-glass opacities in the peripheral regions of both lungs, some of which exhibited solid density with unclear boundaries. In some lesions, thickened intralobular septa was observed; (d–f) on the follow-up CT scan after 4 days, the lesions are in the progressive stage. The involvement in the right upper lobe has decreased, but the density has increased. The lesions in the right middle lobe and lower lobe have expanded and merged, with some areas showing solid density. Additionally, new patchy areas of increased density have appeared in the left lower lobe.

Case 3: The patient is a 56-year-old female who had a fever for 2 days, with the highest temperature reaching 37.7°C. She also experienced chills but no shivering, and there are no symptoms of cough, sputum production, sore throat, or fatigue. She has a history of myocarditis for 7 years. There is no history of contacting with COVID-19 confirmed patients or traveling to epidemic areas, but before the onset of the disease, the patient had traveled to the southern region. Laboratory Examinations: White blood cell count $5.95 \times 10^9/L$, neutrophil percentage 76.6%, lymphocyte percentage 17.3%, C-reactive protein 18.8 mg/L, highly sensitive C-reactive protein >5 mg/L. See Fig. 11.22 for imaging manifestations.

11.5.5.4 Diagnostic key points

(1) Within 14 days of the onset, there is a history of close contact with a confirmed COVID-19 patient or individuals from epidemic areas.

(2) The patient is accompanied by underlying diseases, exhibits typical clinical symptoms such as fever and cough. And the white blood cell count is not elevated, but the C-reactive protein level is increased.

(3) CT Manifestations: In the early stage, the lesions are mainly characterized by ground-glass opacities, distributed in the peripheral and subpleural regions, accompanied by thickening of intralobular septa. In the progressive stage, the extent of the lesions increases, and the density becomes higher, with some areas showing consolidations, along with thickening of intralobular and interlobular septa. In the recovery stage, the lesion scope decreases, and the density decreases as well, with some areas showing fibrous stripes.

(4) Real-time fluorescence RT-PCR testing is positive for SARS-CoV-2 nucleic acid.

11.6 Imaging manifestation of re-positive cases of viral nucleic acid testing

11.6.1 Clinical characteristics

Currently, there have been instances that patients with COVID-19 recovered and discharged from the hospital showed re-positive nucleic acid test. The duration of re-positive results varies, and cases of patients testing re-positive have been reported both at designated isolation facilities and after returning home from isolation. Unlike the initial onset of the disease, where fever, cough, and sore throat are the main symptoms, re-positive cases often exhibit little or no obvious discomfort, or may only

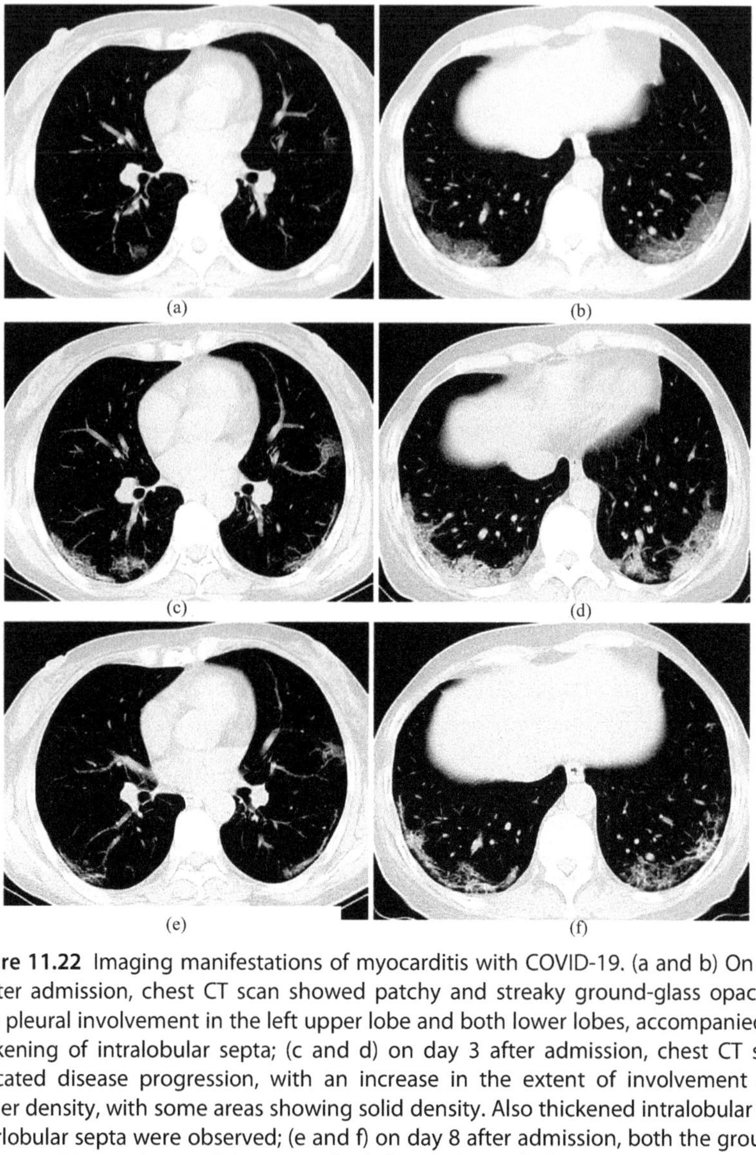

Figure 11.22 Imaging manifestations of myocarditis with COVID-19. (a and b) On day 1 after admission, chest CT scan showed patchy and streaky ground-glass opacities with pleural involvement in the left upper lobe and both lower lobes, accompanied by thickening of intralobular septa; (c and d) on day 3 after admission, chest CT scan indicated disease progression, with an increase in the extent of involvement and higher density, with some areas showing solid density. Also thickened intralobular and interlobular septa were observed; (e and f) on day 8 after admission, both the ground-glass opacities and consolidations in both lungs showed absorption and decreased density. Fibrous stripes were visible, indicating improvement.

present with symptoms such as cough, dry throat, and sore throat. Laboratory examinations showed that indicators such as white blood cell count, C-reactive protein, procalcitonin, lactate, and lactate dehydrogenase have mostly returned to normal levels. When comparing the T lymphocytes at the convalescence stage between re-positive cases and non-re-positive cases, there was no statistically significant difference ($P > .05$). Comparing the CT scans of patients before and after re-positive cases, the majority show lesion absorption and improvement, or no changes, with very few cases showing CT manifestations worsen. The above results suggest that re-positive patients are still in the continuous recovery stage. In addition, the duration of hospitalization for re-positive cases is shorter than that of their initial hospitalization.

Currently, there have been no reports of human-to-human transmission in re-positive cases. It is speculated that this may be due to a significant time gap between the re-positive results and the initial onset of symptoms, during which the SARS-CoV-2 virus may have lost its activity and pathogenicity. As a result, re-positive patients show no clinical symptoms, and no instances of human-to-human transmission have been observed.

The reasons for the re-positive cases in recovered patients may be related to the following points: ① Inadequate adherence to discharge criteria, where the discharge standards outlined in the COVID-19 Diagnosis and Treatment Program were not strictly followed. ② False-negative nucleic acid testing at the time of discharge may be related to the performance of the reagents, sampling and testing methods, and the location of the samples taken. ③ The patient's immune system may be weakened. ④ Discontinuation of antiviral medications beyond the limits of the body's immune function may lead to re-positive results in some patients.

11.6.2 Imaging manifestations

In most cases, the CT manifestations of patients before and after re-positive results show lesion absorption, improvement, or no changes. Rarely, there are cases with CT manifestations showing worsened conditions. Chest CT may show the coexistence of multiple lesions or only present with a single lesion. Ground-glass opacities can be observed in the lungs, appearing as small patchy or nodular opacities. The small bronchi exhibit traction dilation, with increased and thickened bronchovascular bundles. There is also thickening of intralobular septa, along with the presence of fibrous stripes. The pleura is thickened, and there may be pleural effusion present. Huang JiaMei et al. reported a case of exacerbated lesions in a re-positive patient. Before re-positive results, the CT scan showed a few small nodular opacities in the right lower lobe and scattered tiny nodules in both lungs. After re-positive results, the CT scan exhibited a significant increase in the thickening and proliferation of bronchovascular bundles in both lungs compared with before. The lung transparency decreased, showing mild ground-glass opacities.

11.6.3 Case description

Case: Patient, female, 54 years old. He had fever for 6 days, and the SARS-CoV-2 nucleic acid test turned positive 2 days ago. On January 24, 2020, the patient experienced fever, with the highest temperature reaching 38.6°C, accompanied by a slight amount of cough and sputum production, but no other symptoms were present. Self-administration of antipyretic medication resulted in slight improvement. On January 27, 2020, the patient experienced fever again, accompanied by fatigue and intermittent headaches, and sought medical attention at an external hospital. The patient had a history of living in an epidemic area. Laboratory Examinations: White blood cell count 3.77×10^9/L, neutrophil percentage 57%; negative for influenza A virus antigen. On January 28, 2020, and January 30, 2020, the SARS-CoV-2 nucleic acid tests turned out positive, and the patients is considered a confirmed COVID-19 case. From January 30, 2020, to February 18, 2020, the patient was hospitalized in our hospital and treated with antiviral, antipyretic, potassium supplementation, and liver protection therapies, three consecutive SARS-CoV-2 nucleic acid tests returned negative results. The chest CT scan indicated significant improvement in lung lesions, meeting the discharge criteria outlined in The COVID-19 Diagnosis and Treatment Program After discharge, the patient was quarantined at a special site, with normal body temperature and no obvious discomfort. On March 6, 2020, the patient was readmitted to the hospital due to a positive result in the reexamination of SARS-CoV-2 nucleic acid. The patient has no discomfort or symptoms. See Fig. 11.23 for imaging manifestations.

11.6.4 Diagnostic key points

(1) In this case, the imaging manifestations in the acute stage were typical, mainly ground glass shadow and consolidation, and the lesion absorption was slow in the convalescence stage.

(2) The re-positive SARS-CoV-2 nucleic acid results are not correlated with clinical symptoms and imaging manifestations. At the time of re-positive results, the pneumonia lesions have already been almost completely absorbed.

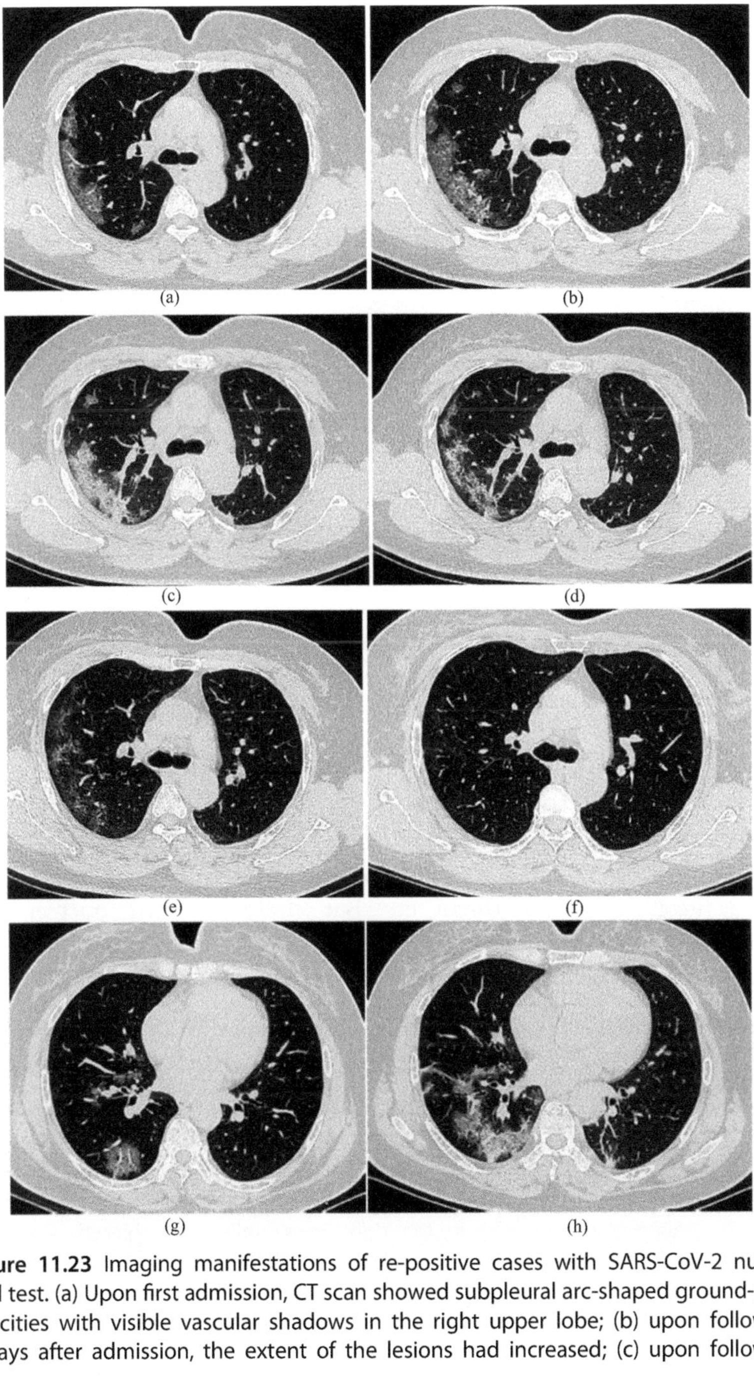

Figure 11.23 Imaging manifestations of re-positive cases with SARS-CoV-2 nucleic acid test. (a) Upon first admission, CT scan showed subpleural arc-shaped ground-glass opacities with visible vascular shadows in the right upper lobe; (b) upon follow-up 4 days after admission, the extent of the lesions had increased; (c) upon follow-up

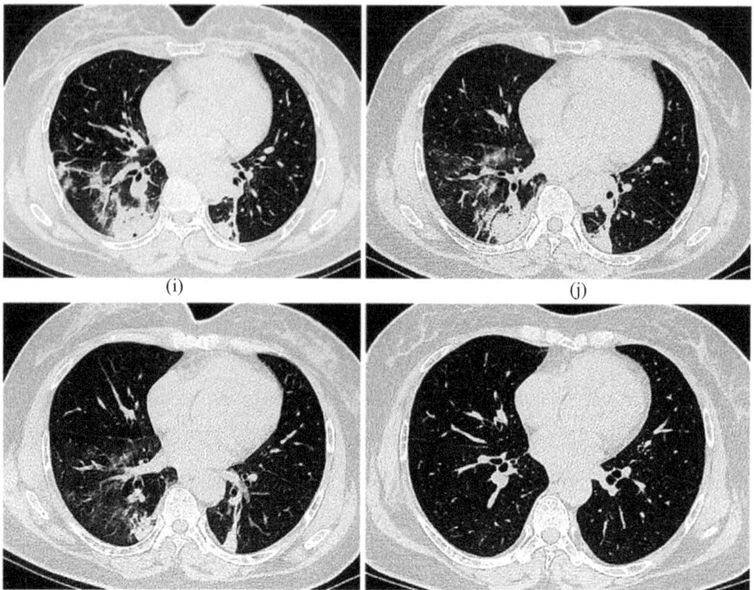

Figure 11.23 Cont'd

11.7 Imaging manifestations of asymptomatic infected persons

11.7.1 Clinical characteristics

Asymptomatic infected persons of COVID-19 refer to individuals who do not have any relevant clinical symptoms or self-perceived recognizable signs such as fever, cough, sore throat, etc., but whose respiratory tract or other samples test positive for SARS-CoV-2 pathogen. Asymptomatic infected persons of COVID-19 can be classified into two situations: ① Individuals

7 days after admission, some lesions showed consolidation, while others had started to be absorbed; (d and e) upon follow-up 11 and 15 days after admission, the lesions had significantly absorbed; (f) after re-admission, CT reexamination showed multiple faint patchy shadows at the original lesion site; (g) upon first admission, CT scan showed multiple ground-glass opacities in the right lower lobe with thickened vascular shadows within; (h–k) upon follow-up (at the same time points as above), the lesions showed increased volume, density, and consolidation, but ultimately showed significant absorption; (l) after re-admission, CT reexamination showed the lesions were almost completely absorbed.

who test positive for nucleic acid but remain asymptomatic throughout the 14-day incubation period without any self-perceived or clinically recognizable symptoms or signs, maintaining the state of Asymptomatic infected persons; ② Individuals who test positive for nucleic acid without any self-perceived or clinically recognizable symptoms or signs at the time of sampling, but subsequently develop certain clinical manifestations, indicating that they are in the state of "Asymptomatic infected persons" during the incubation period.

11.7.2 Imaging manifestations of asymptomatic infected persons

(1) Some asymptomatic infected persons of COVID-19 may show a negative chest CT scan.

(2) Some patients may show patchy or nodular ground-glass opacities, primarily distributed in the subpleural or peribronchial areas, and involving fewer lung lobes. Thickened blood vessels can be seen traversing the lesions, with or without thickening of the interlobular septa. As the disease progresses, the lesion range may expand from the lung periphery to the central part of the lobes. The density of the lesions may transition from ground-glass opacities to a mixture of ground-glass opacities and consolidations or predominantly consolidations. Within the consolidation areas, air bronchograms may be visible. Additionally, there may be thickening of the interlobular septa and intralobular septa, presenting as a "paving stone" sign.

(3) After treatment, the lesions have been absorbed than before with reduced extent and decreased density. Some lesions may be completely absorbed, while others may only present as high-density stripe shadows.

(4) Enlarged mediastinal lymph nodes and pleural effusion are rare.

(5) Compared with COVID-19 patients, asymptomatic infected persons have smaller lesion involvement and fewer lesion numbers. After treatment, these lesions tend to absorb and improve quickly, with some lesions even being absorbed completely.

11.7.3 Case description

Case 1: The patient is a 69-year-old male who was admitted to the hospital due to the discovery of pulmonary shadow for 3 days and a positive result in the SARS-CoV-2 nucleic acid test for 1 day. The patient's son is a confirmed COVID-19 case, and the patient is a close contact. They are under isolation at a local hospital. The patient has a history of good health, with no fever, fatigue, cough, or obvious discomfort. Upon admission, the patient's white blood cell count was decreased, while the counts of neutrophils and lymphocytes, as well as D-dimer levels, were normal. However, C-reactive protein was elevated. See Fig. 11.24 for imaging manifestations.

Case 2: The patient is a 33-year-old male who is a close contact of a confirmed COVID-19 case, his father. He was isolated at a local hospital on

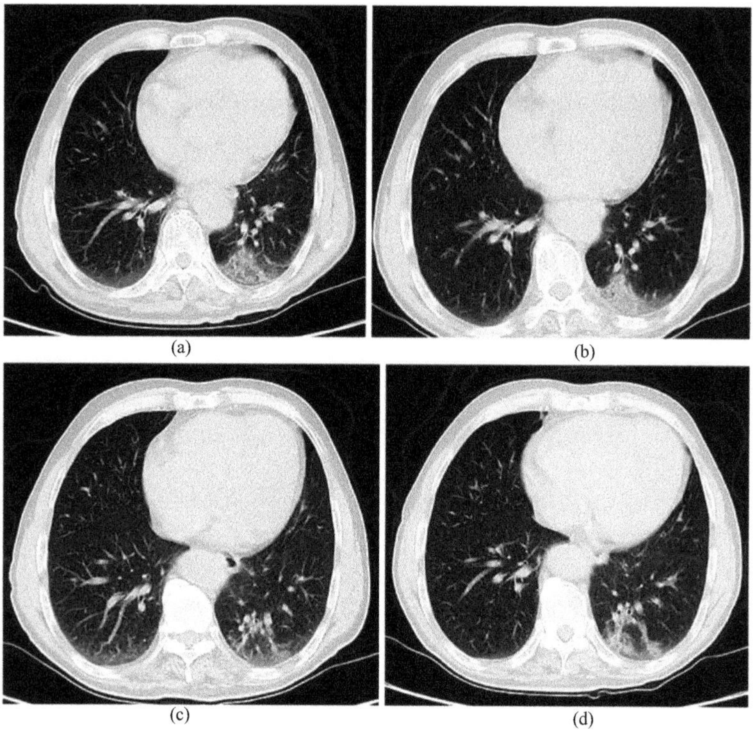

Figure 11.24 Chest imaging manifestations of asymptomatic infected persons of COVID-19 (I). (a) On day 2 after admission, chest CT showed patchy high-density shadows in the left lower lobe, with consolidation at the center of the lesion and ground-glass opacities surrounding it. No obvious abnormalities were observed in the rest of both lungs. This case was classified as a moderate type; (b) on day 4 after admission, chest CT showed increased density in the lesion of the left lower lobe, transitioning from consolidation with ground-glass opacities to consolidation; (c) on day 6 after admission, chest CT showed a reduction in the range and density of the lesion in the left lower lobe; (d) on day 8 after admission, chest CT indicated obvious absorption of the lesion in the left lower lobe, with reduced density and a smaller extent. (This case is provided by Baoding People's Hospital).

January 30, 2020. He had a history of good health and did not experience any obvious discomfort during the isolation period. On February 5, 2020, he developed a fever without any obvious inducement, with the highest body temperature reaching 38.9°C. As a result, he was admitted to the hospital for treatment on February 7, 2020. Upon admission, the patient's white blood cell count, neutrophil count, and lymphocyte count were all within normal range, but the C-reactive protein level was elevated. See Fig. 11.25 for imaging manifestations.

11.7.4 Diagnostic key points

(1) Within 14 days before the onset of the disease, there was a history of contact with confirmed COVID-19 patients.

(2) Real-time fluorescence RT-PCR testing for SARS-CoV-2 nucleic acid is positive, with no obvious clinical manifestations or significant discomfort, indicating an Asymptomatic infected person's state.

(3) The chest CT manifestations in this case demonstrate a pattern consistent with the progression from early stage to progressive stage, peak stage, and recovery stage.

11.8 Imaging manifestations of COVID-19 complications

11.8.1 Clinical characteristics

COVID-19 is primarily characterized by symptoms such as fever, dry cough, and fatigue. Some patients may also experience upper respiratory tract and gastrointestinal symptoms, including nasal congestion, runny nose, and diarrhea. Severe cases of COVID-19 often develop respiratory distress around 1 week after onset. In severe cases, there is rapid progression to acute respiratory distress syndrome (ARDS), viral septic shock, refractory metabolic acidosis, coagulation dysfunction, and multiple organ failure that is difficult to correct. The prognosis for most patients is good, while a small number of patients may have a severe condition that involves multiple systems such as respiratory, neurological, cardiovascular, and digestive systems, which can lead to death.

11.8.2 Imaging manifestations of COVID-19 complications

11.8.2.1 Respiratory complications

In addition to typical chest CT manifestations, severe COVID-19 patients may experience sudden increases in alveolar pressure due to diffuse alveolar

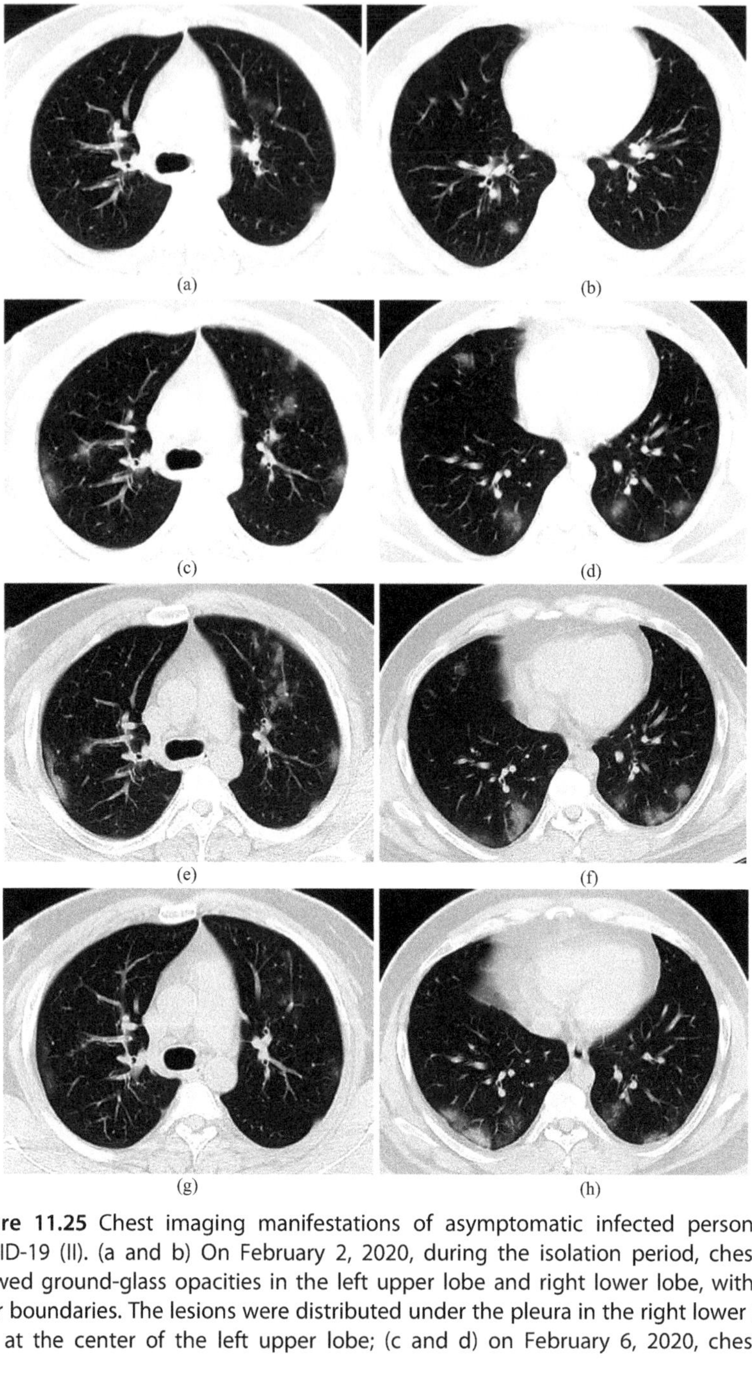

Figure 11.25 Chest imaging manifestations of asymptomatic infected persons of COVID-19 (II). (a and b) On February 2, 2020, during the isolation period, chest CT showed ground-glass opacities in the left upper lobe and right lower lobe, with unclear boundaries. The lesions were distributed under the pleura in the right lower lobe and at the center of the left upper lobe; (c and d) on February 6, 2020, chest CT

damage (DAD), leading to alveolar rupture. This can also result in mediastinal emphysema, giant pulmonary bullae, and pneumothorax. Furthermore, ARDS caused by viral infection can lead to secondary complications, such as invasive pulmonary aspergillosis. In patients with COVID-19-related invasive pulmonary aspergillosis, chest CT manifestations typically show bilateral ground-glass opacities, peripherally distributed nodular infiltrates, and cavitary lesions. Some of these lesions may exhibit an "air crescent" sign.

11.8.2.2 Circulatory complications

COVID-19 patients may experience various complications, including acute myocardial injury, various cardiac arrhythmias, heart failure, myocardial infarction, arterial and venous thromboembolic disorders, and disseminated intravascular coagulation. Among them, acute myocardial injury is the most common cardiovascular complication in COVID-19. MRI T_2-WI signal is increased, and early myocardial enhancement is observed on MRI contrast-enhanced scans; In affected myocardium, T_2-WI signal is increased, and there is a reduction in ventricular wall motion, indicating myocardial infarction; CT angiography reveals partial intraluminal filling defects and pulmonary artery stenosis, suggesting acute pulmonary embolism.

11.8.2.3 Digestive complications

Digestive complications can occur due to drug-related effects, systemic inflammatory response, and multiorgan dysfunction resulting in secondary liver injury. Pathologically, these complications manifest as acute hepatocyte swelling, degeneration, necrosis, and fragmentation, with decreased hepatic blood perfusion. On CT scans, the liver density appears diffusely or focally decreased.

showed an increase in the size and density of the lesions in the left upper lobe and right lower lobe. New ground-glass opacities were observed in the upper lobes of both lungs, the right middle lobe, and the lower lobe of the left lung; (e and f) on February 7, 2020, the first day of admission, the lesions in the left upper lobe and lower lobe had increased, and the density of the lesions in the right lung had increased. The lesions had transformed from ground-glass opacities to consolidations; (g and h) on the ninth day of admission, some lesions in the right upper lobe had already been absorbed, and the lesions in both upper and lower lobes of the lungs had significantly absorbed, with a reduction in size and a decrease in density.

11.8.2.4 Neurological complications

Only a small proportion of COVID-19 patients experience neurological involvement, which may manifest as sudden onset of consciousness disorders, dizziness, and other symptoms after the onset of COVID-19. CT and MRI findings suggest changes indicative of viral encephalitis, encephalomyelitis, or acute necrotizing encephalopathy.

11.8.3 Case description

Case 1: The patient is a 38 years old male with a history of traveling to an epidemic area, and is admitted to the hospital due to bilateral hearing loss and tinnitus for 2 months. On the day of admission, the otoscopic examination of the tympanic cavity was negative; On the second day of admission, the patient had a body temperature of 38.1°C, and Laboratory examinations showed a normal neutrophil count and a decrease in lymphocyte percentage. Multiple respiratory pathogen screenings were all negative. Throat swab nucleic acid testing showed a positive RT-PCR reaction for SARS-CoV-2. On the 10th day of admission, the patient developed high fever and cough, and started receiving high-flow nasal cannula (HFNC) oxygen therapy. On the 20th day of admission, the patient's respiratory distress worsened, and they experienced severe hypoxemia (oxygen saturation below 70%). Subsequently, the patient continued to receive HFNC oxygen therapy until the last CT follow-up showed improvement of the lesions and met the requirements for extubation. See Fig. 11.26 for imaging manifestations.

Cases 2 to 6: In addition to symptoms such as fever, dry cough, and fatigue, all patients experienced severe acute respiratory distress syndrome (ARDS) and tested positive for SARS-CoV-2 and human metapneumovirus (hMPV) by polymerase chain reaction (PCR). Bronchoalveolar lavage fluid culture revealed Aspergillus fumigatus, and the serum galactomannan test was positive. See Fig. 11.27 for imaging manifestations.

11.8.4 Diagnostic key points

(1) In addition to symptoms such as fever, dry cough, and fatigue, the patient exhibited severe acute respiratory distress syndrome (ARDS).

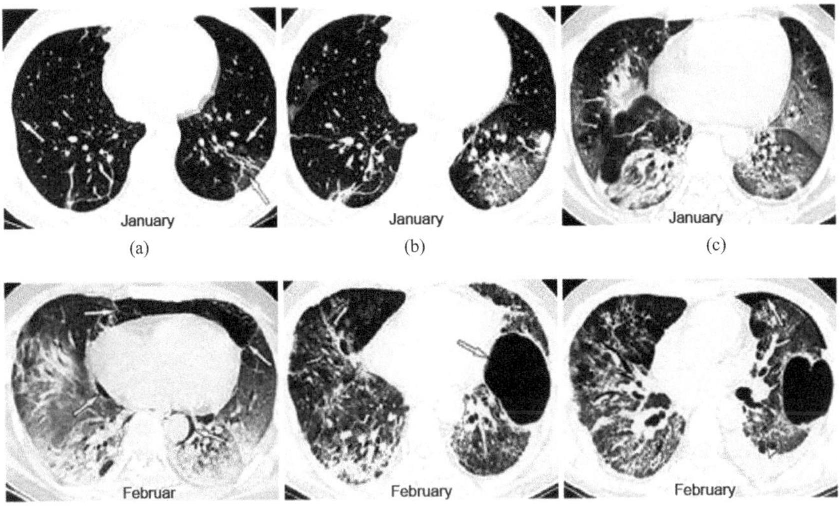

(a) (b) (c)
(d) (e) (f)

Figure 11.26 Imaging manifestations of chest lesion with COVID-19 at progressive and recovery stage. Upon admission, chest CT scan (a) showed multifocal ground-glass opacities (*arrows*) along the bronchovascular bundles and subpleural regions in both lower lobes of the lungs. On the third day of admission (b) and the seventh day of admission (c), chest CT scans showed rapid progression of ground-glass opacities and consolidations in both lower lobes of the lungs. On the 11th day of admission (d), chest CT scan showed consolidations in the subpleural regions of both lower lobes of the lungs and mediastinal emphysema (*arrows*). After a series of supportive treatments, on the 26th day (e), chest CT scan showed improvement in lung lesions and mediastinal emphysema, but a giant pulmonary bulla was observed in the left lung (*arrow*). On the 34th day (f), the final follow-up CT scan showed left-sided pneumothorax (*arrow*) and pleural effusion (*arrow*).

(2) SARS-CoV-2 and hMPV tested positive by RT-PCR. Bronchoalveolar lavage fluid culture showed the growth of Aspergillus fumigatus, and the serum galactomannan test was positive.

(3) Even in the absence of immunodeficiency, invasive pulmonary aspergillosis can occur, especially on the basis of viral infection. Chest CT scan showed ground-glass opacities, interstitial changes, nodular infiltrations, and cavitary lesions, with some lesions exhibiting a crescent sign. Severe COVID-19 patients experienced significant coughing, leading to diffuse alveolar damage (DAD and making the alveoli prone to rupture. A sudden increase in alveolar

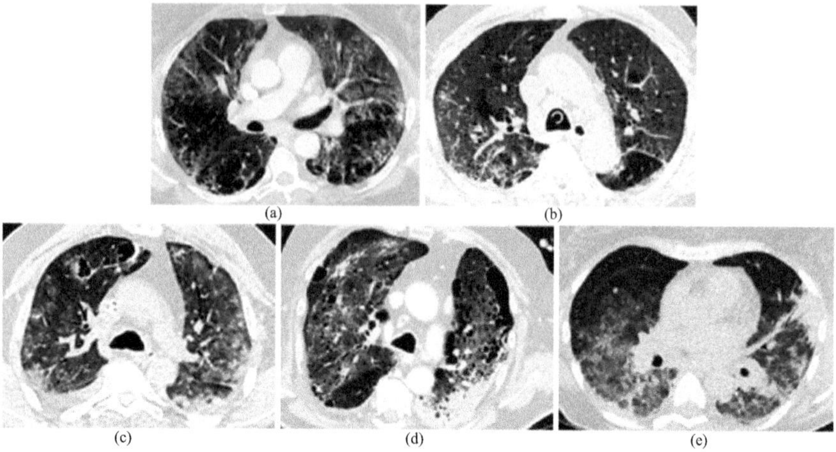

Figure 11.27 Imaging manifestations of COVID-19-related invasive pulmonary aspergillosis. Case 2 (a) Multiple ground-glass opacities with extensive "paving stone" sign and peripherally distributed nodular lesions in both lungs; case 3 (b) multiple ground-glass opacities with infiltrative small nodules, some of which have a diameter of up to 1 cm in both lungs; case 4 (c) multiple ground-glass opacities, diffuse nodular infiltrations, and cavitary lesions, with some showing a crescent sign in both lungs; case 5 (d) giant bullous emphysema and ground-glass opacities in both lungs, with interstitial changes and nodular infiltrations; case 6. (e) Multiple ground-glass opacities in both lungs, with small areas of "paving stone" sign and smaller infiltrative nodules.

pressure can cause mediastinal emphysema, which may be accompanied by interstitial pulmonary emphysema. Giant pulmonary bullae, as well as multiple bullae in the subpleural lung regions, are prone to rupture and can lead to pneumothorax.

11.9 Imaging manifestations of heart disease with COVID-19

11.9.1 Clinical characteristics

COVID-19 primarily affects the lungs but can also involve the heart, reproductive system, and central nervous system, etc. Currently, several studies have shown that a certain proportion of patients with COVID-19 exhibit elevated cardiac injury biomarkers, indicating the presence of myocardial injury. Furthermore, patients with myocardial injury have a higher risk of mortality. As early as late January 2020, researchers noticed that among the first 41 confirmed COVID-19 cases in China, 5 cases (12%) showed evidence of acute myocardial injury. Subsequent larger-scale single-center cohort studies revealed that among 416 hospitalized patients, 82 cases (19.7%) exhibited evidence of myocardial injury. Regardless of the time of admission or symptom onset, patients with myocardial injury had a significantly higher risk of mortality compared with those without myocardial injury, with hazard ratios of 4.26 and 3.41, respectively. In addition to changes in cardiac injury biomarkers, the cardiac involvement in COVID-19 includes abnormal electrocardiograms (e.g., ST-segment elevation, resembling ischemic symptom), cardiogenic shock, heart failure, pericardial effusion, cardiac tamponade, and fulminant myocarditis. Some patients' endomyocardial biopsy results showed myocardial T lymphocyte infiltration.

Currently, apart from endomyocardial biopsy, cardiac magnetic resonance imaging is an important diagnostic tool for myocarditis. Based on the updated 2018 Expert Consensus of Cardiovascular Magnetic Resonance Examination for Nonischemic Myocardial Inflammation (Lake Louise Criteria), the CMR diagnostic criteria for myocarditis are as follows: ① Focal or diffuse T_2-WI signal intensity elevation or T_2 prolongation; ② Focal or diffuse T_1 prolongation on precontrast imaging or increased extracellular volume, or nonischemic pattern of delayed enhancement. The diagnosis of myocarditis can be made when both of the above criteria are met simultaneously. In the current case reports focusing on COVID-19-induced myocarditis, most of them show diffuse or focal T_2, T_2 mapping, and T_1 mapping changes, while delayed enhancement is not prominent. Some cases also exhibit diffuse delayed enhancement with pericardial effusion. These changes differ from the typical manifestations of myocarditis caused by other reasons, where there is often a more prominent delayed enhancement.

In conclusion, cardiac involvement in COVID-19 is relatively common, and cardiac magnetic resonance imaging can detect inflammatory changes in the heart, which is of significant value for disease diagnosis and management.

11.9.2 MR manifestations of heart diseases

(1) COVID-19 involves the heart with myocardial damage, mainly manifested as myocardial edema.

(2) Cardiac Magnetic Resonance Evaluation using T2-WI: Localized T2 STIR signal elevation or the ratio of a global T2-STIR signal to skeletal muscle signal >2, or increased T2 mapping relaxation time.

(3) Cardiac Magnetic Resonance Evaluation using T1 WI: Localized or global T1 and ECV increase, or LGE characterized by nonischemic myocardial injury enhancement.

(4) A diagnosis of myocarditis can be made if any two of the criteria mentioned in (2) and (3) are met on cardiac magnetic resonance imaging.

11.9.3 Case description

Case 1: Patient, female, 24 years old. She had cough for 7 days, and SARS-CoV-2 nucleic acid tested positive for 5 days. She had a history of traveling abroad. She suffered from palpitations and chest pain for 3 days, with a troponin level of 0.56 ng/mL, and she had no history of previous heart-related diseases. See Fig. 11.28 for imaging manifestations.

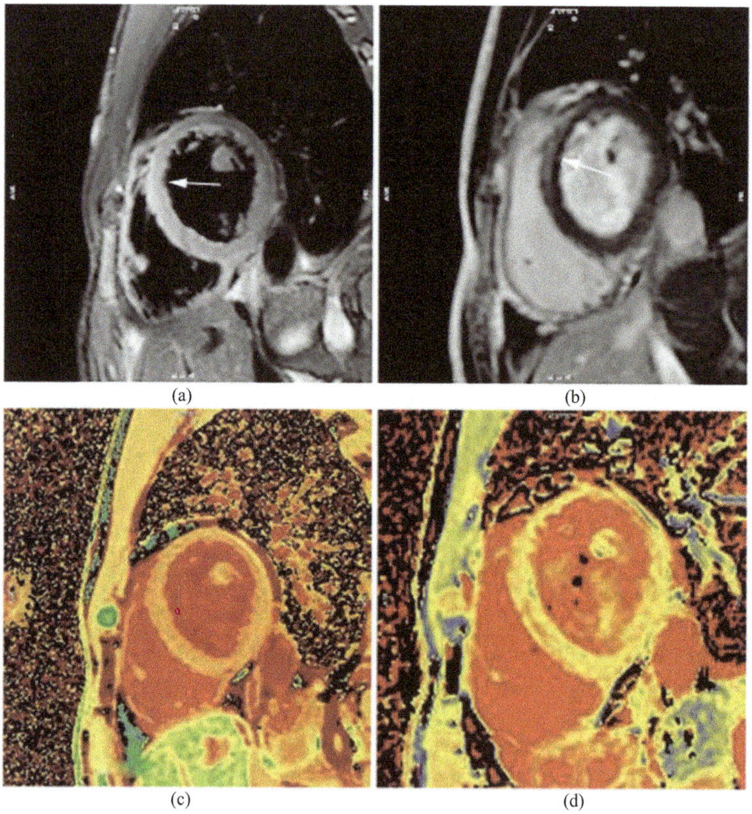

Figure 11.28 CMR imaging manifestations of acute myocardial involvement in COVID-19 patients (I). (a) T2-STIR showed high signal in the interventricular septum; (b) LGE indicated suspicious linear high signal in the mid-myocardium; (c) T1 mapping showed widespread and diffuse high signal; (d) T2 mapping showed widespread and diffuse high signal.

Case 2: Patient, male, 22 years old. He had cough for 7 days, and SARS-CoV-2 nucleic acid tested positive for 5 days. She had a history of traveling abroad. There was no cardiac symptoms or positive signs. The troponin level was 0.25 ng/mL, and the cardiac ultrasound examination was negative. See Fig. 11.29 for imaging manifestations.

11.9.4 Diagnostic key points

(1) The patient had a confirmed history of COVID-19 with a positive nucleic acid test.

(2) The patient showed a mild elevation in troponin levels.

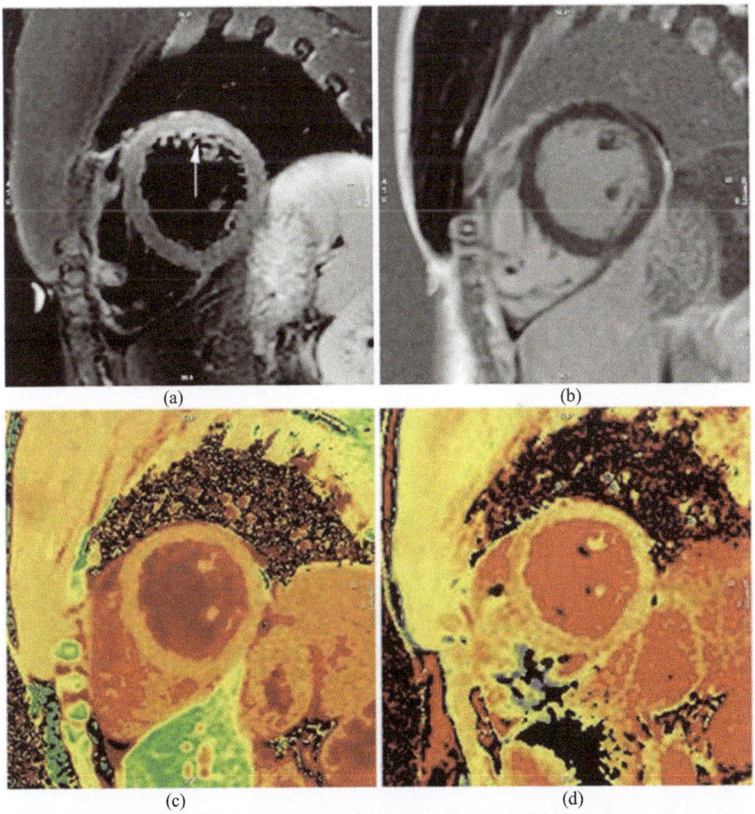

Figure 11.29 CMR imaging manifestations of acute myocardial involvement in COVID-19 patients (II). (a) T2-STIR shows high signal intensity in the anterior wall and anterior lateral wall; (b) LGE negative; (c) T1 mapping indicates extensive and diffuse high signal intensity; (d) T2 mapping indicates extensive and diffuse high signal intensity.

(3) Cardiac magnetic resonance imaging showed localized T2 STIR signal elevation, along with varying degrees of increased T1 values and extracellular volume (ECV) values, indicating changes in myocardial inflammation.

11.10 Imaging manifestations of neurological lesions with COVID-19

11.10.1 Clinical characteristics

SARS-CoV-2 infection is not always limited to the respiratory tract; in certain cases, it can invade the central nervous system and cause neurological lesions. COVID-19 patients may experience neurological symptoms such as headaches, dizziness, muscle pain, and loss of smell, as well as uncommon neurological manifestations, including SARS-CoV-2 infection-related encephalopathy, encephalitis, necrotizing hemorrhagic encephalopathy, stroke, seizures, rhabdomyolysis, and Guillain-Barré syndrome. The incidence of SARS-CoV-2 neurological complications is not yet clear. However, what can be determined is that severe COVID-19 patients are more likely to experience neurological symptoms compared with mild cases. Patients who died from COVID-19 typically undergo brain edema and neurodegeneration.

11.10.2 Imaging manifestations of neurological lesions

Only a small proportion of COVID-19 patients experience neurological abnormalities, and their CT or MRI scans may appear normal or show atypical imaging manifestations. For example, in magnetic resonance diffusion-weighted imaging with high *b*-values, high signal intensity along the ventricular walls suggests ventriculitis; Bilateral cerebral lesions with low signal intensity on T_1-WI and high signal intensity on T_2-WI and T_2-fluid-attenuated inversion recovery (FLAIR) may indicate viral encephalitis; High-density lesions within the brain and ventricles on CT scans suggest acute intracerebral hemorrhage; On magnetic resonance imaging, low signal intensity on T1-WI and diffuse, multiple high signal intensity lesions on T2-WI, often appearing round, oval, or cotton-ball-like, may indicate acute disseminated encephalomyelitis.

11.10.3 Case description

Case 1: The patient is a 51-year-old male, confirmed COVID-19 case, and admitted to the hospital due to sudden onset of altered consciousness. He has no history of cerebrovascular disease, hypertension, coronary heart disease, or other underlying medical conditions. See Fig. 11.30 for imaging manifestations.

Case 2: The patient is a 54-year-old male, confirmed COVID-19 case, and admitted to the hospital due to dizziness after sedation. He has no history of cerebrovascular disease, hypertension, coronary heart disease, or other underlying medical conditions. See Fig. 11.31 for imaging manifestations.

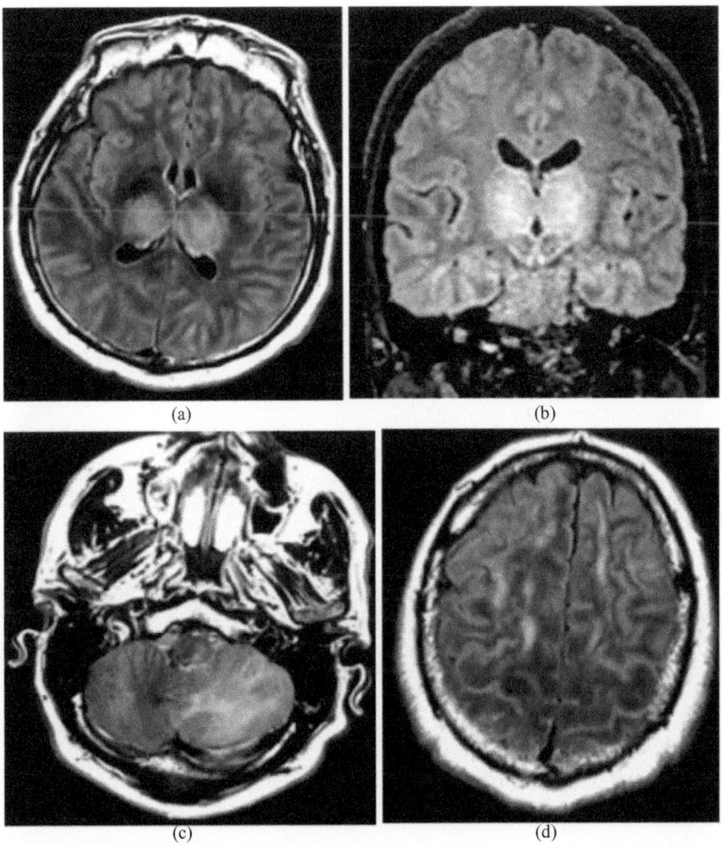

Figure 11.30 Imaging manifestations of acute necrotizing encephalopathy in COVID-19 patients. (a) Axial and (b) coronal T2-FLAIR images showed high signal intensity lesions in the bilateral thalami; with involvement of the bilateral cerebellar hemispheres (c) and cerebral white matter (d).

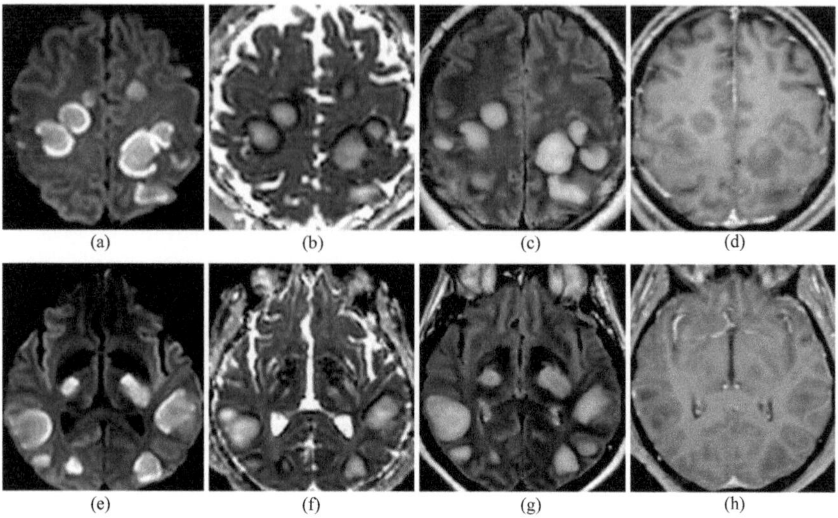

Figure 11.31 Imaging manifestations of nonfusional multiple leukodystrophy in COVID-19 patients. FLAIR and diffusion-weighted imaging showed noncontiguous multifocal high-signal cerebral white matter lesions with variable enhancement. (a and b) are axial diffusion-weighted images; (c and d) are apparent diffusion coefficient (ADC) maps; (e and f) are axial contrast-enhanced FLAIR images; (g and h) are enhanced T1-weighted MRI scans; diffusion-weighted imaging and FLAIR show multiple nodular high signal lesions in the subcortical and cortical spinal tract, with very mild mass effect on adjacent structures; (c and d) demonstrate increased ADC values in the central area corresponding to vasogenic edema and decreased ADC values in the surrounding region corresponding to cytotoxic edema; (g and h) show small areas of mild enhancement after contrast injection.

11.10.4 Diagnostic key points

(1) The patient had a history of traveling to an epidemic area or contacting with individuals from an epidemic area 14 days before the onset of symptoms.

(2) The patient presented with neurological symptoms or exhibited neurological manifestations on imaging.

(3) SARS-CoV-2 was detected in cerebrospinal fluid or brain tissue through nucleic acid testing, or there was evidence of specific intrathecal antibodies against SARS-CoV-2.

(4) SARS-CoV-2 nucleic acid testing was positive, and there were neurological symptoms, with other neurological diseases being ruled out.

11.11 Imaging manifestations of digestive system lesions with COVID-19

11.11.1 Clinical characteristics

SARS-CoV-2 infection-induced gastrointestinal lesions can involve the liver, gallbladder, and gastrointestinal tract, and common symptoms in patients include diarrhea, nausea, vomiting, abdominal pain, and loss of appetite, etc. The autopsy confirmed the SARS-CoV-2 RNA and viral particles can be tested in the liver, gallbladder, gastrointestinal tract, and oral mucosa of the patient. SARS-CoV-2 infection leads to liver injury, mainly manifested by elevated transaminases. Liver function and abnormal transaminase levels are positively correlated with the severity of COVID-19. The proportion of severe cases with gastrointestinal symptoms is higher than that of nonsevere cases. Digestive symptoms in COVID-19 patients typically appear later than respiratory symptoms. Approximately 50% of COVID-19 patients have a positive fecal viral nucleic acid test. In some patients, fecal viral nucleic acid testing remains positive even after a negative result in throat swab virus testing.

11.11.2 Imaging manifestations of digestive system lesions

(1) COVID-19 involvement in the gastrointestinal tract manifests as enteritis and mesenteric inflammatory changes.

(2) The presentation includes segmental intestinal wall swelling and thickening, indistinct perienteric fat space, enhanced mucosal enhancement of the intestinal wall on contrast-enhanced scans, uneven thickening of the mesenteric vasculature, and blurred edges. Mesenteric lymph nodes are increased, exudates from the mesenteric fat spaces are changed and the edges are blurred.

(3) There have been reports of some patients experiencing small bowel and large bowel dilation, increased fluid accumulation, circumferential enhancement of the intestinal wall, and even manifestations of strangulating intestinal obstruction.

(4) Compared with the pulmonary lesions in COVID-19 patients, most of the gastrointestinal lesions involved in COVID-19 can be completely resolved after treatment.

11.11.3 Case description

Case 1: Patient, female, 54 years old. She was confirmed with COVID-19 2 months ago, received symptomatic treatment in a mobile cabin hospital and was discharged after two consecutive negative nucleic acid tests. The follow-up nucleic acid test showed a positive result. In the past 2 days, the patient experienced the new onset of paroxysmal dull pain in middle and upper abdomen, with associated nausea and back pain. There was no nausea, diarrhea, chills or fever. Examination after Admission: C-reactive protein 5 mg/L and white blood cell count 19.6 × 109/L. See Fig. 11.32 for imaging manifestations.

Case 2: Patient, male, 29 years old. He was confirmed with COVID-19 for 17 days, and experienced recurrent fever and diarrhea for 2 days. Examination after Admission: C-reactive protein 5 mg/L, white blood cell count 7.45×10^9/L and neutrophilic granulocyte percentage 77.2%. See Fig. 11.33 for imaging manifestations.

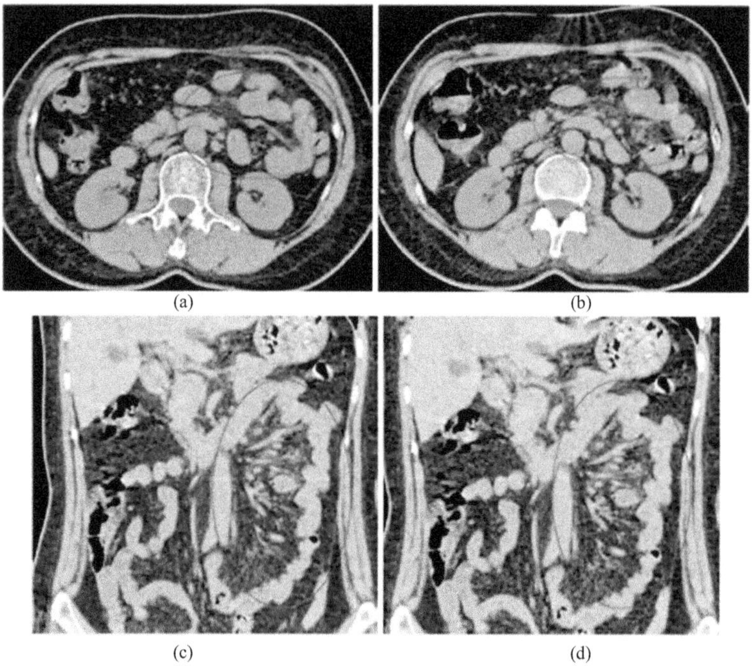

Figure 11.32 Abdominal imaging manifestations of patients with COVID-19. (a–d) show slight swelling and thickening of the intestine wall in the left upper abdomen, even thickening of the mesenteric vasculature, indistinct perienteric fat spaces, and increased number of mesenteric lymph nodes.

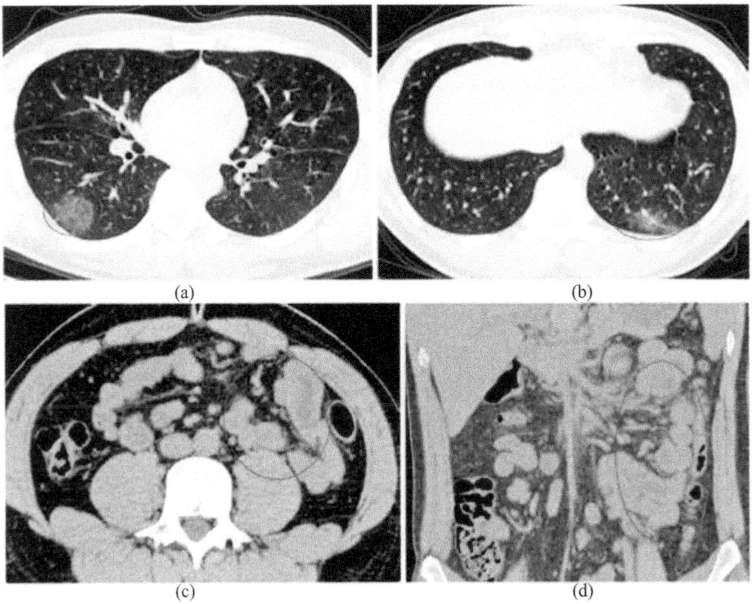

Figure 11.33 Imaging manifestations of chest and abdomen in patients with COVID-19. (a and b) Patchy ground-glass opacities in the subpleural regions of the lower lobes of both lungs, with slightly dilated vessels and slightly blurred vessel edges within the lesion in the left lower lobe; (c and d) localized thickening of the intestinal wall in the left abdominal area, with a small amount of fluid accumulation in the local intestinal lumen, indistinct perienteric fat spaces, and an increased number of mesenteric lymph nodes.

11.11.4 Diagnostic key points

(1) COVID patients with gastrointestinal lesions haD a confirmed history of COVID-19 infection, with or without pulmonary manifestations, and a positive SARS-CoV-2 nucleic acid test.

(2) The patient experienced gastrointestinal symptoms such as abdominal pain and diarrhea, etc.

(3) Abdominal CT Scan: Thickening of the intestinal wall and slight fluid accumulation, mesenteric vascular dilation and blurring, perienteric exudation, and increased number of mesenteric lymph nodes, indicating inflammatory changes in the intestine and mesentery.

Bibliography

[1] Wang W, Yunzhuang X, Qi S, et al. Association of early virus nucleic acid in throat swabs and clinical characteristics in COVID-19 patients early in their illness. J Chongqing Med Univ 2020;45(7):925–8.
[2] Professional Committee on Radiology of Infection and Inflammation of the Chinese Research Hospital Association, (Infectious Disease) Imaging Working Committee of Chinese Association of STD and AIDS Prevention and Control and Infectious Disease Group of Chinese Society of Radiology. Guideline for medical imaging in auxiliary diagnosis of coronavirus disease. Chin J Med Imaging Technol 2020;36(3):321–31.
[3] Zheng Q, Lu Y, Tan L, et al. Advances in clinical and imaging studies of novel coronavirus pneumonia. Electron J Emerg Infect Dis 2020;5(2):140–4.
[4] Specialized Committee for Radiology of Infection and Inflammation of Chinese Research Hospital Association, Professional Committee for Infection Imaging of Chinese Medical Doctor Association Radiologist Branch, Infectious Disease Group of Chinese Society of Radiology, (Infectious Disease) Imaging Working Committee of Chinese Association of STD and AIDS Prevention and Control. Guidelines for imaging diagnosis of novel coronavirus pneumonia (2020 second edition, simple version). J Cap Med Univ 2020;41(2):168–73.
[5] Chen L, Yang X, Zheng N, et al. Clinical characteristics of 67 discharged cases of corona virus disease 2019. Chin J Nosocomiol 2020;30(13):1942–6.
[6] General Office of National Health Committee, State Administration of Traditional Chinese Medicine Office. The COVID-19 Diagnosis and Treatment Program (Trial Fifth Revised Edition) [S/OL]. http://www.gov.cn/zhengce/zhengceku/2020-02/09/5476407/files/765d1e65b7d1443081053c29ad37fb07.pdf.
[7] Wang C, Jianhong L, Yang D, et al. Value of CT features of multiphase lesions for early clinical efficacy prediction of patients with 2019-nCoV pneumonia. Radiol Practice 2020;35(6):711–4.
[8] Ming T, Xiaoyuan F, Liu S, et al. Guidance on imaging diagnosis and infection Control of novel coronavirus pneumonia. Chin Comput Med Imag 2020. https://doi.org/10.19627/j.cnki.cn31-1700/th.20200309.001.
[9] Hong P. Value of imaging in diagnosis and treatment of novel coronavirus. Practic J Clin Med 2020;17(2):13–5.
[10] Huang D, Zhang L, Qingxin G, et al. Analysis of dynamic changes of chest imaging of moderate type of novel coronavirus pneumonia. Guangdong Med J 2020;(12):1–4.
[11] Special Committee on Radiology of Infection and Inflammation of Chinese Research Hospital Association. Committee of the infectious diseases radiology Group of Chinese Society of Radiology; Committee of the infectious diseases Radiology Section of Chinese medical Doctor association; infectious disease imaging Group. Guideline for medical imaging in auxiliary diagnosis of coronavirus disease 2019. Chin J Med Imag Technol 2020;36(3):1–11.
[12] Wu Z, Mcgoogan JM. Characteristics of and important lessons from the coronavirus disease 2019 (COVID-19) outbreak in China: summary of a report of 72 314 cases from the Chinese center for disease Control and prevention. JAMA 2020;323(13):1239–40.
[13] Li X, Wang Y, Bai Y, et al. PET/MR and PET/CT in a severe COVID-19 patient. Eur J Nucl Med Mol Imag 2020;47(10):2478–9.
[14] Jaffe AS, Cleland JGF, Katus HA. Myocardial injury in severe COVID-19 infection. Eur Heart J 2020;41(22):2080–2.
[15] Yang Q, Liu Q, Xu HB, et al. Imaging of coronavirus disease 2019: a Chinese expert consensus statement. Europ J Radiol 2020;127:1–6.
[16] Li K, Wu J, Wu F, et al. The clinical and chest CT features associated with severe and critical COVID-19 pneumonia. Invest Radiol 2020;55(6):327–31.

[17] Zheng Q, Lu Y, Lure F, et al. Clinical and radiological features of novel coronavirus pneumonia. J X Ray Sci Technol 2020;28(3):391−404.

[18] Valette X, Du Cheyron D, Goursaud S. Mediastinal lymphadenopathy in patients with severe COVID-19. Lancet Infect Dis 2020;S1473−3099(20):30310−8.

[19] Chen HX, Ai L, Lu H, et al. Clinical and imaging features of COVID-19. Radiol Infect Dis 2020;7(2):43−50.

[20] Chan JF, Yuan S, Kok KH, et al. A familial cluster of pneumonia associated with the 2019 novel coronavirus indicating person-to-person transmission: a study of a family cluster. Lancet 2020;395(10223):514−23.

[21] Chinese Society of Imaging Technology. Expert consensus on radiological examination scheme and infection prevention and control of COVID-19 (trial first edition). Electron J Emerg Infect Dis 2020;5(2):65−73.

[22] Lu Y, Zhou J, Yimei Mo, et al. Clinical and CT manifestations of 31 cases of coronavirus disease 2019. Electron J Emerg Infect Dis 2020;5(2):79−82.

[23] Hanxiong G, Xiong Y, Shen N, et al. Clinical imaging features of novel coronavirus pneumonia (COVID-19). Radiol Practice 2020;35(2):125−30.

[24] Song L, Zeng Y, Xiaoming G, et al. Imaging manifestations and differential diagnosis of novel coronavirus pneumonia. Electron J Emerg Infect Dis 2020;5(2):83−6.

[25] Lei J, Li J, Li X, et al. CT imaging of the 2019 novel coronavirus (2019-nCoV) Pneumonia. Radiology 2020;295(1):18.

[26] Guo Y, Liu S. Radiological diagnosis of COVID-19: expert recommendation from the Chinese society of radiology. Chin J Radiol 2020;4:279−85. 1st ed.

[27] Yi J, Lu X, Jin R, et al. Diagnosis, treatment and prevention of severe acute respiratory syndrome coronavirus 2 infection in children: experts' consensus statement. Chin J Appl Clin Pediat 2020;(2):143−50. 2nd ed.

[28] Ai X, Fu X, Lin L, et al. Characteristics of 30 re-hospitalized patients with re-positive of 2019-nCoV nucleic acid. Chin J Infect Control 2020;19(7):591−6.

[29] Song S, Shi Y, Tang H, et al. Analysis of the causes and countermeasures for positive recurrence of SARS CoV 2 RNA in discharged patients with COVID-19. Chin J Microecol 2020;32(7):786−8.

[30] Huang J, Yuan C, Huang D. CT manifestations and clinical features of COVID-19 patients with nucleic acid conversion in guangzhou. J Practic Med 2020. https://kns.cnki.net/kcms/detail/44.1193.R.20200628.0851.002.html.

[31] General Office of National Health Commission of the People's Republic of China. Questions and answers on the prevention and control of asymptomatic COVID-19 patients [EB/OL]. [2020-03-31], http://www.nhc.gov.Cn/jkj/s3578/202003718c79c96f3e46409dd49303d41a00ef.shtml.

[32] Wang Y, Chen J, Xiang W, et al. Analysis on CT imaging of asymptomatic COVID-19 patients. Med J Wuhan Univ 2020;41(3):353−6.

[33] Hongjun L. Guideline for medical imaging in auxiliary diagnosis of coronavirus disease 2019. Chin J Med Imag Technol 2020;36(3):321−31.

[34] Lu H, Han R, Yu P, et al. A correlation study of CT and clinical features of different clinical types of COVID-19. Chin J Radiol 2020;54(4):300−4.

[35] Meng H, Xiong R, He R, et al. CT imaging and clinical course of asymptomatic cases with COVID-19 pneumonia at admission in Wuhan [J]. China. J Infect 2020;81(1):33−9.

[36] Shi H, Han X, Jiang N, et al. Radiological findings from 81 patients with COVID-19 pneumonia in Wuhan, China: a descriptive study. Lancet Infect Dis 2020;20(4):425−34.

[37] Bernheim A, Mei X, Huang M, et al. Chest CT findings in Coronavirus Disease-19 (COVID-19): relationship to duration of infection. Radiology 2020;295(3):200463.

[38] Hu Z, Song C, Xu C, et al. Clinical characteristics of 24 asymptomatic infections with COVID-19 screened among close contacts in Nanjing, China. Sci China Life Sci 2020;63(5):706−11.

[39] Sun R, Liu H, Wang X. Mediastinal emphysema, giant bulla, and pneumothorax developed during the course of COVID-19 pneumonia. Korean J Radiol 2020;21(5):541−4.

[40] Koehler P, Cornely OA, Böttiger BW, et al. COVID-19 associated pulmonary aspergillosis. Mycoses 2020;63(6):528−34.

[41] Ellul MA, Benjamin L, Singh B, et al. Neurological associations of COVID-19. Lancet Neurol 2020;19(9):767−83.

[42] Bikdeli B, Madhavan MV, Jimenez D, et al. COVID-19 and thrombotic or thromboembolic disease: implications for prevention, antithrombotic therapy, and follow-up: JACC State-of-the-Art Review. J Am Coll Cardiol 2020;75(23):2950−73.

[43] Akhmerov A, Marbán E. COVID-19 and the heart. Circ Res 2020;126(10):1443−55.

[44] Cha MH, Regueiro M, Sandhu DS. Gastrointestinal and hepatic manifestations of COVID-19: a comprehensive review. World J Gastroenterol 2020;26(19):2323−32.

[45] Sala S, Peretto G, Gramegna M, et al. Acute myocarditis presenting as a reverse Tako-Tsubo syndrome in a patient with SARS-CoV-2 respiratory infection. Eur Heart J 2020;41(19):1861−2.

[46] Inciardi RM, Lupi L, Zaccone G, et al. Cardiac involvement in a patient with coronavirus disease 2019 (COVID-19). JAMA Cardiol 2020;5(7):1−6. 27.

[47] Huang C, Wang Y, Li X, et al. Clinical features of patients infected with 2019 novel coronavirus in Wuhan, China. Lancet 2020;395(10223):497−506.

[48] Shi S, Qin M, Shen B, et al. Association of cardiac injury with mortality in hospitalized patients with COVID-19 in Wuhan, China. JAMA Cardiol 2020;5(7):802−10.

[49] Hu D, Liu K, Li B, et al. Large intracardiac thrombus in a COVID-19 patient treated with prolonged extracorporeal membrane oxygenation implantation. Eur Heart J 2020;41(32):3104−5.

[50] Ferreira VM, Schulz-Menger J, Holmvang G, et al. Cardiovascular magnetic resonance in nonischemic myocardial inflammation: expert recommendations. J Am Coll Cardiol 2018;72(24):3158−76.

[51] Carod-Artal FJ. Neurological complications of coronavirus and COVID-19. Rev Neurol 2020;70(9):311−22.

[52] Kremer S, Lersy F, DE Sèze J, et al. Brain MRI findings in severe COVID-19: a retrospective observational study. Radiology 2020;297(2):242−51.

[53] Kandemirli SG, Dogan L, Sarikaya ZT, et al. Brain MRI findings in patients in the intensive care unit with COVID-19 infection. Radiology 2020;297(1):232−5.

[54] Behzad S, Aghaghazvini L, Radmard AR, et al. Extrapulmonary manifestations of COVID-19: radiologic and clinical overview. Clin Imag 2020;66:35−41.

[55] Ma C, Cong Y, Zhang H. COVID-19 and the digestive system. Am J Gastroenterol 2020;115(7):1003−6.

[56] Aroniadis OC, Dimaio CJ, Dixon RE, et al. Current knowledge and research priorities in the digestive manifestations of COVID-19. Clin Gastroenterol Hepatol 2020;18(8):1682−4.

[57] Han C, Duan C, Zhang S, et al. Digestive symptoms in COVID-19 patients with mild disease severity: clinical presentation, stool viral RNA testing, and outcomes. Am J Gastroenterol 2020;115(6):916−23.

[58] Ng SC, Tilg H. COVID-19 and the gastrointestinal tract: more than meets the eye. Gut 2020;69(6):973−4.

CHAPTER 12

Application of AI in COVID-19

Li Li[1], Lv Zhehao[2], Ren Meiji[1], Guo Jiachen[1], Zhou Zhen[3] and Shi Yanbin[4]
[1]Beijing YouAn Hospital, Capital Medical University, Beijing, China; [2]The First Affiliated Hospital of Harbin Medical University, Harbin, China; [3]Beijing Deepwise Science and Technology Co., Ltd., Beijing, China; [4]The Sixth People Hospital of Zhengzhou, Zhengzhou, China

12.1 Application of AI in radiology

Since the 1950s, AI has developed rapidly and played a significant role in various fields. In 2020, the Artificial Intelligence Standardization White Paper (2020 Edition) in China provided a relatively comprehensive definition of artificial intelligence. Artificial Intelligence (AI) is defined as the theory, method, technique, and application system that utilize digital computers or digital computer-controlled machines to simulate, extend, and expand human intelligence, perceive the environment, acquire and utilize the knowledge to obtain the best results. In recent years, AI has rapidly developed with the technical support of big data, high-performance computers and the internet, which has received significant attention in the field of medical imaging and gained recognition and support from clinical physicians. Compared with the traditional imaging workflows that are highly dependent on human labor, AI can provide safer, more accurate, and efficient imaging solutions. Before the COVID-19 epidemic, AI was primarily applied in three aspects: lesion detection, clinical diagnosis, and disease monitoring.

"Early detection, early diagnosis, early isolation, and early treatment" are the key principles for the prevention and control of COVID-19 epidemic. On February 4, 2020, the Ministry of Industry and Information Technology of China issued an initiative to encourage increased efforts in scientific research and fully utilize the power of AI to accelerate the use of AI to address technological deficiencies in epidemic prevention and control, and to collectively combat the COVID-19. During this epidemic, imaging diagnosis has played a crucial and irreplaceable role in various

The Pathology and Imaging of Coronavirus Pneumonia
ISBN 978-0-443-32930-2
https://doi.org/10.1016/B978-0-443-32930-2.00018-0

aspects of COVID-19 diagnosis and treatment. The timely development and application of COVID-19 imaging AI products not only help alleviate the strain on medical resources and improve the efficiency of radiologists, but also reduce the risk of in-hospital infections and enhance the response capabilities of healthcare systems in vulnerable areas.

12.1.1 Application of the "Tianyan AI Platform"

The intelligent positioning assistance and adaptive positioning frame functions of the "Tianyan AI Platform" have played a significant role, not only greatly reducing the risk of in-hospital infection between patients and frontline technicians, but also minimizing cross-infection between patients. Furthermore, the adaptive positioning frame function can also reduce errors in lung scans and minimize unnecessary radiation exposure for patients due to inadequate or excessive scan frames caused by human factors.

12.1.2 AI-assisted COVID-19 diagnosis

See Section 12.2 of this Chapter for details.

12.1.3 Remote AI diagnosis platform for COVID-19 based on expert system

Building a remote diagnosis platform is an effective approach to enhance the response capabilities of vulnerable areas to COVID-19, especially in situations where there is a limited availability of professional medical teams. Remote healthcare systems can not only address the challenges of medical access in regions with limited medical capacity, but can also be applied to healthcare institutions with better facilities to facilitate early triage and noncontact diagnostic services, thereby avoiding patient overcrowding and the risk of new clusters of infections.

12.2 AI-assisted COVID-19 diagnosis

Due to the relatively high rate of missed diagnosis of COVID-19 in X-ray chest imaging, chest CT examination has become the preferred imaging method for screening and assessing the severity of COVID-19. This section mainly introduces the "CT + AI" assisted diagnosis for COVID-19.

12.2.1 Earning screening of suspected COVID-19 patients

Currently, there has been some progress in deep learning research on chest imaging. Classic networks such as U-Net, UNet++, and V-Net are the mainstream in the research for COVID-19 screening. Researchers utilized 51 CT images from COVID-19 patients and 55 CT images from non-COVID-19 patients to train a U-Net-based segmentation model, and the segmentation of lesions was used as classification labels (COVID-19 or non-COVID-19). Evaluation Results: Accuracy of 95.2%, sensitivity of 100%, and specificity of 93.6%. In an additional test on a dataset of 16 cases of viral pneumonia and 11 cases of nonpneumonia, the model has correctly identified all viral pneumonia patients and nine nonpneumonia patients. As a result of AI, the reading time for the radiologists has been reduced by 65%. Another model based on U-Net, using 3D CNN as input model (DeCoVNet), achieved a sensitivity of 90.7%, specificity of 91.1%, and an area under the curve (AUC) of 0.959 in the chest CT images from 313 COVID-19 patients and 229 control subjects. Jin et al. used a combined model of UNet++ and ResNet50 and achieved a sensitivity of 97.4% and specificity of 92.2% in the chest CT images from 1136 subjects (723 COVID-19 patients and 413 control subjects). Mei et al. applied AI technology to combine chest CT images with clinical symptoms, exposure history, and Laboratory examinations for the rapid diagnosis of COVID-19 positive patients.

12.2.2 Quantitative analysis of COVID-19 pulmonary lesions

The "CT + AI" pneumonia-assisted diagnostic system facilitates quantitative analysis of pulmonary lesions, which is beneficial for evaluating the severity of a patient's conditions, determining the disease type and stage, predicting lesion progression, and assessing prognosis.

Regarding the typing of COVID-19, researchers compared radiologist's visual scores with quantified assessment results generated by computer software. The correlation analysis showed that AI demonstrated a moderate to high correlation with radiologists' scores in terms of lesion percentage assessment, a moderate correlation in comparing the proportion of ground-glass density lesions and the average lesion density, as well as a moderate correlation in comparing the proportion of consolidation lesions and the average lesion density. Researchers divided patients into severe and non-severe groups, compared their clinical and radiological data upon admission, and established a predictive model for COVID-19 severity. The results showed that the AI-based clinical-imaging model outperformed CT semiquantitative analysis parameters and commonly used clinical pneumonia severity indices in predicting the severity of COVID-19.

12.2.3 Differential diagnosis with non-COVID-19

Researchers proposed a convolutional neural network model for the classification of COVID-19 and other viral pneumonia. They used chest CT images from 99 patients (44 COVID-19 patients and 55 cases of typical viral pneumonia) as input for the 3D convolutional neural network, featuring the accuracy of 73.1%, specificity of 67%, and sensitivity of 74%. Researchers proposed a deep learning CT diagnostic system with ResNet-50 as the backbone to screen for COVID-19 and bacterial pneumonia patients from the healthy population. The model was trained and tested using 88 COVID-19 cases, 101 bacterial pneumonia cases, and 86 healthy subjects, with the accuracy of the model in pneumonia classification and diagnosis of 86.0% and 94.0%, respectively. Researchers, for the first time, utilized a V-Net -based deep learning model to segment candidate infection regions, then sent features such as infection regions and relative infection distance into the ResNet-18 network for classification, and increased the number of patients in the experiments to achieve a more balanced distribution of training data: 219 COVID-19 patients, 224 H1N1 influenza patients, and 175 healthy subjects. The overall accuracy of the model was 86.7%. Researchers further expanded the CT dataset to include 4356 images from 3322 patients (1296 COVID-19 images, 1735 community-acquired pneumonia images, and 1325 nonpneumonia images). They used 2D slices as inputs with shared weights for the ResNet-50 model (COVNet), which has effectively distinguished between COVID-19 and community-acquired pneumonia, achieving a sensitivity of 90%, specificity of 96%, and an AUC of 0.96.

In summary, the functions of AI mainly include the following aspects: ① Precise Early Screening: AI technology can rapidly perform automatic lesion detection, segmentation, accurate quantification, and generate formatted reports. This helps in timely identifying hidden lesions when patients' symptoms are not obvious or when nucleic acid tests are negative, thereby improving the accuracy of diagnosis with the assistance of radiologists. ② Quantitative Evaluation: AI can perform comprehensive quantification of individual lesions, lobar lesions and total lung lesions. The auxiliary diagnosis and differential diagnosis were made based on the distribution, shape and density of the lesions. ③ Intelligent Follow-up: AI enables quantitative comparison and analysis of lesion count, lesion volume changes, and lesion density changes in multiple image series, to assess the pneumonia severity and predict t disease progression trends.

Although multiple studies have confirmed that AI technology does improve the efficiency and diagnostic accuracy of radiologists, the reasons why AI technology has not been widely adopted in the postpandemic period may be attributed to the following aspects:① Lack of understanding among radiologists about AI technology, leading to doubts about the credibility of AI-assisted diagnosis. ② Lack of unified standards for image scanning protocols and quality, as nonstandardized imaging data can impact the accuracy of AI models. ③ False positive or false negative results in AI software's detection of lesions due to nonstandardized annotations and other factors, leading to misdiagnosis and missed diagnosis occurrences. ④ Complete differentiation between COVID-19 and other types of pneumonia cannot be achieved at present.

Due to the relatively short development time of COVID-19 "CT + AI" software, its maturity still needs to be further improved in the following aspects.

①Data Set Construction and Quality Control: To achieve the research and development goals, it is essential to build an appropriate training data set, the coverage of regions, equipment and scanning parameters, population characteristics and imaging staging into account. ② Data Annotation Rules: Data annotation is a crucial step in the implementation of medical imaging AI algorithms. Currently, there is a lack of unified industry annotation consensus. The training and its effect of annotators also need to be strengthened. ③ Integration and Modeling of Imaging, Clinical, and Laboratory Examinations: Assist physicians in achieving rapid diagnosis during epidemic screening, avoid misdiagnosis, and enable accurate stratification and comparison, provide efficacy evaluation and disease course monitoring, and thus provide more precise information for clinical treatment. ④ Algorithm Breakthroughs: Medical imaging, and medicine as a whole, are multidimensional data resources, and the processing and modeling of multimodal, multitask dynamic medical data depend on algorithm breakthroughs.

The currently developed "CT + AI" software mainly focuses on COVID-19 detection and quantitative evaluation, assisting radiologists in conducting large-scale medical image screening and diagnosis of suspected cases. Its main advantage lies in improving the radiologists' diagnosis efficiency and reducing diagnosis time. However, the product still has its limitations in etiological analysis, treatment guidance, and genetic testing. In the future, AI shall be further integrated with diverse features such as clinical information, epidemiology, radiology, and prognosis related to

COVID-19. By establishing multimodal models, we aim to comprehensively and intelligently assess the disease's pathogenesis, provide early screening warnings, and offer scientific evidence for guiding precise clinical diagnosis and treatment.

Bibliography

[1] Qiang X, Zhang Q, Lu G. Current and challenge in clinical transformation of the new generation of artificial intelligence in medical imaging. Chin J Radiol 2019;53(11):913–5.
[2] Wang Y, Wang Y, Shan H, et al. Progress in the application of AI in novel coronavirus pneumonia. Inf Commun 2020;(6):1–4.
[3] Cao Y, Xu Z, Feng J, et al. Longitudinal assessment of COVID-19 using a deep learning-based quantitative CT pipeline: illustration of two cases. Radiol Cardiothorac Imaging 2020;2(2):e200082.
[4] Huang L, Han R, Ai T, et al. Serial quantitative chest CT assessment of COVID-19: deep-learning approach. Radiol Cardiothorac Imaging 2020;2(2):e200075.
[5] Mei X, Lee HC, Diao KY, et al. Artificial intelligence-enabled rapid diagnosis of patients with COVID-19. Nat Med 2020;26(8):1224–8.
[6] Shen C, Yu N, Cai S, et al. Quantitative computed tomography analysis for stratifying the severity of Coronavirus Disease 2019. J Pharm Anal 2020;10:123–9.
[7] Bowen L, Tang R, Zhiyan Z, et al. Research on the application value of MSCT pixel index quantitative analysis in COVID-19 assessment. Elect J Emerg Infect Dis 2021;6(1):22–6.
[8] Li L, Qin L, Xu Z, et al. Artificial intelligence distinguishes COVID-19 from community acquired pneumonia on chest CT. Radiology 2020:200905.
[9] Yi X, Liu S. The value of medical image artificial intelligence in diagnosis and treatment of novel coronavirus pneumonia. China Med Equip 2020;35(6):59–62.
[10] Xinling L, Ying W. Application and development of artificial intelligence in detection and diagnosis of pulmonary nodules. Elect J Emerg Infect Dis 2019;4(3):185–9.

CHAPTER 13

Imaging differential diagnosis

Lu Yibo[1], Zheng Guangping[2], Fu Chaohong[3], Lu Puxuan[4], Zhao Dawei[5], Gong Lianggeng[6], He Yulin[7], Lv Zhehao[8], Qiao Guoqing[9] and Zheng Qiuting[4]

[1]HIV/AIDS Clinical Treatment Center of Guangxi (Nanning), The Fourth People's Hospital of Nanning, Nanning, Guangxi Zhuang Autonomous Region, China; [2]The Third People's Hospital of Shenzhen, Shenzhen, China; [3]The Sixth People Hospital of Zhengzhou, Zhengzhou, China; [4]Shenzhen Center for Chronic Disease Control, Shenzhen, China; [5]Beijing YouAn Hospital, Capital Medical University, Beijing, China; [6]The Second Affiliated Hospital of Nanchang University, Nanchang, China; [7]The First Affiliated Hospital of Nanchang University, Nanchang, China; [8]The First Affiliated Hospital of Harbin Medical University, Harbin, China; [9]The First Hospital of China Medical University, Shenyang, China

13.1 Imaging differential diagnosis of SARS

13.1.1 Differentiation of SARS from other pneumonia

Based on the different pathogenesis and clinical manifestations of pneumonia, there are distinct points of emphasis in distinguishing SARS from pneumonia in patients with normal immune function and those with impaired immune function.

13.1.1.1 Differentiation of SARS from pneumonia in patients with normal immune functions

The most common type of pneumonia in patients with normal immune function is bacterial or *Mycoplasma pneumonia*. The common imaging manifestations of SARS can be helpful in diagnosing this disease, but some of them are less frequently observed, which can aid in diagnosis by exclusion.

13.1.1.1.1 Common imaging manifestations of SARS

Ground-glass opacities are relatively common signs in SARS, appearing as small patchy lesions with a circular or rounded shape in the early stage, with an incidence rate of 80%. The lesions can rapidly progress to diffuse ground-glass opacities, accounting for 16.7% of the entire course of the disease. Consolidation of the lungs occurs in 67% of cases, and the "paving stone" sign is observed in 6.7% of cases.

The Pathology and Imaging of Coronavirus Pneumonia
ISBN 978-0-443-32930-2
https://doi.org/10.1016/B978-0-443-32930-2.00001-5

In terms of dynamic changes, most SARS cases show a rapid progression from the early stage of small patchy lesions to multiple and diffuse lesions involving one or both lungs. Generally, bacterial pneumonia or *Mycoplasma pneumonia* often exhibits more localized lesions, and many cases do not go through a stage of multiple or diffuse lung involvement before absorption of the lesions.

13.1.1.1.2 Signs less common in SARS

Single lobe consolidation without accompanying ground-glass opacities is rare in SARS, which is the characteristic of bacterial pneumonia. Localized distribution of alveolar consolidation is seen not only in pneumococcal pneumonia but also in infections such as gram-negative bacterial pneumonia and aspiration pneumonia. During the early and progressive stages of the disease (within 1−14 days of the onset), cavities are generally not observed in the images. However, they may appear later in the disease course, particularly when complications of infection occur. Large amounts of pleural effusion are generally rare.

13.1.1.2 Differentiation of SARS from pneumonia in patients with impaired immune functions

Patients with impaired immune functions are prone to develop opportunistic lung infections, which are observed in patients with AIDS and those who have undergone organ transplantation and are on long-term immunosuppressive therapy. Opportunistic pneumonia is commonly seen in Pneumocystis pneumonia and cytomegalovirus pneumonia. Due to impaired immune function, chest imaging often shows multiple and diffuse lesions, with rapid progression, which has many similarities with the imaging features of SARS. Imaging-based differential diagnosis shall focus on specific medical history, laboratory examinations, and differences observed in imaging.

Pneumocystis pneumonia commonly occurs in patients with HIV/AIDS, especially when the CD_4 cell count is less than 200 cells/mm^3. Chest X-ray findings typically show diffuse reticular and ground-glass opacities symmetrically distributed around the bilateral hilar regions or involving both lungs. On CT imaging, there is widespread ground-glass opacities, and cystic lesions may be observed in the upper lung fields (accounting for 10%−34% of cases).

Cytomegalovirus (CMV) pneumonia is the most common viral infection in the lungs of AIDS patients and occurs in the late stage of immunosuppression when the CD_4 count is less than 100 cells/mm^3. It is also observed in patients after organ transplantation. X-ray chest and CT scans show multiple ground-glass opacities, often occurring in the middle and lower lung fields, and may be accompanied by interstitial thickening.

In human immunodeficiency virus (HIV) positive patients, bacterial pneumonia can present as multiple consolidations distributed according to lung lobes and segments, or as diffuse bilateral opacities. The progression speed of the lesions is faster compared with pneumonia in the general population.

13.1.2 Differentiation of SARS from noninflammatory diseases

Although some diffuse alveolar and interstitial lesions may not present with clinical manifestations of acute inflammation, they still need to be distinguished from SARS based on imaging. Common diseases with acute onset include acute pulmonary edema, allergic pneumonitis, diffuse alveolar hemorrhage, and noninfectious acute interstitial pneumonia, etc. When distinguishing, it is important to note that these diseases lack inflammatory manifestations and require considering their respective imaging features.

Acute cardiogenic pulmonary edema is characterized by an enlarged heart shadow, interstitial pulmonary edema manifestations, and a butterfly sign. Allergic pneumonia is characterized by a history of allergies and corresponding clinical manifestations. CT scans often show ground-glass opacities combined with diffuse micronodules. Diffuse pulmonary hemorrhage is caused by various factors and presents with clinical manifestations such as hemoptysis and anemia, along with imaging features of pulmonary interstitial thickening.

Common chronic noninflammatory diseases include pulmonary alveolar proteinosis, chronic eosinophilic pneumonia, and bronchiolitis obliterans organizing pneumonia. When distinguishing these diseases, it is important to note that they lack acute inflammatory manifestations and exhibit certain imaging characteristics. Pulmonary alveolar proteinosis often presents with the "gravel road" sign, and the imaging of lung lesions commonly exhibits a "map-like" distribution. In the early stage of solitary peripheral small-sized bronchioloalveolar carcinoma, it may appear as rounded ground-glass opacities, known as ground-glass density nodules. Diffuse bronchioloalveolar carcinoma can present with multiple ground-glass density nodules involving multiple lung lobes.

13.1.3 Differentiation of SARS from ARDS

ARDS (Acute Respiratory Distress Syndrome) can be caused by various factors such as infection, bleeding, and allergies, with the most common imaging manifestation of diffuse lung disease, and clinically, it presents with progressive hypoxemic respiratory distress. Autopsy studies have shown that ARDS is one of the causes of SARS death. It usually occurs after the second week of the disease course. On X-ray, it presents with diffuse bilateral opacities or "white lung." Therefore, it is necessary to differentiate ARDS from SARS. Due to the late stage of SARS, it may be complicated by infections caused by various pathogens, leading to abnormalities in blood routine, which is similar to ARDS caused by infections.

During the process of differential diagnosis, it shall be noted that in the early stages of ARDS, lung changes may be mild, and chest X-rays may appear normal or only show small patchy shadows. However, this condition can cause a decrease in PaO_2 and lead to symptoms such as rapid breathing. As the disease progresses, the number of lung shadows increases, and the progressive worsening of hypoxia becomes evident. The clinical manifestation of hypoxia in SARS generally occurs after the appearance of diffuse lung shadows on imaging. After using steroids, lung imaging and clinical manifestations can significantly improve.

The principle of differential diagnosis for SARS is to closely combine imaging manifestations with medical history, clinical manifestations, and laboratory examinations. For general pneumonia patients, it is important to pay attention to the clinical and imaging characteristics of each disease. In the differentiation from pneumonia in patients with impaired immune function, attention shall be paid to relevant medical history. In the differentiation from diseases other than pneumonia, the presence or absence of inflammatory clinical manifestations is crucial. The dynamic changes in clinical and imaging manifestations can provide diagnostic clues for challenging cases of differentiation. The imaging and pathological manifestations of SARS patients with ARDS are shown in Fig. 13.1.

13.2 Imaging differential diagnosis of MERS

MERS shall primarily be differentiated from SARS, human-infected H7N9 avian influenza, or pneumocystis pneumonia. Belonging to the same coronavirus family, SARS and MERS share approximately 70%–80% homology in the nucleotide sequence of the same PCR fragment, and also

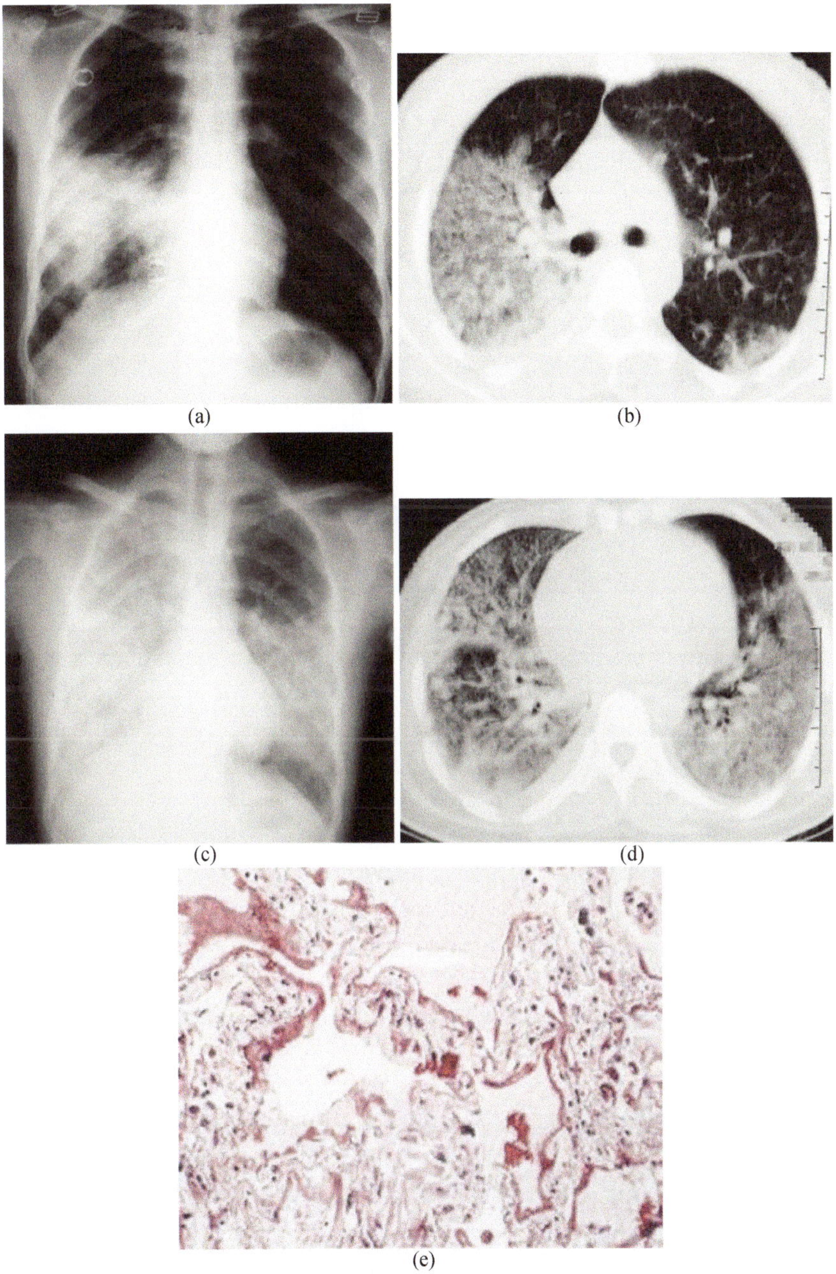

Figure 13.1 Imaging and pathological manifestations of SARS patients with ARDS. (a) On the ninth day of the disease course, the chest X-ray showed large patchy opacities in the right middle and lower lung fields; (b) CT scan showed ground-glass opacities in the posterior segment of the right lung and left lower lung, more pronounced in the right lung; (c) on the 11th day of the disease course, the chest X-ray showed "white lung" appearance in the right lung and left middle and lower lung fields; (d) CT scan showed diffuse ground-glass opacities and consolidation in both lungs; (e) pathological examination revealed the alveolar cavities filled with uniformly pale-staining eosinophilic exudate, which appeared as serous or fibrinous fluid. Concentrated exudate formed hyaline membranes adhering to the alveolar walls (HE $\times$ 200).

have some common imaging manifestations. Both of them can present with ground-glass opacities and lung consolidation in the middle and lower lung fields, accompanied by rapid disease progression.

13.2.1 Imaging differentiation of MERS from SARS

The main clinical features of SARS are sudden onset with fever as the initial symptom, accompanied by headache, body aches, malaise, fatigue, and other systemic symptoms, as well as respiratory symptoms such as cough, chest tightness, and difficulty breathing. In a few cases, it may progress to ARDS. Peripheral blood white cell count is normal or decreased. As of September 26, 2003, WHO reported that there were 32 countries and regions affected by SARS, with a global total of 8098 confirmed cases and 774 deaths, resulting in a fatality rate of 9.5%. In mainland China, there were a total of 5327 confirmed cases and 349 deaths from SARS.

Pulmonary imaging examination is a crucial basis for diagnosing SARS, and consecutive imaging examinations can reveal the dynamic changes in the lesions. ① Early Stage: The lesions are mostly localized, and may be unilateral, but more commonly bilateral changes are observed. The CT examination shows small patchy ground-glass opacities, which are mostly single but can also be round in shape. A small number of lesions present as single small patchy consolidations, multiple small patchy or larger patchy opacities. Larger lesions may appear as ground-glass opacities and consolidations, which can involve an entire lung segment (Fig. 13.2a). ② Progressive Stage: In most patients, the lesions worsen and progress within 14 days after the onset of the disease. The small patchy shadows in the early stage of the disease can transform into larger, multiple, or diffuse lesions within 3–7 days. The lesions progress from one lung to both lungs and from one lung field to multiple lung fields. Most patients reach the peak of lung infiltration or "critical stage" with the most severe lesions between 8 and 14 days. The CT manifestations still show ground-glass opacities combined with lung consolidation. Lesions with a diameter of more than 3 cm account for over 90%. The lesions often become multiple and show a diffuse distribution in both lungs. Most of the patients have lesions distributed in a mixed pattern of central and peripheral lung regions, while central distribution of lesions is rarely seen (Fig. 13.2b–g). ③ Convalescent Stage: The lesions usually begin to be absorbed around 2–3 weeks after the onset of the disease with decreased shadow size, and the density is gradually reduced and absorbed. In some patients, even though the clinical symptoms

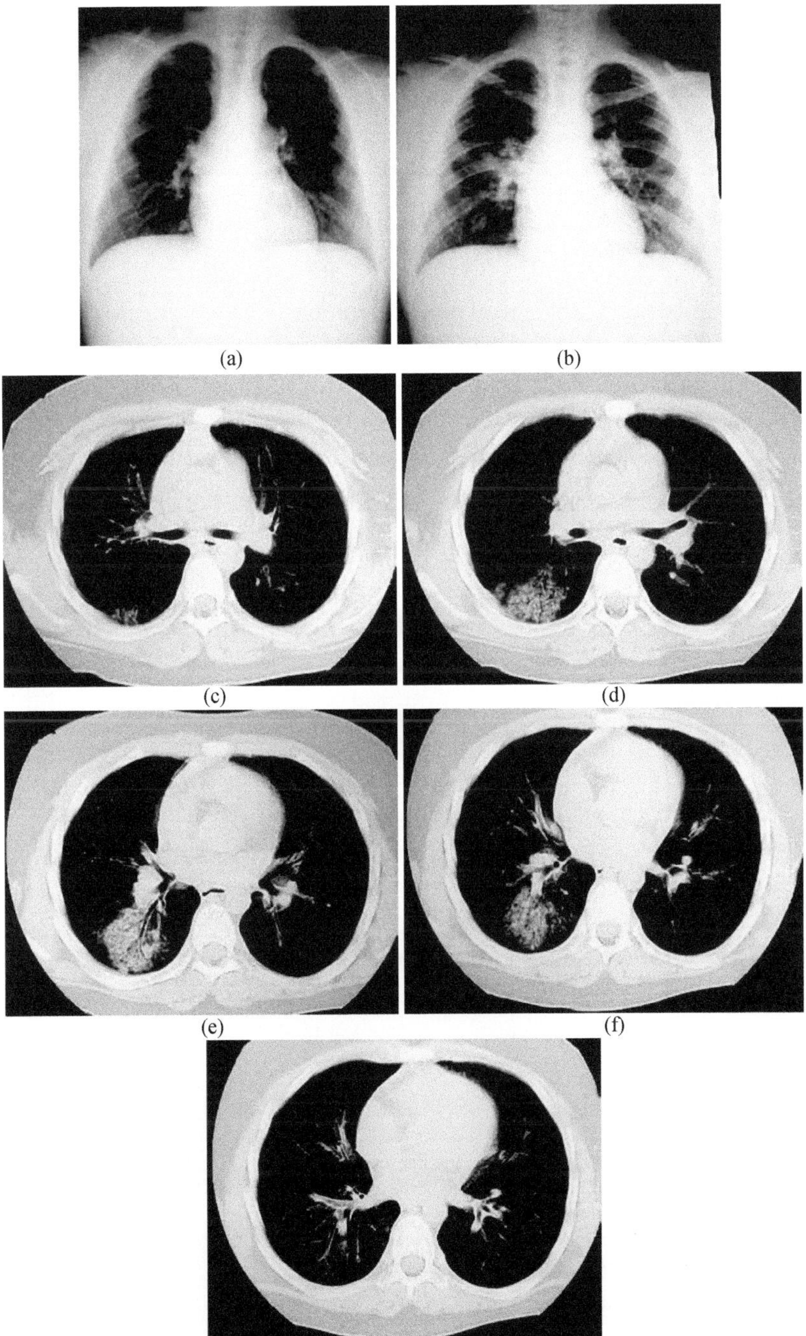

Figure 13.2 SARS imaging manifestations. (a) Chest X-ray on March 25, 2003 showed a basically normal appearance; (b) chest X-ray on March 27, 2003 showed unclear structures in the right hilum, with patchy opacities in the hilar region; (c–g) CT scans on March 29, 2003: (c) CT lung window showed scattered patchy ground-glass opacities in the right lower lung dorsal segment; (d–f) CT lung window showed consolidation with air bronchogram in the right lower lung; (g) scattered small patchy ground-glass opacities in the right lower lung.

improve and disappear, CT follow-up may reveal the presence of pulmonary interstitial hyperplasia during the absorption process of lung lesions. With dynamic observation, these lesions may be gradually absorbed, but in some cases, they may progress to pulmonary interstitial fibrosis. The imaging manifestations of SARS on CT scans are relatively characteristic and can be summarized as follows: ① Small patchy ground-glass opacities in the lung periphery; ② Early-stage lesions are often single, rapidly progressing to multiple lobes or bilateral lungs with diffuse ground-glass opacities or consolidations along with ground-glass opacities; ③ Enlarged lymph nodes in the hilum or mediastinum are often absent; ④ Lesions usually do not present with cavities, air-filled spaces, or calcifications on imaging; ⑤ Pleural effusion or pneumothorax is uncommon in the early stages, but some patients may develop a small amount of pleural effusion or pneumothorax in the later stages; ⑥ In the convalescent stage, the absorption of lesions may be slow, and imaging changes such as reticular opacities and interstitial hyperplasia can be observed.

SARS and MERS share similarities in their imaging characteristics, as both can present with bilateral ground-glass opacities and lung consolidations. However, the main manifestation of the ground-glass opacities of MERS in the early stage is mainly distributed in the subpleural lung under multiple small flaky pieces. On the other hand, in SARS, early-stage lesions may also present as small patchy ground-glass opacities, but they do not show a distinct subpleural distribution pattern, and single lesions are more common in the early stages. SARS is more commonly associated with patchy and large consolidated lung opacities compared with MERS. Both SARS and MERS are caused by beta-coronaviruses, with their receptors of ACE2. ACE2 is mainly located on the alveolar type II epithelial cells deep in the human lungs, which makes individuals susceptible to severe respiratory diseases such as pneumonia. Due to their similar pathogenesis, the imaging manifestations of SARS and MERS are essentially alike, which poses certain challenges in radiological differential diagnosis.

13.2.2 Imaging differentiation of MERS from human-infected H7N9 avian influenza

Human-infected H7N9 avian influenza virus caused by H7N9 subtype avian influenza virus is an acute respiratory infectious disease. Since the first detection of human-infected H7N9 avian influenza virus in February 2013, outbreaks of h-infected H7N9 avian influenza virus have occurred in over 10 provinces (autonomous regions and municipalities) in China. The epidemiological characteristics include a clear history of contact with live poultry or their secretions or excreta in live poultry markets, with chickens being the main source of infection.

In the early stage of human-infected H7N9 avian influenza pneumonia, it can manifest as localized or scattered small patchy shadows, with a predilection for the middle and lower lung fields (Fig. 13.3a). During the progressive stage (Day 3–6), CT examination showed patchy or large patchy ground-glass opacities primarily distributed along the bronchi in both lungs. Additionally, pulmonary consolidation, which is more common, can be observed, and the sign of bronchial inflation may be observed (Fig. 13.3b–o). Severe cases may present with pleural effusion. Hilar and mediastinal lymphadenectasis is rare. The absorption stage is mainly characterized by fibrotic bands and small patchy consolidation shadows (Fig. 13.3p and q).

Human-infected H7N9 avian influenza pneumonia is mainly characterized by ground-glass opacities and lung consolidation. However, the distribution of ground-glass opacities in H7N9 avian influenza pneumonia is mainly centered around the bronchial bundles, which differs from MERS, where ground-glass opacities are mainly distributed under the pleura. Lung consolidations are more commonly located in the lower lobes of both lungs in human-infected H7N9 avian influenza pneumonia, and there is not much difference in this aspect between H7N9 and MERS. The imaging features of the two diseases partly overlap, which makes their differentiation somewhat challenging.

13.2.3 Differentiation of MERS from pneumocystis jiroveci pneumonia

Pneumocystis jiroveci pneumonia (PJP), formerly known as Pneumocystis carinii pneumonia, is one of the most common opportunistic infections in AIDS patients. The Pneumocystis family enters the lung interstitium or alveolar spaces through endogenous or exogenous infection pathways, leading to diffuse alveolar exudation and foamy changes. The lesions involve consolidations, and the interstitial tissue shows lymphocytes and plasma cells. There is thickening of the alveolar septa, and the lesions are mostly distributed diffusely. In the late stage, fibrosis may occur. Patients mainly present with progressive dyspnea, fever, dry cough, and hypoxemia. Other manifestations may include chills, chest pain, and weight loss.

The imaging manifestations of Pneumocystis jiroveci pneumonia (PJP) include: ① Ground-glass opacities (GGO) in the lungs, which are the most

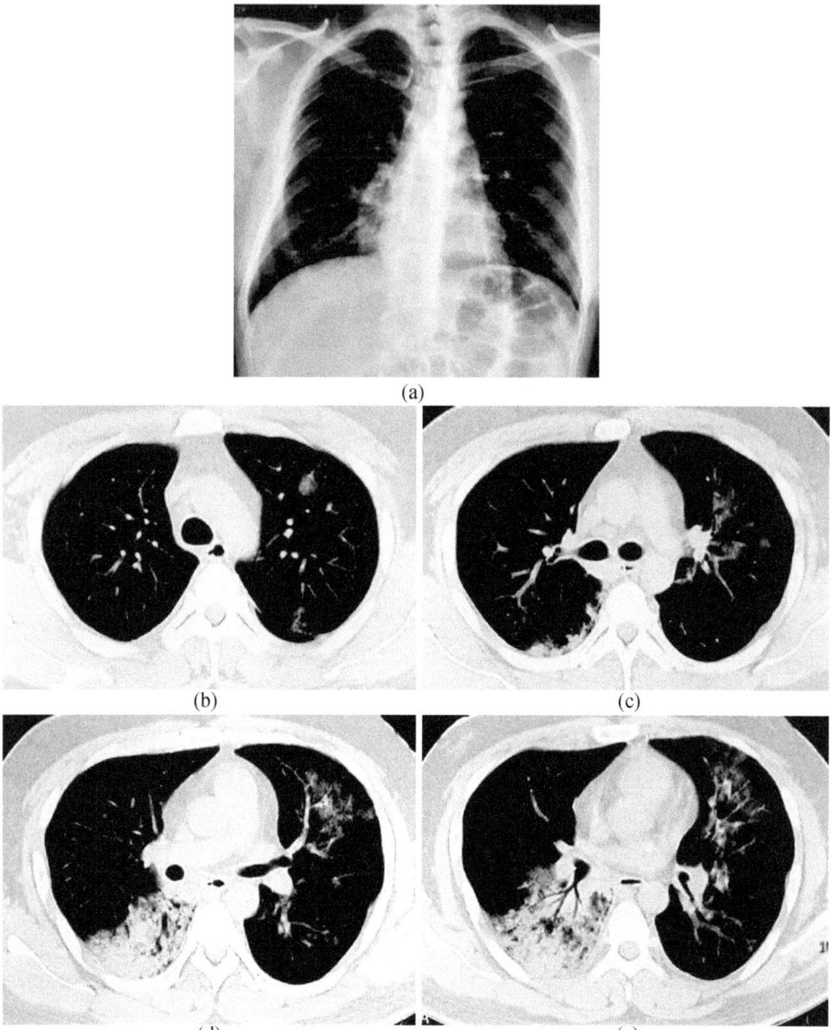

Figure 13.3 Imaging manifestations of human-infected H7N9 avian influenza pneumonia. On December 30, 2013, the patient presented with fever without any obvious inducement; (a) X-ray chest examination on January 2, 2014; (b–o) CT re-examination on January 4, 2014; (b–j) CT lung window showed patchy ground-glass opacities in both upper lungs and left lower lung, mainly distributed along the bronchovascular bundles. In the right lower lung, there were large patchy consolidation shadows with bronchial inflation signs, and in the left lower lung, small patchy consolidation shadows were observed. (k) CT mediastinal window showed irregular high-density consolidation shadow in the right lower lung, and no pleural effusion was observed; (l–o) coronal reconstruction displayed ground-glass opacities and consolidation shadows in both lungs; (p and q) CT re-examination on January 18, 2014, showed that the ground-glass opacities and lung consolidation in both lungs were mostly absorbed. CT lung window (p) and coronal reconstruction (q) showed fibrous strip shadows and patchy consolidation shadows in both lungs.

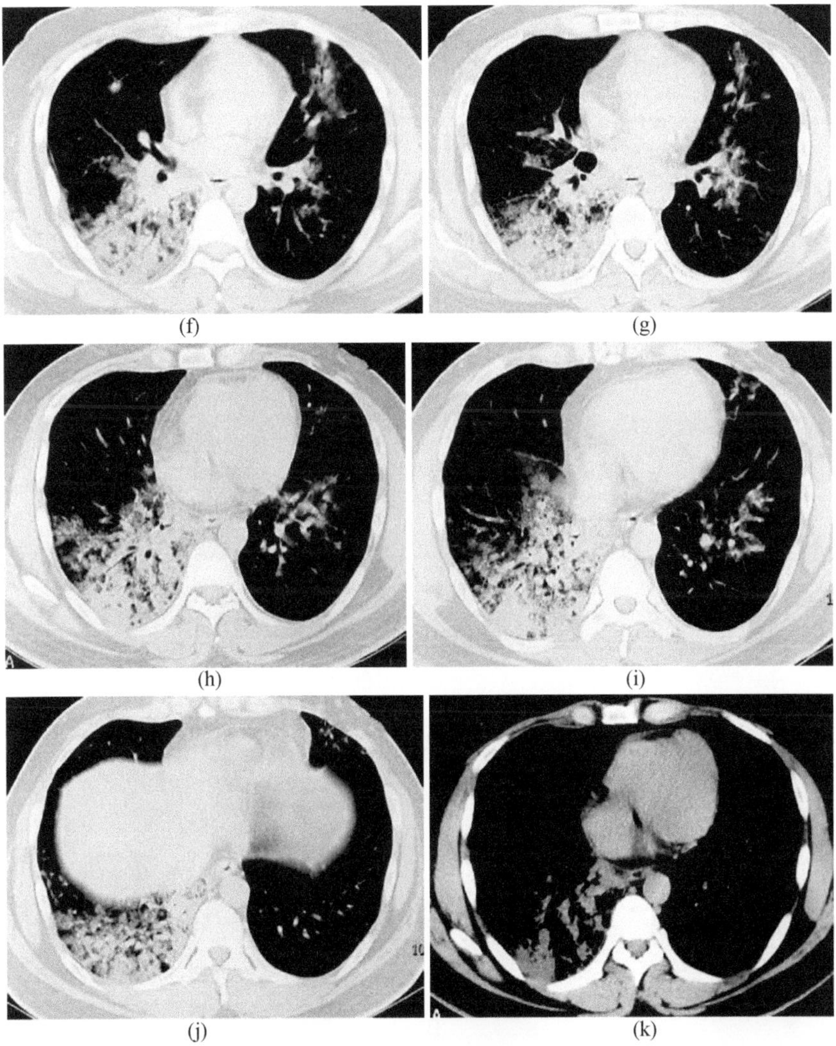

Figure 13.3 Cont'd

characteristic imaging manifestations. With the progression of the disease, GGOs may develop into consolidations, diffuse reticular nodular opacities, or symmetric reticular opacities (Fig. 13.4). The lesions are often centered around the hila, and as the disease progresses, they spread outward from the hila. They may also gradually spread from the lower to the upper lungs, resulting in diffuse, patchy, and map-like changes. Normal lung areas and affected areas may intersect and merge. ② Due to inflammation and fibrosis

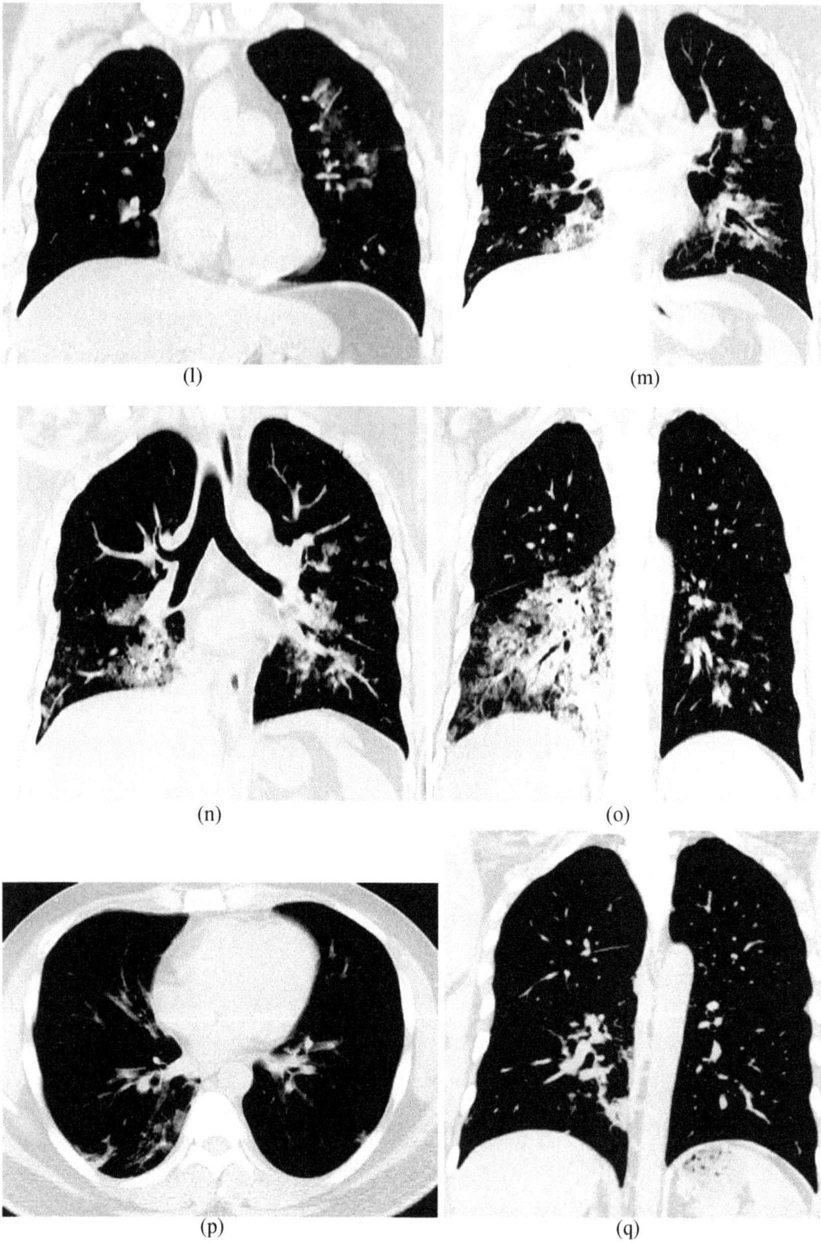

Figure 13.3 Cont'd

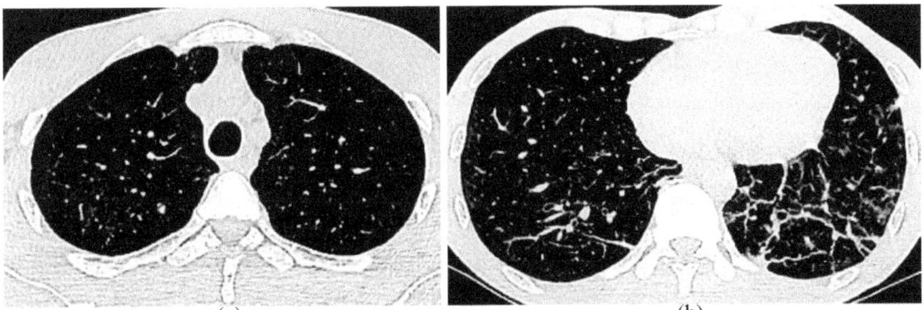

(a) (b)

Figure 13.4 Imaging manifestations of pneumocystis jiroveci pneumonia (I). The patient is a 38-year-old male with AIDS. He has had recurrent hemoptysis for 2 months, and worsened with chest pain for the past 10 days. (a–b) diffuse patchy ground-glass opacities with confluent lesions and unclear borders are distributed throughout both lungs. Scattered stripe opacities with increased density can be observed in the lower lobes of both lungs.

of the alveoli and interstitium, the lung tissue undergoes repeated reconstruction, leading to the formation of lung cysts (Fig. 13.5a). These cysts are often found in the upper lobes of both lungs. In some cases, the lung cysts may rupture, leading to pneumothorax. ③ In the early or progressive stages of the disease, a clear crescent or arch-shaped area can be observed in the lung tissue beneath the pleura, which is known as the "crescent sign" (Fig. 13.5b).

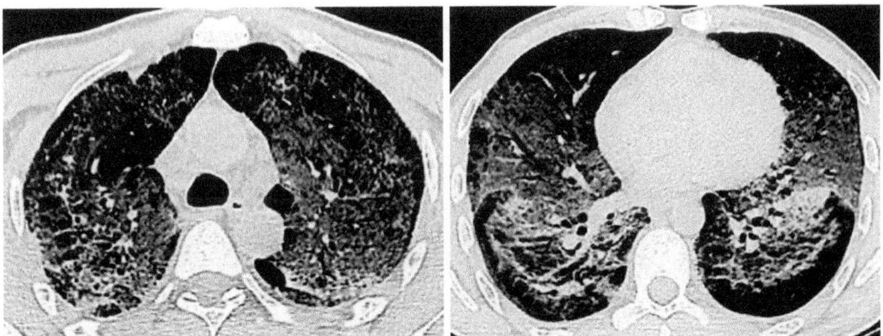

(a) (b)

Figure 13.5 Imaging manifestations of pneumocystis jiroveci pneumonia (II). The patient is a 42-year-old male with AIDS and has experienced dyspnea for 14 days. (a and b) Diffuse distribution of ground-glass opacities and reticular opacities in both lungs, scattered pulmonary cysts below the pleura in the upper lobes, and focal lesions concentrated near the hila in the lower lobes, with clear crescent-shaped areas under the pleura.

Comparison between Pneumocystis jiroveci pneumonia (PJP) and Middle East Respiratory Syndrome (MERS): MERS is generally more susceptible in the population, with most patients having a history of exposure to the epidemic area or contact with infected individuals. The clinical manifestations are mainly characterized by upper respiratory symptoms such as fever, dry cough, and fatigue. On the other hand, PJP rarely infects individuals with normal immune function and mainly affects patients with impaired immune function, malnourished individuals, or those with tumor-related immune damage or organ transplant recipients, leading to opportunistic infection. The clinical manifestations include progressive dyspnea, fever, dry cough, and hypoxemia. In terms of imaging manifestations, both primarily present with ground-glass opacities. In MERS, the lesions are mainly distributed in the periphery of both lungs, which are localized in the early stages, but with disease progression, they may coalesce and result in consolidation. On the other hand, PJP typically shows a symmetric distribution centered around the hila, and the lesions tend to be more diffuse throughout the lungs. Differentiation can be conducted by combining the clinical history, the patient's immune status, and nucleic acid testing.

13.3 Imaging differential diagnosis of COVID-19

13.3.1 Differentiation of COVID-19 from influenza A (H3N2) virus

Influenza, commonly known as "the flu", is an acute respiratory infectious disease caused by the influenza virus. The main types of influenza viruses that infect humans are influenza A and influenza B. Among them, the influenza A subtype H3N2 has been actively circulating since the global pandemic of 1968 and continues to cause new outbreaks in populations. The H3N2 virus can evade host immune recognition through antigenic drift and antigenic shift. It has consistently been one of the main subtypes of influenza A virus that causes seasonal epidemics in populations, posing a serious threat to human health during influenza outbreaks. The main clinical manifestations of patients include fever (with a temperature above 38°C), cough, sputum production, nasal congestion, dizziness, fatigue, sore throat, headache, chills, shortness of breath, runny nose, and muscle pain. Digestive symptoms such as constipation and jaundice may also accompany the illness. In most patients, chest X-rays and/or chest CT scans show interstitial pneumonia and lobar pneumonia as the main features, with accompanying pulmonary edema. CT images often reveal large or small patchy ground-glass opacities, consolidation, thickening of the interlobular septa, reticular opacities, centrilobular nodules, central nodular tree-in-bud sign of pulmonary lobule, air trapping, and fibrous stripes, etc. The proportion of abnormal findings on chest X-rays or CT scans in patients with influenza A H3N2 is significantly lower than those in other influenza patients. Abnormal lung auscultation includes dry and wet crackles, wheezes, and fine rales, etc.

Case 1: Patient, male, 20 years old. Four days ago, the patient traveled to an epidemic area for business and caught a cold after being exposed to cold weather. He developed a fever with a maximum temperature of 38.6°C, sometimes accompanied by chills, headache, and fatigue. He also experienced coughing with yellowish sticky sputum, which was not easily expectorated and in small amounts. His appetite decreased, but he did not have any symptoms of a sore throat or muscle soreness. He self-administered "compound paracetamol and amantadine hydrochloride tablet and Pudilan antiinflammatory tablets" and experienced slight improvement in the above symptoms. One day before seeking medical attention, he returned and added "aspirin" and "Lianhua Qingwen Capsules (granules)" to his medication. He suffered the acute ill appearance, congestion, and swelling in the throat, with bilateral tonsils mildly enlarged (Grade I). Serum Amyloid A Protein 297.80 mg/L, Whole blood procalcitonin: 0.16 ng/mL, White blood cell count 14.57×10^9/L, Neutrophil count 11.94×10^9/L, Lymphocyte count 1.82×10^9/L. The urban center for disease control confirmed that the patient's throat swab tested positive for the influenza A H3N2 virus. See Fig. 13.6 for imaging manifestations.

13.3.2 Differentiation of COVID-9 from mycoplasma pneumoniae pneumonia

Chlamydia is a prokaryotic microorganism, lacking a cell wall and containing ribosomes as cell organelles. Featuring pathogenicity, it is the primary pathogen responsible for pediatric community-acquired pneumonia. It is now widely believed that Chlamydia infection is caused by pathogen-induced damage to the respiratory mucosa, triggering the body's immune response. Due to individual differences and varied immune responses, the CT manifestations of Chlamydia pneumonia are diverse, yet they still possess certain characteristic features. The radiological findings are similar to respiratory viral infections. Mycoplasma Pneumoniae Pneumonia can present as tracheobronchitis, bronchiolitis, bronchopneumonia, and

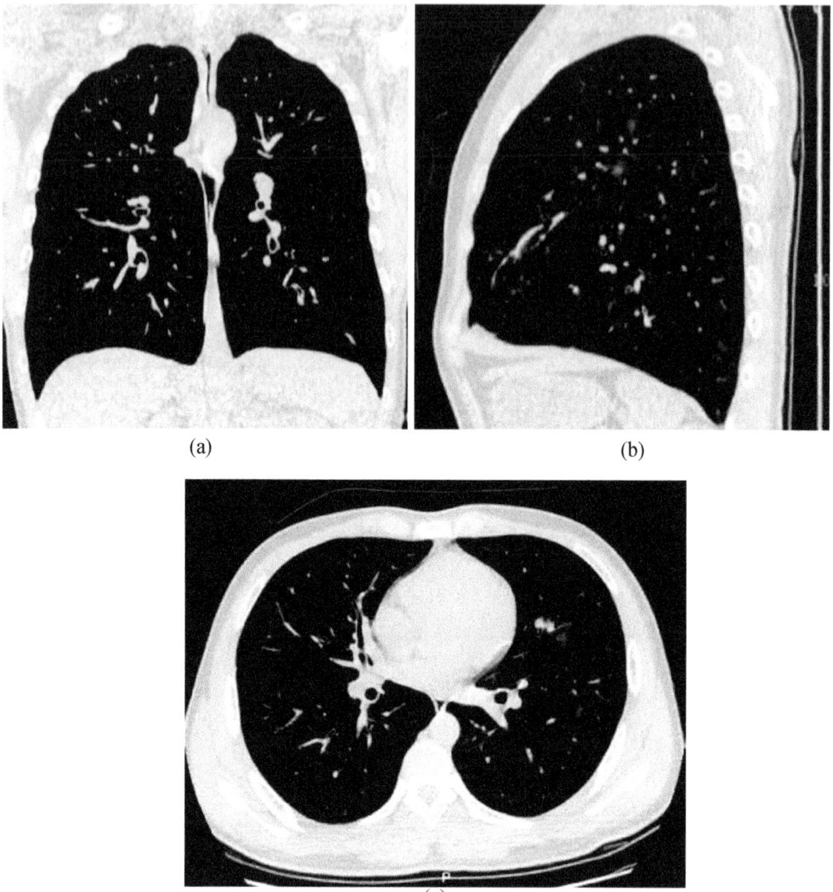

(a) (b) (c)

Figure 13.6 Imaging manifestations of patients with influenza A H3N2 virus. (a) Coronal view; (b) sagittal view; (c) axial view showing multiple small patchy ground-glass opacities in the left lung.

interstitial pneumonia. Its typical pathological changes involve lymphocytic and plasma cell infiltration around the bronchi, with neutrophils and macrophages accumulating around and within the bronchial lumens. The most common X-ray chest imaging manifestations in pediatric mycoplasma pneumoniae pneumonia are lobar or segmental consolidation. On CT scans, it often presents as centrilobular nodules, ground-glass opacities, and consolidations. Other typical features include thickened bronchial walls, tree-in-bud sign, and enlarged hilar lymph nodes. The main CT features of adult mycoplasma pneumoniae pneumonia include extensive lesions that often involve multiple lung lobes, with unilateral or bilateral ground-glass opacities and consolidations. The lower lobes of both lungs are commonly affected, and the distribution is often in a lobular pattern. At the same time, the patient's clinical symptoms do not match the CT changes, meaning that the clinical symptoms improve significantly or disappear, but the lung shadows do not show obvious absorption. Without etiological examination, it is difficult to distinguish it from lobar pneumonia. COVID-19 generally affects both lungs and is located beneath the pleura. Considering the epidemiological history, it is not difficult to distinguish it.

Case 2: Patient, female, 9 years old. The patient was admitted to the hospital after having a fever and cough for 5 days. Outpatient intravenous therapy was administered for 4 days with unsatisfactory results, and the body temperature remained at 38.8°C. The breath sounds in the right lung were slightly diminished, and a few fine crackles can be heard. The quantitative test of mycoplasma pneumoniae DNA was positive, mycoplasma pneumoniae antibodies were positive, mycoplasma pneumoniae antibodies were positive, chlamydia pneumoniae antibody titer was 1:40, human serum amyloid A protein was 236.34 mg/L, C-reactive protein was 201.45 mg/L, erythrocyte sedimentation rate was 34.00 mm/h, white blood cell count was 10.38×10^9/L, and lymphocyte count was 2.27×10^9/L. Respiratory syncytial virus antibodies, adenovirus antibodies, influenza virus type A antibodies, influenza virus type B antibodies, parainfluenza virus antibodies, and *Legionella pneumophila* antibodies were all negative. Sputum smear: White blood cells (+++), no bacteria was seen. See Fig. 13.7 for imaging manifestations.

Case 3: Patient, male, 5 years old. He was admitted to with cough for 1 week and fever for 6 days, his body temperature was 37.9°C, and he had coarse breath sounds in both lungs. Chlamydia pneumoniae antibody titer: 1:640. White blood cell count 9.90×10^9/L and Lymphocyte count 2.46×10^9/L. See Fig. 13.8 for imaging manifestations.

13.3.3 Differentiation of COVID-19 from mycotic pneumonia

Pulmonary aspergillosis infection is a rare infectious disease of the lungs. The incidence of pulmonary aspergillosis infection is increasing due to the use of steroids and anticancer drugs during leukemia chemotherapy. The imaging manifestations show scattered patchy, nodular, or consolidative shadows in the middle and lower parts of the lungs. It can also form cavities, and the lesions may be unilateral or bilateral. In some cases, there may be a small amount of reactive pleural effusion on the affected side.

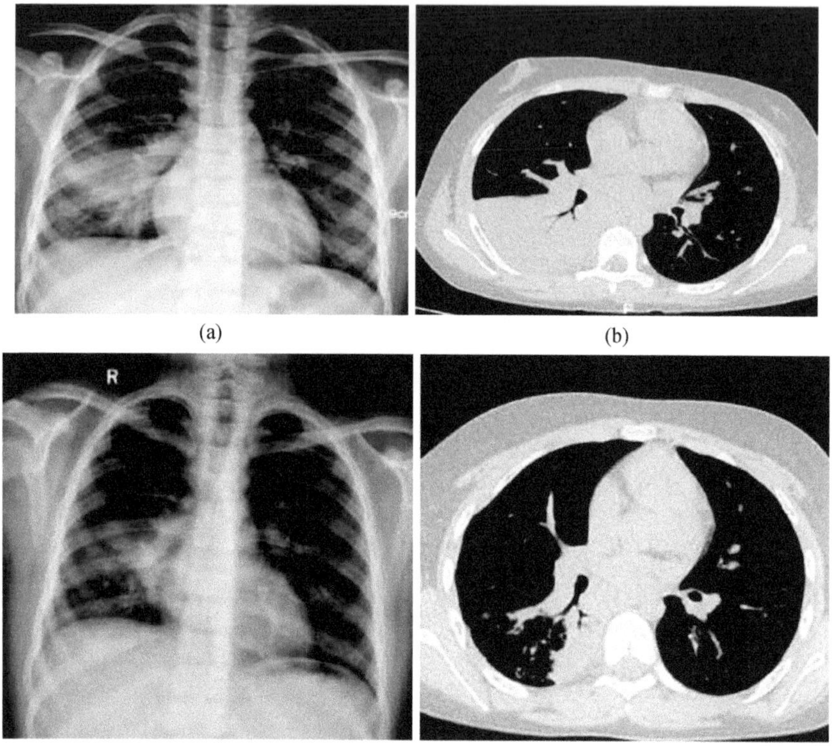

(c) (d)

Figure 13.7 Imaging manifestations of mycoplasma pneumoniae pneumonia (I). (a) Chest X-ray 1 day after the onset; (b) CT scan 6 days after the onset; (c) chest X-ray 11 days after the onset; (d) CT scan 18 days after the onset, showing patchy high-density shadows in the lower lobe of the right lung. After antiinfection treatment, the extent of the lesions decreased.

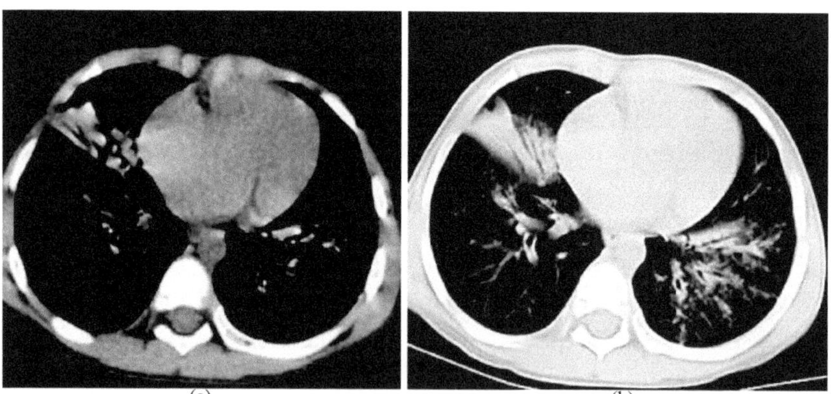

(a) (b)

Figure 13.8 Imaging manifestations of mycoplasma pneumoniae pneumonia (II). (a) Mediastinal window; (b) lung window showing patchy high-density shadows in the right middle lobe and left lower lobe of the lungs.

Case 4: Patient, female, 59 years old. During chemotherapy for acute lymphocytic leukemia, the patient experienced recurrent chest tightness, shortness of breath, accompanied by a bloating sensation in the back, significant changes after urination, and fever, with a maximum body temperature of 38.9°C. White blood cell count 0.20×10^9/L, and lymphocyte count 0.17×10^9/L. Aspergillus galactomannan is tested positive. See Fig. 13.9 for imaging manifestations. In this case, the patient's imaging

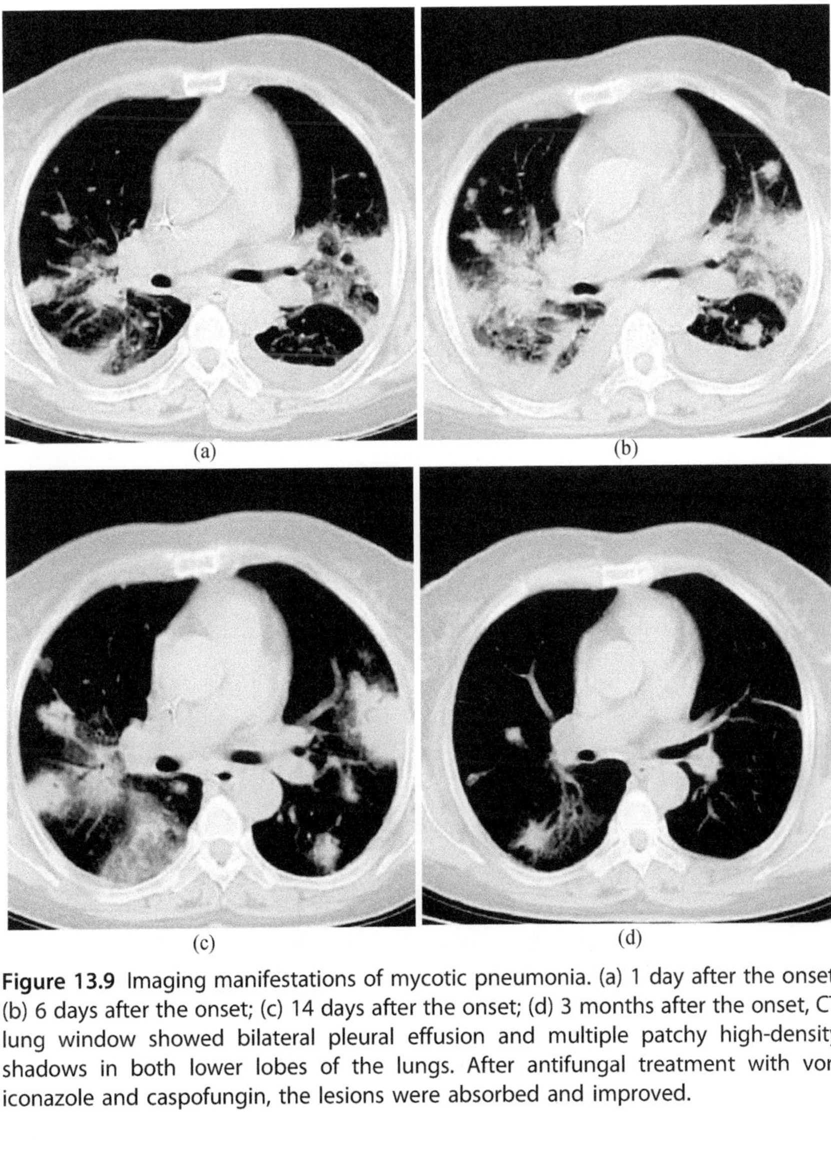

Figure 13.9 Imaging manifestations of mycotic pneumonia. (a) 1 day after the onset; (b) 6 days after the onset; (c) 14 days after the onset; (d) 3 months after the onset, CT lung window showed bilateral pleural effusion and multiple patchy high-density shadows in both lower lobes of the lungs. After antifungal treatment with voriconazole and caspofungin, the lesions were absorbed and improved.

manifestations include multiple patchy shadows in the lower lobes of both lungs, accompanied by pleural effusion. Considering the medical history, it is not difficult to differentiate it from COVID-19.

13.3.4 Differentiation of COVID-19 from lupus pneumonitis

Systemic lupus erythematosus (SLE) predominantly occurs in females and can cause pulmonary infiltrate, leading to lupus pneumonitis. While not very common clinically, its incidence ranges from 5% to 12%. Lupus pneumonitis is a type of interstitial lung disease associated with collagen vascular diseases and autoimmune disorders. In the chest, diffuse pulmonary alveolar infiltrates are observed, particularly prominent at the lung bases.

Case 5: Patient, male, 36 years old. The patient presented with chest pain and shortness of breath for 3 days and had coarse breath sounds in both lungs. Lupus-related antibodies are tested positive. Erythrocyte sedimentation rate 38 mm/h, White blood cell count 5.35×10^9/L, and Lymphocyte count 1.02×10^9/L. See Fig. 13.10 for imaging manifestations. In this case, the patient is a male, which is less commonly seen clinically, but the imaging manifestations are quite typical. Combining the clinical data, epidemiology, and laboratory test results, it is not difficult to distinguish it from COVID-19.

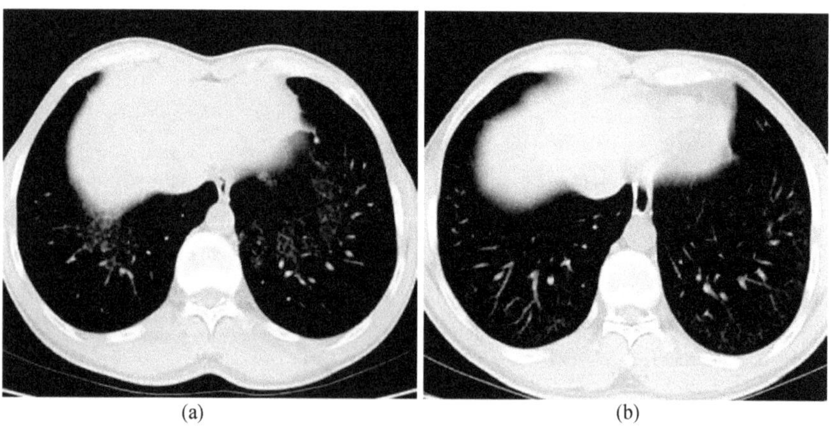

(a) (b)

Figure 13.10 Imaging manifestations of lupus pneumonitis. (a) CT lung window at 3 days after the onset; (b) CT lung window at 1 month after the onset, showing multiple patchy high-density shadows in both lower lobes of the lungs. After treatment, the lesions were mostly absorbed.

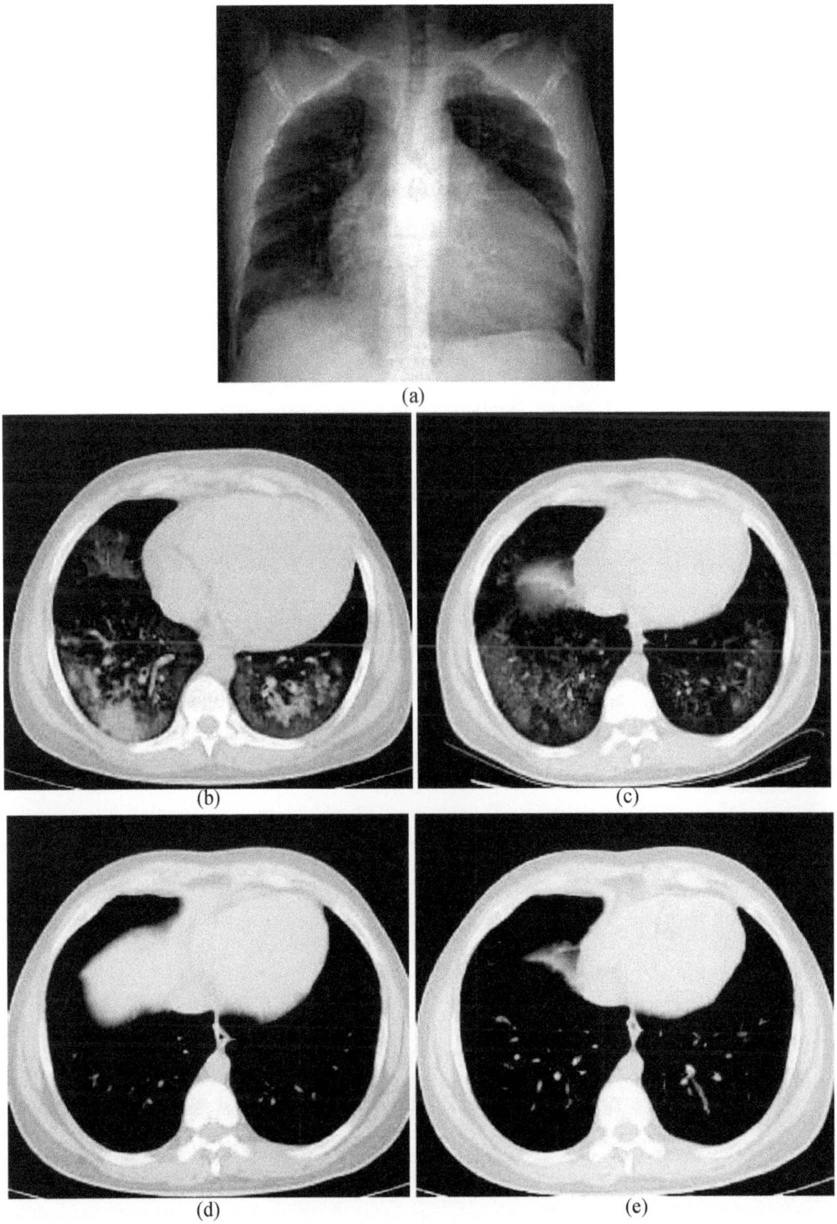

Figure 13.11 Imaging manifestations of cardiogenic pulmonary edema combined with pulmonary infection. (a and b) 3 days after the onset, the localized images showed significantly enlarged cardiac silhouette, and large patchy ground-glass opacities and small patchy high-density shadows in both lower lobes of the lungs; (c) CT lung window 6 days after the onset; (d) 16 days after the onset. (e) 25 days after the onset, after comprehensive treatment including oxygen therapy, antiinfection, sputum expectoration, bronchodilators, cardiac and diuretic therapy for heart failure, anticoagulation, and myocardial nutritional support, the condition was improved with absorption.

13.3.5 Differentiation of COVID-19 from cardiogenic pulmonary edema combined with pulmonary infection

Pulmonary edema is a pathological process where excessive accumulation of fluid occurs in the interstitial tissue and, with or without overflow, enters the alveolar spaces due to factors such as heart failure, resulting in the production of interstitial fluid exceeding its reabsorption.

Case 6: Patient, male, 22 years old. The patient presented with cough, shortness of breath, nausea and vomiting for 3 days and had a body temperature of 36.3°C. Coarse breath sounds were heard in both lungs, and dry and wet rales can be heard with low heart sounds. White blood cell count 12.28×10^9/L, lymphocyte count 1.51×10^9/L, and C-reactive protein 24.10 mg/L. Throat swab SARS-CoV-2 nucleic acid test was negative. Respiratory pathogen spectrum was negative. Myocardial Enzyme: Aspartate aminotransferase 41 U/L, Creatine kinase 43 U/L, Lactate dehydrogenase 371 U/L, Hydroxybutyrate dehydrogenase 257 U/L. See Fig. 13.11 for imaging manifestations. In this case, the patient has pulmonary edema combined with pulmonary infection, and combining the medical history and SARS-CoV-2 nucleic acid test, it can be differentiated.

Bibliography

[1] Chan KS, Zheng JP, Mok YW, et al. SARS: prognosis, outcome and sequelae. Respirology 2003;8(Suppl. l):36−40.

[2] Hui DS, Joynt GM, Wong KT, et al. Impact of severe acute respiratory syndrome (SARS) on pulmonary function, functional capacity and quality of life in a cohort of survivors. Thorax 2005;60(5):401−9.

[3] Venkataraman T, Coleman CM, Frieman MB. Overactive epidermal growth factor receptor signaling leads to increased fibrosis after severe acute respiratory syndrome coronavirus infection. J Virol 2017;91(12):00182−217.

[4] Lu P, Yang G, Liu J, et al. Imaging diagnosis of pneumocystis carinii pneumonia in patients with acquired immunodeficiency syndrome. Chin J Med Imag 2003;11(3):166−8.

[5] Zhou B, Tang X, Lu P. Infectious atypical pneumonia. Beijing: People's Medical Publishing House; 2004. p. 128−88.

[6] Lu P, Zhou B. Diagnostic imaging of emerging infectious diseases. Beijing: People's Medical Publishing House; 2013. p. 37−48.

[7] Lu PX, Zhu BP. Diagnostic imaging of emerging infectious diseases. Berlin: Springer; 2015. p. 5−28.

[8] Lu PX, Zhou BP, Chen XC, et al. Chest X-ray imaging of patients with SARS. Chin Med J 2003;116(7):972−5.

[9] Lu P, Yang G, Weiye Y, et al. Imaging follow-up of SARS patients complicated with pulmonary fibrosis. Chin J Med Imag Technol 2004;20(12):1901−3.

[10] Lu P. The characteristic of epidemiology and medical imaging in human avian influenza outbreak in China. Electron J Emerg Infect Dis 2017;2(2):124−6.

[11] Deng Y, Huang H, Yuan J, et al. Imaging manifestations and therapeutic evaluation of pneumonia caused by human infected H7N9 avian influenza virus. Electron J Emerg Infect Dis 2017;2(1):46–9.
[12] Shi R, Chunzi S, Qian M, et al. Value of clinical imaging in diagnosis and treatment of human-infected H7N9 avian influenza. Electron J Emerg Infect Dis 2016;1(1):50–2.
[13] Jingjing L, Zeng Z, Lu P, et al. Imaging follow-up study of human-infected H7N9 avian influenza virus pneumonia. Radiol Practice 2016;31(3):228–31.
[14] Lu P, Zeng Z, Zheng F, et al. Characteristics of imaging manifestations and dynamic changes in patients with severe pneumonia caused by H7N9 avian influenza virus. Radiol Practice 2014;29(7):740–4.
[15] Huang X, Zeng Z, Lu P, et al. The clinical imaging analyses of the pneumonia of human infection with H7N9 subtype avian influenia virus in 12 cases. Chin J CT MRI 2014;12(2):8–11.
[16] Qiu J, Renhua W. Imaging manifestations of pneumocystis jiroveci pneumonia. Electron J Emerg Infect Dis 2019;4(4):235–9.
[17] Catherinot E, Lanternier F, Bougnoux M, et al. Pneumocystis jirovecii pneumonia. Infect Dis Clin 2010;24(1):107–38.
[18] Jianjian L, Yang S. Research progress on the diagnosis of HIV/AIDS complicated with pneumocystis carinii pneumonia. Chin J Mycol 2009;4(3):186–8.
[19] HIV/AIDS Group, Branch of Tropical Diseases and Parasitology, Chinese Medical Association. Expert consensus on diagnosis and treatment of HIV infection combined with invasive fungal diseases. Chin J Infect Dis 2019;37(10):581–93.
[20] Hardak E, Brook O, Yigla M. Radiological features of pneumocystis jirovecii pneumonia in immune compromised patients with and without AIDS. Lung 2010;188(2):159–63.
[21] Lan BW, Lu PX, Zeng YJ, et al. Clinical imaging research of the fifirst Middle East respiratory syndrome in China. Radiol Infect Dis 2015;(2):173–6.
[22] Yuelong S. Problems and challenges of influenza prevention and control in China. Chin J Prev Med 2013;47(5):391–3.
[23] Korsun N, Angelova S, Trifonova I, et al. Predominance of influenza A (H3N2)viruses during the 2016, 2017 season in Bulgaria. J Med Microbiol 2018;67(2):228–39.
[24] Yan Q, Zhong J, Zhou H, et al. Clinical and epidemiological characteristics of 245 cases of influenza A (H3N2). Chin J Tuberc Respir Dis 2019;42(7):510–4.
[25] Jin X, Zhang Z, Wang D. Applied pathophysiology of the respiratory system. Wuhan: Huazhong University of Science and Technology Press; 2007.
[26] He Y, Liu C, Zheng S, et al. The role of chest digital radiography in children with *Mycoplasma pneumonia*. Radiol Practice 2015;30(1):75–7.
[27] Lyv Z, Chen TT, Mao L, et al. The chest CT imaging characteristics of *Mycoplasma pneumonia* with different age groups of children. Radiol Infect Dis 2017;4(4):150–6.
[28] Gong L, Zhang CL, Zhen Q. Analysis of clinical value of CT in the diagnosis of pediatric pneumonia and *Mycoplasma pneumonia*. Exp Ther Med 2016;11(4):1271–4.
[29] Zhao D, Chen H, Yang Q, Deng L. Value of clinical signs in the identification of *Mycoplasma pneumonia* in community acquired pneumonia in children. Zhonghua Er Ke Za Zhi 2016;54(2):104–10.
[30] Qu D, Lin L, Shengqi L. CT features of adult mycoplasma pneumoniae pneumonia. Chin J Med Imaging Technol 2010;26(2):269–71.
[31] Jin H, Guo Y, Liu B, et al. Clinical analysis of asthma status with Aspergillus lung infection. Guangxi Med J 2005;27(10):1572–3.
[32] Chen Q, Lu P, Wu S. Clinical diagnosis and treatment of pulmonary aspergillosis. Beijing: People's Medical Publishing House; 2017.

[33] Qiu C, Lu PX, Wu SP. Pulmonary aspergillosis: diagnosis and cases. Berlin: Springer; 2019.
[34] Wanzhen Y. Pulmonary manifestations of systemic lupus erythematosus. J Clin Pulm Med 2000;5(1):17–20.
[35] Zheng G, Zhang Q, Tan W, et al. Clinical imaging analysis of inactive tuberculosis complicated by pulmonary aspergillosis. Electron J Emerg Infect Dis 2018;3(1):34–6.
[36] Yanyun D, Huang K, Lan K, et al. Analysis of clinical characteristics of AIDS complicated with bacterial pneumonia. Electron J Emerg Infect Dis 2019;4(1):20–3.

CHAPTER 14

Diagnosis and treatment

Zhang Lina[1], Li Tongzeng[2], Yingmin Ma[2], Sun Lijun[2] and Liu Yang[2]
[1]Department of Radiology, The Fourth Affiliated Hospital of China Medical University, Liaoning, China; [2]Beijing YouAn Hospital, Capital Medical University, Beijing, China

14.1 Diagnostic criteria

14.1.1 Diagnostic criteria of COVID-19

14.1.1.1 Suspected cases

Combining the history of epidemiology and clinical manifestations, the case compliant with any one sign of the history of epidemiology and any two symptoms from the clinical manifestations is judged as the suspected case.

Where there is no specific history of epidemiology, the case compliant with any two symptoms from the clinical manifestations and the SARS-CoV-2-specific IgM antibodies tested positive, or the case compliant with any three symptoms from the clinical manifestations is judged as the suspected case.

14.1.1.1.1 History of epidemiology

(1) There is the travel history or residency in a community with reported cases within 14 days prior to the onset of the disease;

(2) There is the history of contact with a confirmed SARS-CoV-2 infected patient or asymptomatic affected person within 14 days prior to the onset of the disease;

(3) There is the history of contact with patients in fever or respiratory symptoms from a community with reported cases within 14 days the onset of the disease;

(4) Clustered cases (2 or more cases with fever and/or respiratory symptoms occurring within a small area such as a household, workplace, school class, etc. within 2 weeks).

The Pathology and Imaging of Coronavirus Pneumonia
ISBN 978-0-443-32930-2
https://doi.org/10.1016/B978-0-443-32930-2.00010-6

14.1.1.1.2 Clinical manifestations

(1) Fever and/or respiratory symptoms, and other COVID-19 related clinical manifestations;
(2) Having the aforementioned COVID-19 radiological features;
(3) In the early stages of the disease, there may be a normal or decreased white blood cell count and a normal or decreased lymphocyte count.

14.1.1.2 Confirmed cases

Suspected cases with one of the following etiological or serological evidence:

(1) Real-time fluorescent RT-PCR detected SARS-CoV-2 nucleic acid positive;
(2) Virus gene sequencing showed high homology with known SARS-CoV-2;
(3) Positive for SARS-CoV-2 specific IgM and IgG antibodies;
(4) SARS-CoV-2 specific IgG antibodies converted from negative to positive or the IgG antibody titer in the convalescent stage showed a 4-fold or higher increase compared with the acute stage.

14.1.2 Diagnostic criteria of SARS

The diagnosis of SARS can be made by combining the epidemiological history, clinical symptoms and signs, general laboratory examinations, pulmonary X-ray imaging changes, and positive SARS pathogen detection, excluding other diseases with similar manifestations.

The presence of clinical symptoms and the appearance of pulmonary X-ray changes are the fundamental criteria for diagnosing SARS.

Having clear supporting evidence from an epidemiological perspective and being able to exclude other diseases are the most important supporting criteria for making a clinical diagnosis. For cases where clear epidemiological evidence cannot be immediately obtained during the initial visit, rigorous epidemiological follow-up shall be conducted after the visit. Dynamic observation of the changes in the patient's conditions (symptoms, oxygenation status, pulmonary X-ray imaging), the effectiveness of antimicrobial treatment, and the results of specific SARS pathogen detection are of great significance for the diagnosis.

Clinical physicians shall promptly classify and carry out appropriate disposal of relevant individuals based on the following criteria.

14.1.2.1 Medical isolation observer

Those who have had contact with SARS patients or suspected SARS patients within the past 2 weeks but do not exhibit clinical manifestations of SARS shall be categorized as medical isolation observers and undergo medical isolation observation.

14.1.2.2 Suspected cases

For individuals who lack clear epidemiological evidence but possess other supporting evidence for SARS, they can be considered as suspected cases. Further epidemiological follow-up shall be conducted, and arrangements for etiological examination shall be made to confirm the diagnosis. For individuals with epidemiological evidence and clinical symptoms but no pulmonary X-ray changes, they shall also be considered as suspected cases. For such cases, it is necessary to conduct dynamic reexamination of chest X-rays or chest CT scans. Once pulmonary lesions appear, and after excluding other diseases, a clinical diagnosis can be made.

14.1.2.3 Clinical diagnosis and definitive diagnosis

A clinical diagnosis of SARS can be made for individuals who have epidemiological evidence of SARS, corresponding clinical manifestations, pulmonary X-ray imaging changes, and the exclusion of other diseases. Based on clinical diagnosis, a definitive diagnosis can be made if secretion testing for SARS-CoV RNA is positive, or if serum (or plasma) testing for SARS-CoV specific antigen N protein is positive, or if serum SARS-CoV antibodies converts from negative to positive, or if the antibody titer increases by $\geq$4 times.

14.1.2.4 Diagnostic criteria of severe SARS

A diagnosis of severe SARS can be made if any of the following three criteria are met.

(1) Respiratory distress, respiratory rate (RR) $\geq$30 breaths per minute for adults at rest state, and accompanied by any of the following conditions:

① X-ray chest imaging shows multilobar lesions or lesions occupying more than one-third of the total lung area on the frontal chest X-ray;

② Disease progression with an increase in lesion area by over 50% within 48 h, occupying more than one-fourth of the total lung area on the frontal chest X-ray.

(2) The presence of hypoxemia, with the oxygenation index less than 300 mmHg (1 mmHg = 0.133 kPa).
(3) The occurrence of shock or MODS (Multiple Organ Dysfunction Syndrome).

14.1.3 Diagnostic criteria of MERS

14.1.3.1 Suspected cases

The patients who are consistent with the history of epidemiology and clinical manifestations, but have no laboratory evidence are judged as suspected cases.

(1) History of Epidemiology: There is a history of traveling or residing in the Middle East or epidemic outbreak areas within 14 days before the onset of the disease, or a history of close contact with suspected/clinically diagnosed/confirmed cases.
(2) Clinical Manifestations: Fever accompanied by respiratory symptoms that cannot be easily explained by other pathogenic infections.

14.1.3.2 Clinically diagnosed cases

(1) Patients who meet the criteria for suspected cases and whose laboratory screening result (e.g., positive for a single target PCR or single serum antibody) is positive are judged as clinically diagnosed cases.
(2) Patients who meet the criteria for suspected cases and whose laboratory test result is negative or even unable to be determined due to improper collection or handling of a single specimen are judged as clinically diagnosed cases.

14.1.3.3 Confirmed cases

A confirmed laboratory diagnosis of MERS can be made if any of the following four criteria are met:

(1) At least two target PCR tests are positive.
(2) Positive product from a single target PCR, confirmed by gene sequencing.
(3) MERS-CoV is isolated from respiratory specimens.
(4) Seroconversion or a fourfold or higher increase in MERS-CoV antibodies in serum of the convalescence stage compared with that of the acute stage.

14.2 Treatment principle

14.2.1 Determine the treatment place according to the disease conditions

(1) Suspected and confirmed cases shall be isolated and treated in designated hospitals with effective isolation and protective conditions. Each suspected case shall be isolated and treated in a single room, while confirmed cases can be treated in the same ward with multiple patients.

(2) Severe cases shall be admitted to the Intensive Care Unit (ICU) for treatment as early as possible.

(3) For severe and critical cases, in addition to treatment, proactive measures shall be taken to prevent and treat complications, manage underlying conditions, prevent secondary infections, and provide timely organ function support.

14.2.2 Principle of SARS treatment

Although the pathogen of SARS has been largely identified, the pathogenesis is still not fully understood, and currently, there is a lack of specific treatments targeting the etiology. Based on the above understanding, clinical management shall primarily focus on symptomatic supportive care and treatment of complications. Blind use of medication shall be avoided, especially the prolonged and high-dose combination of multiple drugs (such as antibiotics, antiviral drugs, immunomodulators, corticosteroids, etc.).

14.2.3 Principle of MERS treatment

According to the severity of the condition, the treatment location shall be assessed as follows: Suspected, clinically diagnosed, and confirmed cases shall be isolated and treated in hospitals with effective isolation and protective conditions. Severe cases shall be promptly admitted to the Intensive Care Unit (ICU) for treatment. Strict isolation and protective measures shall be implemented during the transshipment process.

14.3 General treatment and supporting treatment

14.3.1 General treatment

(1) Stay in bed, give intensive supportive treatment and ensure adequate energy intake; monitor water and electrolyte balance, maintain internal environment stability; closely monitor vital signs, oxygen saturation, etc.

(2) Monitor blood routine, urine routine, C-reactive protein, biochemical indicators (liver enzymes, cardiac enzymes, renal function, etc.), coagulation function, arterial blood gas analysis, and chest imaging according to the conditions. If conditions permit, cytokine testing can be performed.
(3) Timely administer effective oxygen therapy measures, including nasal cannula, oxygen mask, and high-flow nasal cannula oxygen therapy. If conditions permit, hydrogen-oxygen mixture inhalation therapy (H_2/O_2: 66.6/33.3) can be used for treatment.
(4) Antibacterial Treatment: Avoid blind or inappropriate use of antibacterial drugs, especially the combination of broad-spectrum antibacterial drugs.

14.3.2 Supportive treatment

14.3.2.1 Respiratory support

(1) Nasal Cannula or Mask Oxygen Therapy: Severe patients with a PaO_2/FiO_2 ratio lower than 300 mmHg shall receive oxygen therapy immediately. After receiving nasal cannula or mask oxygen therapy, closely observe the patient for a short period (1−2 h). If respiratory distress and/or hypoxemia do not relief, high-flow nasal cannula oxygen therapy (HFNC) or noninvasive mechanical ventilation (NIV) shall be used.
(2) High-flow nasal cannula oxygen therapy (HFNC) or noninvasive ventilation (NIV): HFNC or NIV shall be administered when the PaO_2/FiO_2 ratio is below 200 mmHg. For patients receiving HFNC or NIV, and in the absence of contraindications, it is recommended to implement prone position ventilation, also known as awake prone position. The duration of prone position treatment shall be greater than 12 h.
(3) Some patients undergoing HFNC or NIV therapy have a high risk of treatment failure, and it is essential to closely monitor their symptoms and signs. If there is no improvement in the patient's condition
(4) After a short period (1−2 h) of treatment, especially after receiving prone position therapy and hypoxemia remains unchanged, or if there are signs of increased respiratory rate, excessive tidal volume, or strong inspiratory effort, it often indicates poor effectiveness of HFNC or NIV treatment. In such cases, invasive mechanical ventilation shall be initiated promptly.

(5) Invasive Mechanical Ventilation: In general, when the PaO_2/FiO_2 ratio is below 150 mmHg, tracheal intubation and invasive mechanical ventilation shall be considered. However, considering that the clinical manifestations of hypoxemia in critical COVID-19 patients may not be typical, the decision to perform tracheal intubation and invasive mechanical ventilation shall not be based solely on whether the PaO_2/FiO_2 ratio meets the standard. Instead, real-time evaluation shall be conducted by taking into account the patient's clinical manifestations and organ functions. It is worth noting that the potential harm caused by delaying tracheal intubation may be greater.

(6) Early and appropriate use of invasive mechanical ventilation is a crucial treatment approach for critical patients. Implementing mechanical lung-protective ventilation strategies, lung recruitment therapy can be used for patients with moderate to severe acute respiratory distress syndrome or those undergoing invasive mechanical ventilation with FiO_2 higher than 50%. Furthermore, it is decided whether to perform lung recruitment techniques according to the reactivity of lung recruitment. It shall be noted that some COVID-19 patients may have poor lung recruitable volume, and excessive positive end-expiratory pressure (PEEP) shall be avoided to prevent barotrauma.

(7) Airway Management: Strengthen airway humidification, and recommend using active heated humidifiers. If conditions permit, use circuit heated wires to ensure effective humidification; Closed suction shall be used, and if necessary, tracheal suction with a bronchoscope shall be performed; Active airway clearance therapies such as vibration sputum drainage, high-frequency chest wall oscillation, and postural drainage shall be carried out; Early passive and active mobilization shall be initiated to promote sputum drainage and lung rehabilitation, provided that oxygenation and hemodynamics are stable.

(8) ECMO Initiation Criteria: ECMO shall be considered for evaluation as early as possible when the following conditions are met, in addition to being under optimal mechanical ventilation conditions ($FiO_2 \geq 80\%$, tidal volume of 6 mL/kg of ideal body weight, PEEP $\geq$5 cmH_2O, and no contraindications), and when protective ventilation and prone position have not yielded satisfactory results: ① $PaO_2/FiO_2 < 50$ mmHg for more than 3 h; ② $PaO_2/FiO_2 < 80$ mmHg for more than 6 h; ③ Arterial blood pH < 7.25, with $PaCO_2 > 60$ mmHg for more than 6 h, and RR > 35 breaths per minute; ④ RR > 35 breaths per minute,

with arterial blood pH < 7.2 and plateau pressure >30 cmH_2O; ⑤ Presence of concomitant cardiogenic shock or cardiac arrest.

Critical patients who meet the indications for ECMO and have no contraindications shall initiate ECMO treatment as early as possible. Delaying may result in poor patient outcomes.

The ECMO mode selection is as follows: For respiratory support only, the veno-venous ECMO (VV-ECMO) is most commonly used; for simultaneous respiratory and circulatory support, veno-arterial ECMO (VA-ECMO) is chosen; In case of hypoxia in the head and upper extremities during VA-ECMO, Veno-Arterio-Venous ECMO (VAV-ECMO) mode can be applied. After implementing ECMO, strict adherence to a lung-protective ventilation strategy shall be carried out. Recommended Initial Settings: Tidal volume < 4–6 mL/kg of ideal body weight, plateau pressure ≤ 25 cmH_2O, driving pressure <15 cmH_2O, PEEP (Positive End-Expiratory Pressure) set at 5–15 cmH_2O, respiratory rate (RR) set at 4–10 breaths per minute, and FiO_2 (Fraction of Inspired Oxygen) < 50%. For patients with difficult-to-maintain oxygenation, excessive inspiratory efforts, evident consolidations in dependent lung areas, or requiring active airway secretion drainage, prone position ventilation can be employed.

Children have weaker cardiopulmonary compensatory abilities compared with adults and are more sensitive to hypoxia. Therefore, they require more aggressive oxygen therapy and ventilatory support strategies. Indications for such treatments shall be appropriately broadened.

14.3.2.2 Circulatory support

In critical patients who may have concurrent shock, appropriate use of vasopressor drugs shall be administered on the basis of adequate fluid resuscitation. It is essential to closely monitor the patient's blood pressure, heart rate, urine output, as well as lactate and base excess levels, is essential. Hemodynamic monitoring shall be conducted when necessary to guide fluid administration and the use of vasopressor medications, and thus to improve tissue perfusion.

14.3.2.3 Anticoagulation treatment

Severe or critical patients have a higher risk of developing thromboembolism. For patients without contraindications to anticoagulation and with a significant elevation of *D*-dimer levels, it is recommended to use anticoagulant medication for prophylaxis. In the event of a thromboembolism, anticoagulation treatment shall be conducted according to the corresponding guidelines.

14.3.2.4 Acute kidney injury and renal replacement therapy

Critical patients may experience concomitant acute kidney injury, and efforts shall be made to actively identify the underlying causes, such as factors related to low perfusion and medications. While actively correcting the underlying causes, it is essential to pay attention to maintain water, electrolyte, and acid-base balance. The indications for continuous renal replacement therapy (CRRT) include: ① Hyperkalemia; ② Severe acidosis; ③ Pulmonary edema or excessive fluid overload refractory to diuretics.

14.3.2.5 Blood purification treatment

The blood purification system includes techniques such as plasma exchange, adsorption, perfusion, and blood/plasma filtration. It can clear inflammatory factors, block the "cytokine storm," and thereby reduce the damage caused by the inflammatory response to the body. It can be used for the early and middle-stage treatment of cytokine storms in severe and critical patients.

14.3.2.6 Multisystem inflammatory syndrome in children

The treatment principle involves multidisciplinary collaboration, early antiinflammatory interventions, correction of shock and coagulation dysfunction, organ function support, and, if necessary, antimicrobial therapy. Patients with typical or atypical Kawasaki disease manifestations shall receive treatment similar to the classic Kawasaki disease treatment protocol. The main treatment involves intravenous immunoglobulin, corticosteroids, and oral aspirin therapy.

14.3.2.7 Other treatment measures

Xuebijing treatment can be considered; Probiotics can be used as a gut microbiota regulator to maintain gut microbial balance and prevent secondary bacterial infections; In severe and critical pediatric cases, intravenous immunoglobulin may be considered as appropriate.

Pregnant patients with severe or critical conditions shall actively consider terminating the pregnancy, with cesarean section being the preferred method. Patients often experience anxiety and fear, and as such, psychological counseling shall be strengthened, and if necessary, supplemented with medication treatment.

14.3.3 Treatment of SARS

14.3.3.1 General treatment and disease monitoring

Stay in bed, pay attention to maintain water and electrolyte balance, and avoid exertion and vigorous coughing. Closely observe changes in the patient's conditions (as many patients may be in the progressive stage during the 2–3 weeks following the onset of the disease). In general, continuous nasal cannula oxygen therapy is provided in the early stage (with an oxygen concentration typically at 13 L/min).

According to the patient's conditions, pulse volume blood oxygen saturation shall be monitored at scheduled intervals or continuously every day. Regularly review blood routine, urine routine, blood electrolytes, liver and kidney functions, myocardial enzyme, T lymphocyte subpopulations (when possible), and chest X-ray, etc.

14.3.3.2 Symptomatic treatment

(1) If the body temperature is above 38.5°C or there is significant body aches, antipyretic and analgesic medications can be used. For individuals with high fever, physical cooling measures such as applying cold packs, alcohol sponge baths, and cooling blankets are recommended. Salicylate-containing antipyretic and analgesic drugs are prohibited for use in children.

(2) Patients with cough and sputum can be given antitussive and expectorant medications.

(3) For patients with impaired organ functions such as heart, liver, or kidney, appropriate treatment measures shall be taken.

(4) Patients with diarrhea shall pay attention to rehydration and correction of water and electrolyte imbalances.

14.3.3.3 Use of glucocorticoids

The purpose of using glucocorticoids is to suppress abnormal immune pathological reactions, alleviate severe systemic inflammatory response, and prevent or reduce late-stage pulmonary fibrosis. Glucocorticoid therapy can be considered when any of the following indications is present: ① Severe toxic symptoms, persistent high fever that does not subside, and the highest body temperature remains above 39°C after symptomatic treatment for more than 5 days; ② Chest X-ray showing multiple or large opacities with rapid progression, and within 48 h, the lesion area increases by more than 50% and occupies more than 1/4 of the total lung area on the frontal chest X-ray; ③ Compliant with the diagnostic criteria for acute lung injury or ARDS (Acute Respiratory Distress Syndrome).

14.3.3.4 Antiviral treatment

Currently, specific drugs targeting SARS-CoV have not been discovered. Clinical retrospective analysis data shows that commonly used antiviral drugs such as ribavirin are ineffective against SARS. The efficacy of protease inhibitor drugs such as lopinavir/ritonavir is yet to be verified.

14.3.3.5 Immunotherapy

The efficacy of nonspecific immunostimulants such as thymosin, interferon, and intravenous immunoglobulin in treating SARS has not been confirmed, and their routine use is not recommended. The clinical efficacy of convalescent serum for SARS has not been confirmed. However, it can be cautiously tried in critical patients with a definite diagnosis under close observation.

14.3.3.6 Use of antimicrobial agents

The use of antimicrobial agents serves primarily two purposes: first, for empirical therapy in suspected patients to aid in the diagnosis; second, for the treatment and control of secondary bacterial and fungal infections.

14.3.3.7 Psychotherapy

For suspected cases, appropriate admission conditions shall be arranged to reduce patients' concerns about nosocomial cross-infection; for confirmed cases, it is essential to provide increased care and explanations, guiding patients to deepen their understanding of the self-limiting nature and curability of the disease.

14.3.3.8 Treatment principles of severe SARS

Although the conditions of the majority of SARS patients can naturally improve, approximately 30% of cases are classified as severe, among which some may progress to acute respiratory distress syndrome (ARDS) or even death. Therefore, for severe patients, it is necessary to conduct close and dynamic observation, enhance monitoring and timely provide respiratory support, reasonably use glucocorticoids, and strengthen nutritional support and organ function protection. Besides, pay attention to water, electrolyte, and acid-base balance, prevent and treat secondary infections and promptly manage complications.

14.3.3.8.1 General treatment and disease monitoring

The general treatment and disease monitoring for nonsevere patients are essentially the same, but for severe patients, it is important to strengthen the monitoring of vital signs, intake and output of fluids, electrocardiogram, and blood glucose levels. When the blood glucose level is elevated above the normal level, the application of insulin to bring them under control within the normal range may help reduce complications.

14.3.3.8.2 Respiratory Support Treatment

Severe SARS patients shall be regularly monitored for changes in SpO_2. A decrease in SpO_2 after exercise is an early indication of respiratory failure and shall be promptly addressed and treated.

(1) Oxygen Therapy: For severe cases, even if there are no signs of hypoxia at rest, continuous nasal cannula oxygen therapy shall also be administered. For individuals with hypoxemia, it is usually necessary to administer higher oxygen flow rates to maintain SpO_2 at 93% or above. If needed, an oxygen mask can be used for oxygen supplementation. Try to avoid activities that require discontinuing oxygen therapy, such as going to the restroom or undergoing medical examinations. If the flow rate of oxygen inhalation $\geq$5 L/min (or inhaled oxygen concentration $\geq$40%), the SpO_2 remains below 93%, or the SpO_2 can be maintained at 93% or above even after sufficient oxygen therapy, but the respiratory rate is still 30 breaths per minute or higher, and the respiratory load remains at a high level, noninvasive positive pressure ventilation (NIPPV) shall be considered promptly.

(2) Noninvasive positive pressure ventilation (NIPPV): NIPV can improve symptoms of respiratory distress and enhance lung oxygenation, aiding patients in overcoming critical conditions and potentially reducing the need for invasive ventilation. The indications for NIPV application: ① Respiratory rate (RR) > 30 breaths per minute; ② SpO_2 < 93% under the oxygen flow rate of 5 L/min. Contraindications: ① Life-threatening situations requiring urgent endotracheal intubation; altered level of consciousness; ② Vomiting, upper gastrointestinal bleeding; ③ Excessive airway secretions and impaired ability to clear sputum; ④ Inability to cooperate with NIPPV treatment; ⑤ Hemodynamic instability and multiorgan dysfunction.

The commonly used modes and corresponding parameters for NIPPV are as follows: ① Continuous Positive Airway Pressure (CPAP): The commonly used pressure level is generally set between 4 and 10 cmH_2O (1 cmH_2O = 0.098 kPa); ② Pressure Support Ventilation (PSV) + Positive End-Expiratory Pressure (PEEP): The PEEP level is typically set between 4 and 10 cmH_2O, and the inspiratory pressure level is generally set between 10 and 18 cmH_2O. When FiO_2 is less than 0.6, the PaO_2 shall maintain at or above 70 mmHg, or SpO_2 at or above 93%.

When applying NIPPV, the following points shall be noted: Choose an appropriate, well-sealed nasal mask or full-face mask; It shall be used continuously throughout the day (including during sleep), and any intermittent breaks shall be shorter than 30 min. When applying NIPV, the pressure level shall begin at a low setting (e.g., 4 cmH_2O) and gradually increase to the predetermined level; If severe coughing occurs, temporarily disconnecting the ventilator tubing shall be considered to avoid barotrauma; If the expected results ($SpO_2 \geq 93\%$, improvement in dyspnea) are not achieved after 2 h of NIPV application, invasive ventilation may be considered as an alternative.

(3) Invasive Positive Pressure Ventilation: The indications for implementing invasive positive pressure ventilation in SARS patients are as follows: ① NIPV treatment is intolerable, or there is no improvement in respiratory distress, unsatisfactory improvement in oxygenation function, $PaO_2 < 70$ mmHg, and a deteriorating trend in the conditions. ② Presence of life-threatening clinical manifestations or multiple organ failure, requiring urgent endotracheal intubation for resuscitation.

The approach and method for establishing an artificial airway shall be chosen based on each hospital's experience and the specific conditions of the patient. To shorten the procedure time and reduce the chance of cross-infection among medical staff, the use of oral endotracheal intubation or fiberoptic bronchoscopy-guided nasal intubation can be considered under strict protective measures. Tracheostomy shall only be performed after other artificial airway methods have been established beforehand to ensure the safety.

The specific ventilation mode for invasive positive pressure ventilation can be selected based on the hospital's equipment and the clinical experience of the medical staff. Generally, pressure-limited ventilation modes can be selected. For example, in the early stage, pressure-regulated volume control ventilation (PRVCV) + PEEP, pressure control (PC), or volume control (VC) + PEEP can be selected. After improvement, it can be switched to synchronize intermittent mandatory ventilation (SIMV + PSV) + PEEP, and before extubation, PSV + PEEP can be used.

Ventilation parameters shall be set according to the principles of "lung-protective ventilation strategy": ① Use small tidal volumes (6–8 mL/kg), appropriately increase the ventilation rate, and limit the inspiratory plateau pressure to less than 35 cmH_2O. ② Apply appropriate levels of PEEP to maintain alveolar recruitment, allowing collapsed alveoli to reopen and avoiding lung injury caused by repeated opening and closing of alveoli during tidal breathing. The therapeutic range for PEEP is 5–20 cmH_2O, with an average around 10 cmH_2O. At the same time, attention shall be paid to the impact of increased PEEP on the circulatory system.

During the ventilation process, patients with respiratory dyscoordination and anxiety shall receive adequate sedation, and if necessary, muscle relaxants shall be administered to prevent a decline in oxygenation function. The following sedative drugs can be used: ① Midazolam maleate: Initially, administer 3–5 mg through intravenous injection, followed by a maintenance dose of 0.05–0.20 mg/(kg·h). ② Propofol: Initially, administer 1 mg/kg through intravenous injection, followed by a maintenance dose of 1–4 mg/(kg·h). Based on this, opioid drugs can be intermittently used as needed, and muscle relaxants can be added if necessary. Vecuronium bromide (4 mg) can be used as muscle relaxant, which is administered through intravenous injection, and it can be repeated if necessary.

14.3.3.8.3 Application of glucocorticoids

For severe cases that meet the criteria for acute respiratory distress syndrome (ARDS), corticosteroids shall be administered timely and regularly to reduce pulmonary exudation, injury, and subsequent pulmonary fibrosis, as well as to improve lung oxygenation function. Currently, the adult dosage used in most hospitals is equivalent to 80–320 mg of methylprednisolone per day. The specific dosage can be adjusted based on the patient's conditions and individual differences. In some critical patients, short-term (3–5 days) methylprednisolone pulse therapy can be considered, with a daily dose of 500 mg. After the condition improves or X-ray chest images show absorption of the lesions, the dose of methylprednisolone can be gradually reduced and eventually stopped. Typically, a reduction of 1/3 of the dose every 3–5 days can be chosen.

14.3.3.8.4 Clinical nutrition support

Due to the presence of malnutrition in most critical patients, eating easily digestible food at early stage shall therefore be encouraged. When the condition worsens and the patient is unable to eat normally, timely clinical nutrition support shall be provided, using a combination of enteral nutrition and parenteral nutrition. Nonprotein energy shall be 105–126 kJ

(25–30 kcal) per kilogram per day. The proportion of fat can be increased appropriately to reduce the burden on the lungs. Medium/long-chain mixed lipid emulsions have minimal impact on liver function and immune function. The protein intake shall be 1.0–1.5 g/(kg·d), as excessive intake may have adverse effects on liver and kidney function. Water-soluble and fat-soluble vitamins shall be supplemented. Plasma albumin level shall be maintained within the normal range.

14.3.3.8.5 Prevention and treatment of secondary infections

Critical patients often have a weakened immune system, requiring close monitoring and prompt treatment of secondary infections. If necessary, prophylactic antimicrobial therapy shall be cautiously considered.

14.3.4 Treatment of MERS

14.3.4.1 General treatment and close monitoring

(1) Stay in bed, maintain water and electrolyte balance, and closely monitoring the changes in the conditions.
(2) Regularly review blood routine test, urine test, arterial blood gas analysis, blood biochemistry, and chest imaging.
(3) According to changes in oxygen saturation, take effective oxygen therapy measures timely, including nasal cannula, oxygen mask, and if necessary, noninvasive or invasive ventilation.

14.3.4.2 Antiviral treatment

Currently, there are no specific and effective drugs for MERS-CoV. In vitro experiments have shown that the combination therapy of ribavirin and interferon-α has some antiviral effects. However, the clinical research results are still uncertain. Antiviral therapy can be tried in the early stages of the disease, but caution shall be exercised regarding the adverse reactions of the medications during the treatment process.

14.3.4.3 Antibiotics treatment

Avoid using antibiotics blindly or inappropriately, strengthen bacteriology monitoring, and apply antibiotics when secondary bacterial infections occur.

14.3.4.4 TCM treatment

Based on literature data and combined with the experience of traditional Chinese medicine (TCM) in treating diseases like "febrile diseases" and "wind-heat lung heat", treatment shall be conducted through TCM syndrome differentiation and treatment under the guidance of a TCM physician.

14.3.4.5 Treatment suggestions for severe cases

The treatment principles for severe and critical cases involve symptomatic treatment, prevention and management of complications, and providing effective organ function support. Implement effective respiratory support (including oxygen therapy, noninvasive/invasive mechanical ventilation), circulatory support, liver support, and kidney support, etc. For critical patients with poor response to invasive mechanical ventilation, hospitals with the necessary capabilities can consider implementing extracorporeal membrane oxygenation (ECMO) support technology. Maintain the gastrointestinal function of critical patients and apply microecological modulator when appropriate to regulate gut health.

14.4 Antiviral treatment

At present, there are no approved antiviral drugs for treating COVID-19 in China. On October 23, 2020, remdesivir, an antiviral drug from Gilead Science was approved by the United States Food and Drug Administration (FDA) for marketing. It is indicated for use in COVID-19 patients aged 12 years and above, who weigh at least 40 kg and require hospitalization. Earlier, remdesivir has already received Emergency Use Authorization (EUA) from the FDA. On November 20, 2020, the United States FDA approved the Emergency Use Authorization for the combination therapy of Eli Lilly and Company's rheumatoid arthritis drug, baricitinib, with remdesivir as a treatment for SARS-CoV-2. This drug was authorized for use in hospitalized adults and pediatric patients aged 2 years and above, who require oxygen support, and are either suspected or confirmed COVID-19 patients in the lab. According to FDA data, in a clinical trial conducted on hospitalized COVID-19 patients, the combination therapy of baricitinib and remdesivir was found to reduce the time to recovery within 29 days after the start of treatment, as compared with patients who received remdesivir with a placebo. Currently, baricitinib has not been approved as a standalone treatment for COVID-19.

On August 19, 2020, the National Health Commission of China issued The COVID-19 Diagnosis and Treatment Program (Trial Eighth Version), which does neither recommend the use of lopinavir/ritonavir or ribavirin as standalone treatment, nor the use of hydroxychloroquine or the combination of hydroxychloroquine with azithromycin. The following drugs can continue to be used and further evaluate their efficacy in clinical practice: 1. α-interferon: Adult dose is 5 million units or an equivalent dose, diluted in

2 mL of sterile water for injection, and administered through nebulization twice daily. The treatment course shall not exceed 10 days. Ribavirin: It is recommended to use ribavirin in combination with interferon (dose same as above) or lopinavir/ritonavir (adult dose is 200 mg/50 mg per tablet, take 2 tablets at a time, twice daily). The adult dose of ribavirin alone is 500 mg per dose, administered through intravenous infusion 2–3 times daily. The treatment course shall not exceed 10 days. Chloroquine phosphate (used for adults aged 18–65 years, with body weight greater than 50 kg) is administered at 500 mg per dose, twice daily, for a treatment course of 7 days; For adults with body weight less than 50 kg, the dosage is 500 mg twice daily for the first 2 days, followed by 500 mg once daily from the third to the seventh day. Arbidol (adult dose is 200 mg) is administered three times daily for a treatment course not exceeding 10 days.

Lopinavir/ritonavir may cause adverse reactions such as diarrhea, nausea, vomiting, and liver function impairment. It is important to be cautious about potential drug interactions with other medications. Simultaneous use of three or more antiviral drugs is not recommended. If intolerable toxic side effects occur, relevant medications shall be discontinued.

Several clinical studies have been conducted during the emergency clinical trials of antiviral drugs. Although no antiviral drugs have been proven effective through rigorous "randomized, double-blind, placebo-controlled studies," some medications have shown potential therapeutic effects based on clinical observation studies. Currently, a more consistent opinion suggests that drugs with potential antiviral effects shall be used in the early stages of the disease. It is recommended to prioritize their use in patients with severe high-risk factors or those showing a tendency toward severe illness.

In the absence of sufficient evidence from evidence-based medicine, antiviral treatment for COVID-19 patients can only draw on the experience of treating other coronaviruses in the past. However, as clinical trial results and treatment experiences regarding COVID-19 antiviral drugs continue to be updated and accumulated, the antiviral treatment protocols for COVID-19 patients are also being further updated. Currently, there is no definitive research confirming that drugs such as ribavirin, arbidol, and oseltamivir can shorten the time for virus detection, nor have they been found to have a positive effect on improving the prognosis of COVID-19 patients.

As of August 6, 2020, apart from drug clinical trials, most international clinical guidelines no longer recommend the use of lopinavir/ritonavir (or other HIV protease inhibitors), chloroquine phosphate, hydroxychloroquine (regardless of whether they are combined with azithromycin or not), and tocilizumab for COVID-19 patients. Currently, the only antiviral drug recommended for clinical treatment of COVID-19 patients in clinical guidelines in Europe, the United States, and other regions is remdesivir. Remdesivir is a nucleotide analog prodrug that can inhibit RNA-dependent RNA polymerase. It has demonstrated promising activity against MERS and SARS in cell culture and animal models and has been used in clinical trials for the treatment of Ebola virus. Cell culture studies have shown that this drug can inhibit SARS-CoV-2 in cell lines. The first COVID-19 patient in the United States who received this medication experienced immediate and remarkable improvement, and subsequent research has found that remdesivir can reduce the mortality rate in severe patients.

The Infectious Diseases Society of America (IDSA) recommends the use of remdesivir for COVID-19 patients who require oxygen therapy but do not need high-flow oxygen, noninvasive ventilation, or ECMO support. The recommended treatment duration is 5 days, and for patients who do not show clinical improvement after 5 days of treatment, the duration can be extended to 10 days. If a patient's condition worsens during remdesivir treatment and requires high-flow oxygen or mechanical ventilation, the treatment course shall be completed. However, for COVID-19 patients who need high-flow oxygen, noninvasive ventilation, mechanical ventilation, or ECMO support, as well as those with mild or moderate cases, there is insufficient data to confirm the benefit of this drug, and experts do not provide a recommendation for its use.

14.4.1 Antiviral treatment of SARS

During the outbreak of SARS, no effective antiviral treatment was found. Clinical data reviews and analyses did not reveal any significant therapeutic effect of ribavirin as an antiviral treatment. The Diagnosis and Treatment Protocol for Severe Acute Respiratory Syndrome (SARS) issued in 2003 suggested the use of protease inhibitors for treatment. In vitro studies have shown that lopinavir-ritonavir may have some antiviral activity, but its efficacy has not been confirmed in clinical trials. Similarly, remdesivir has been found to be effective against SARS in in vitro studies, but its therapeutic efficacy has not been confirmed through clinical practice.

14.4.2 Treatment of MERS

The Diagnosis and Treatment Protocol for Middle East Respiratory Syndrome (MERS) Cases issued in China in 2015 stated that there were no specific and effective antiviral drugs for MERS-CoV currently. In vitro experiments have shown that the combination therapy of Ribavirin and Interferon-α exhibits some antiviral effects, but the clinical research results are still uncertain. Antiviral therapy can be tried in the early stages of the disease, and during its use, attention shall be paid to the side effects of the medication. Observational studies have found that interferon combined with ribavirin may improve the 14-day survival rate of MERS patients, but it does not have an impact on the 28-day survival rate.

14.5 Use of glucocorticoids and immunomodulators

According to the diagnosis and treatment protocol of COVID-19 in China, for patients with progressive deterioration of oxygenation index, rapid radiological progression, and excessive activation of the body's inflammatory response, short-term use of glucocorticoids is recommended (generally 3–5 days, not exceeding 10 days). The suggested dosage is equivalent to 0.5–1.0 mg/(kg·d) methylprednisolone. It shall be noted that high-dose glucocorticoids may delay the clearance of the virus due to their immunosuppressive effects. The Recommendations on the Use of Glucocorticoids in Patients with Novel Coronavirus Pneumonia suggests starting methylprednisolone at a low dosage. ① For patients with a body weight <80 kg, the recommended dosage of methylprednisolone is as follows: Day 1: 20 mg per dose, twice daily. Day 2: If the body temperature is <38°C, 20 mg per dose, twice daily, and continue for 6 days; During the treatment period, if the body temperature is ≥38°C on any day, the dose for that day can be increased to 40 mg per dose, and given twice daily. ② For patients with a body weight ≥80 kg, the recommended dosage of methylprednisolone is as follows: Day 1: 40 mg per dose, twice daily. Day 2: If the body temperature is <38°C, 40 mg per dose, twice daily, and continue for 6 days; during the treatment period, if the body temperature is ≥38°C on any day, the dose for that day can be increased to 60–80 mg per dose, and given twice daily. The indications for using glucocorticoids are as follows: ① Adults (age ≥18 years); ② Confirmed SARS-CoV-2 infection through polymerase chain reaction (PCR) or serum antibody testing; ③ Symptomatic patients (including fever, cough, or other relevant infection symptoms) within 10 days of symptom onset, with confirmed pneumonia and rapid disease progression on imaging; ④ Patients with resting SpO_2 (oxygen saturation) ≤93% or respiratory distress (respiratory rate ≥30 times/minute) or PaO_2/FiO_2 ratio (oxygenation index) ≤300 mmHg without oxygen supplementation.

Precautions for the use of glucocorticoids are similar to those for other conditions, and include the following: ① Patients with diabetes who are receiving oral medications or insulin treatment; ② Known allergies to methylprednisolone, hydrocortisone, dexamethasone, or other excipients; ③ Refractory hypertension; ④ Epilepsy or delirium; ⑤ Glaucoma; ⑥ History of active gastrointestinal bleeding within the past 3 months; ⑦ Difficult-to-correct hypokalemia; ⑧ Known secondary bacterial or fungal infections; ⑨ Immunosuppressed state (e.g., chemotherapy, radiation therapy, or HIV infection within the past 1 month); ⑩ Severe lymphocytopenia (absolute lymphocyte count <300/μL) in peripheral blood.

In the Consensus on Diagnosis, Treatment and Management of Severe COVID-19 jointly issued by Wuhan Tongji Hospital and several other hospitals, it is recommended to administer methylprednisolone to patients with moderate symptoms who undergo high fever that have not brought down and difficulty in breathing. The recommended dosage is 40 mg per day, and gradually reduced after 2–3 days. The treatment course is 5 days. For severe cases, methylprednisolone shall be given at a dosage of 40–80 mg per day, which are divided into two doses, and then the dosage shall be reduced every 2–3 days until it is discontinued. The treatment course is 7–10 days. For critical patients, the recommended dosage of methylprednisolone is 80–160 mg per day, which is divided into two doses, and then the dosage shall be reduced every 2–3 days with the treatment course of 7–10 days.

A large-scale randomized controlled experiment from the UK Recovery Trial consisted of 2014 patients who received a daily dosage of 6 mg of dexamethasone for a treatment duration of 10 days. The control group, on the other hand, comprised 4321 patients who received standard care. The results showed that compared with the control group, the experimental group had a decreased mortality rate and an increased 28-day discharge rate. Scholars conducted a systematic analysis of 23 randomized controlled trials on COVID-19 treatment, and the results showed that glucocorticoids can reduce the mortality rate and the incidence of the need for tracheal intubation and mechanical ventilation in COVID-19 patients.

Although there is inconsistency in the dosage and duration of steroid treatment for COVID-19, current data indicates that steroids are effective for severe patients. Clinicians can individualize steroid therapy based on each patient's conditions.

Other immunotherapies include blood products of patients with COVID-19, such as convalescent plasma, immunoglobulins, and neutralizing monoclonal antibodies, all of which are in the clinical trial stage. Additionally, targeted antiinflammatory agents such as interleukin inhibitors, interferons, and kinase inhibitors are also under investigation.

According to the Chinese COVID-19 treatment guidelines, for patients with extensive bilateral lung lesions or severe cases, and those with elevated IL-6 levels in laboratory tests, Tocilizumab (TCZ) can be considered for use. Specific Usage: The initial dose is 4—8 mg/kg with the recommended dose of 400 mg, dilute it with 0.9% normal saline to 100 mL, and the infusion time is more than 1 h; If the initial treatment is not effective, a second dose (with the same dosage as the initial dose) may be administered after 12 h.

The maximum cumulative number of doses shall not exceed two with the maximum single dosage of not more than 800 mg. Caution shall be taken for allergic reactions, and it is contraindicated in individuals with tuberculosis or other active infections.

Tocilizumab is a monoclonal antibody against the IL-6 receptor, which can block COVID-19 inflammation storm. Some studies suggest that after SARS-CoV-2 infection, pathogenic T cells are rapidly activated, producing granulocyte-macrophage colony-stimulating factor (GM-CSF) and IL-6, etc. GM-CSF will further activate $CD14^+$ $CD16^+$ inflammatory monocytes, generating larger amounts of IL-6 and other inflammatory cytokines, resulting in the formation of an inflammation storm and severe immune damage to the lungs and other organs. Therefore, researchers believe that IL-6 and GM-CSF are two key inflammatory factors involved in triggering the inflammation storm in COVID-19 patients. Tocilizumab is a recombinant humanized monoclonal antibody targeting the IL-6 receptor and is a first-line treatment for autoimmune diseases with elevated IL-6 levels, such as rheumatoid arthritis.

Clinical studies conducted by research teams in China suggest that this drug is likely to prevent patients from progressing to severe and critical conditions by blocking the inflammation storm, thereby reducing the mortality. Some scholars have used Tocilizumab in patients requiring endotracheal intubation, and compared with the control group, although the Tocilizumab treatment group had a higher incidence of secondary infections, it still reduced the mortality rate. However, the Infectious Diseases Society of America (IDSA) in the United States believes that the effectiveness of Tocilizumab treatment is still uncertain and may have some adverse reactions. Therefore, the use of this drug to treat COVID-19 patients is only recommended for clinical trials.

The treatment guidelines in China recommend the emergency use of intravenous immunoglobulin for COVID-19 patients with rapid disease progression who have moderate or severe symptoms. The recommended dosage is 20 mL for patients with moderate symptoms and 40 mL for patients with severe symptoms. The immunoglobulin is administered intravenously, and based on the patient's improvement, it can be re-administered every other day, with a total of no more than five doses.

Convalescent plasma from recovered patients in Chinese Diagnosis and Treatment Protocol: Suitable for patients with rapid disease progression, severe, and critical cases. Refer to the Clinical Treatment Protocol for Convalescent Plasma of COVID-19 (Trial Version 2) for specific administration and dosage.

14.5.1 Application of glucocorticoids and immunomodulator in SARS treatment

The application indications of steroids recommended in China's SARS diagnosis treatment protocol are as follows: ① Severe toxemia, persistent high fever that have not brought down, and the highest body temperature remains above 39°C for more than 3 days even after symptomatic treatment; ② X-ray chest images showing multiple or large infiltrates, rapidly progressing, with an increase in lesion area exceeding 50% within 48 h and occupying more than 1/4 of the total lung area on frontal chest X-ray. ③ Meet the diagnostic criteria for acute lung injury or ARDS (Acute Respiratory Distress Syndrome). The application is appropriate if any of the above indications are present.

The recommended daily dosage for adults is 80–320 mg of methylprednisolone, which is administered intravenously. The specific dosage can be adjusted based on the patient's conditions and individual differences. When clinical manifestations improve or X-ray chest images show absorption of pulmonary shadows, the dosage shall be gradually reduced and discontinued. Generally, the dosage shall be reduced by 1/3 every 3–5 days. Usually, after 1–2 weeks of intravenous administration, it can be switched to oral prednisone or prednisolone. Generally, the treatment duration shall not exceed 4 weeks, and excessive dosage or prolonged treatment shall be avoided. Antacid and gastric mucosal protective agents shall be used concomitantly. Additionally, caution shall be exercised to prevent secondary infections, including bacterial and/or fungal infections, and potential latent tuberculosis infection shall also be considered.

According to the Chinese SARS diagnosis and treatment protocols, the efficacy of nonspecific immunomodulators such as thymosin, interferon, and intravenous immunoglobulin in the treatment of SARS has not been confirmed, and their routine use is not recommended. The clinical efficacy of convalescent serum for SARS has not been confirmed, but some meta-analysis studies suggest its effectiveness in the treatment of SARS. Therefore, for high-risk patients with confirmed diagnosis, convalescent serum can be considered for trial use under close observation.

14.5.2 Application of glucocorticoids and immunomodulator in MERS treatment

The use of corticosteroids and immunomodulators are not mentioned in the Chinese MERS diagnosis and treatment protocols. Some scholars have found that using steroids may not lead to improved clinical outcomes, so caution shall be exercised when using steroid treatment for MERS patients. Based on the treatment experience with SARS and influenza, the use of convalescent plasma and human immunoglobulin can be considered for severe patients.

14.6 Application of mechanical ventilation and ECMO

According to the diagnosis and treatment protocol of COVID-19 in China, critical cases shall be admitted to the Intensive Care Unit (ICU) as early as possible and receive prompt and effective oxygen therapy, which includes nasal cannula, oxygen mask, and high-flow nasal cannula oxygen therapy.

14.6.1 Nasal cannula and mask oxygen therapy

Patients with severe COVID-19 (PaO_2/FiO_2 less than 300 mmHg) shall be given oxygen therapy immediately.

(1) They shall receive oxygen therapy through nasal cannula or face mask, and promptly assess whether respiratory distress and/or hypoxemia are relieved.

(2) After receiving oxygen therapy through nasal cannula or face mask, they shall be closely observed for a short period (1–2 h). If respiratory distress and/or hypoxemia do not relief, high-flow nasal cannula (HFNC) or noninvasive ventilation (NIV) shall be applied.

14.6.2 HFNC or NIV

For patients with a PaO_2/FiO_2 less than 200 mmHg, high-flow nasal cannula (HFNC) or noninvasive ventilation (NIV) shall be administered.

(1) For patients receiving HFNC or NIV, and without contraindications, it is recommended to perform prone position ventilation while the patient is awake. The duration of prone position treatment shall be greater than 12 h per day whenever possible.

Prone position ventilation is an effective respiratory support method that can significantly improve patient oxygenation and even prognosis, which is to enhance the ventilation-perfusion ratio in patients with ARDS, increases end-expiratory lung volume, and promote more even distribution of ventilation by lung recruitment and altering chest wall elasticity. This approach can not only improve oxygenation but also reduce the likelihood of developing ventilator-associated lung injury (VALI). It is recommended to use prone position ventilation for critical ARDS patients with an oxygenation index of <100 and PEEP >10 cmH_2O. The daily duration of prone position ventilation is advised to be >12 h. Combining electrical impedance tomography (EIT) technology to dynamically assess real-time lung ventilation, implementing personalized respiratory support strategies would be more reasonable.

(2) Some patients undergoing HFNC or NIV therapy have a high risk of treatment failure, and it is essential to closely monitor their symptoms and signs. In case of poor treatment response to HFNC or NIV, timely initiation of invasive mechanical ventilation shall be considered. If the FiO_2 remains at or above 70% during HFNC or NIV therapy, regular monitoring of arterial blood gases shall be intensified to further assess the patient's oxygenation status.

Severe and critical COVID-19 patients require a certain level, or even a high level, of positive end-expiratory pressure (PEEP) to keep the pulmonary alveoli open due to diffuse pulmonary exudation and consolidation, which is difficult to achieve with HFNC. Therefore, caution shall be exercised when applying HFNC to patients with $PaO_2/FiO_2 < 150$ mmHg, especially those without preexisting cardiopulmonary diseases. If continuous HFNC treatment for 24 h fails to significantly improve the oxygenation state, it is recommended to apply noninvasive positive pressure ventilation or even invasive positive pressure ventilation with endotracheal intubation.

(3) NIPV: For patients with $PaO2/FiO_2 \geq 150$ mmHg, NIPV can be considered as an alternative treatment when conventional oxygen therapy and HFNC show inadequate efficacy. It can also be used interchangeably with HFNC. Patients with $150 > PaO_2/FiO_2 \geq 100$ mmHg shall be treated with caution, and be prepared for tracheal intubation if necessary. $PaO_2/FiO_2 < 100$ mmHg is not recommended for routine application.

(4) Invasive positive pressure ventilation (IPPV): In general, when PaO_2/FiO_2 is less than 150 mmHg, endotracheal intubation and invasive mechanical ventilation shall be considered. However, considering that the clinical manifestations of hypoxemia in critical COVID-19 patients may not be typical, the decision to perform tracheal intubation and invasive mechanical ventilation shall not be based solely on whether the PaO_2/FiO_2 ratio meets the standard. Instead, real-time evaluation shall be conducted by taking into account the patient's clinical manifestations and organ functions. It is worth noting that the potential harm caused by delaying tracheal intubation may be greater.

Early and appropriate use of invasive mechanical ventilation is a crucial treatment approach for critical patients. For patients with moderate to severe ARDS, or when FiO_2 is greater than 50% during invasive mechanical ventilation, lung recruitment therapy can be considered. Furthermore, it is decided whether to perform lung recruitment techniques according to the reactivity of lung recruitment. Please note that some COVID-19 patients may have poor lung recruitability, so avoid excessively high PEEP levels to prevent barotrauma.

During IPPV, a lung-protective ventilation strategy shall be employed, which includes using small tidal volumes (4–8 mL/kg of ideal body weight) and low inspiratory pressures (plateau pressure <30 cmH_2O) for mechanical ventilation to reduce the risk of ventilator-induced lung injury. If patients experience patient-ventilator asynchrony, timely administration of sedatives and neuromuscular blocking agents shall be considered.

In critical COVID-19 patients, the lung manifestations exhibit heterogeneity and polymorphic changes. To maintain effective oxygenation ($SpO_2 > 93\%$), a lung protective ventilation strategy is recommended, involving PEEP titration based on esophageal pressure monitoring to maintain the end-expiratory transpulmonary pressure ≥ 0 cmH_2O and end-inspiratory transpulmonary pressure ≤ 20 cmH_2O. Routine lung recruitment therapy is not recommended. If lung recruitment is necessary, it is advised to perform dynamic monitoring of transpulmonary pressure or end-tidal carbon dioxide pressure for increased safety. If esophageal pressure monitoring is not available, the ventilator parameters can be set based on the dynamic assessment of plateau pressure (Pplat) ≤ 30 cmH_2O and driving pressure.

(Pplat-PEEP) < 15 cmH_2O, to achieve a balance between oxygenation and the prevention of ventilator-associated lung injury.

(5) Rescue Therapy: For severe ARDS patients, lung recruitment is recommended. When human resources are sufficient, prone ventilation shall be performed for 12 h or more per day. When prone position ventilation is not effective, if conditions permit, consideration shall be given to ECMO (Extracorporeal Membrane Oxygenation) as soon as possible.

ECMO, also known as "Extracorporeal Membrane Oxygenation", is a form of extracorporeal life support (ECLS) technology used to completely replace the function of a patient's lungs and partially replace the function of the heart, thereby providing continuous extracorporeal respiration and circulation to sustain the patient's life and gain time for the diagnosis and treatment of the underlying disease. In the past 2 years, the clinical application level of ECMO in China has made rapid progress. The latest published literature has shown that during the treatment of severe COVID-19 in Wuhan, some critical patients received ECMO as an auxiliary support, and some of these patients successfully recovered.

The ECMO initiation criteria as proposed in the Chinese COVID-19 diagnosis and treatment guidelines are as follows: ECMO shall be considered for evaluation and implementation as early as possible when the following conditions are met, in addition to being under the optimal mechanical ventilation conditions ($FiO_2 \geq 80\%$, tidal volume 6 mL/kg ideal body weight, PEEP ≥ 5 cmH_2O, and no contraindications), and protective ventilation and prone position have been ineffective: ① $PaO_2/FiO_2 < 50$ mmHg for more than 3 h ② $PaO_2/FiO_2 < 80$ mmHg for more than 6 h ③ Arterial blood pH < 7.25 and $PaCO_2 > 60$ mmHg for more than 6 h, with a respiratory rate (RR) > 35 breaths per minute. ④ When RR > 35 breaths per minute, arterial blood pH < 7.2, and plateau pressure (Pplat) > 30 cmH2O. ⑤ Concurrent cardiogenic shock or cardiac arrest.

For critical patients with severe COVID-19 and no contraindications, early initiation of ECMO treatment is recommended. Delaying the timing of ECMO initiation may lead to poor patient outcomes.

The Extracorporeal Life Support Professional Committee of the Chinese Medical Association proposes the following criteria for ECMO initiation in severe pneumonia caused by SARS-CoV-2 infection: ① Within the first

7 days of invasive ventilation. ② If the PaO_2/FiO_2 is ≤50 mmHg for more than 3 h while in the supine position, or if after attempting prone position, the PaO_2/FiO_2 is ≤80 mmHg (FiO_2 > 80%, PEEP >10 cmH_2O) for more than 6 h; or if the FiO2 is 1.0 and the PaO_2/FiO_2 is < 100 mmHg. ③ If the inspiratory plateau pressure is >25 cmH_2O or the driving pressure is >15 cmH_2O, and there is no significant improvement in oxygenation after 24 h of treatment. ④ If the arterial blood pH is < 7.25 and $PaCO_2$ is >60 mmHg for more than 6 h, with a respiratory rate (RR) > 35 breaths per minute. ⑤ If the RR is >35 breaths per minute, the blood pH is < 7.2, and the plateau pressure is >30 cmH_2O.⑥ In the presence of severe barotrauma (e.g., pneumomediastinum, pneumothorax). ⑦ In cases of severe air leak syndrome. ⑧ Concurrent cardiogenic shock or cardiac arrest.

The application of ECMO does not have absolute contraindications, but it requires an experienced team for operation and maintenance. During the treatment, close monitoring of coagulation function is essential, and proactive measures shall be taken to prevent nosocomial infections. Adverse prognostic factors associated with ECMO can be considered as relative contraindications.

(1) Conditions of irreversible diseases may be combined, such as severe brain dysfunction, severe central nervous system damage, advanced malignancies, etc;
(2) Existence of contraindications to anticoagulation therapy, such as COVID-19-induced liver failure with severe coagulopathy, significant hemorrhage, recent occurrence or enlargement of intracranial hemorrhage, etc;
(3) Mechanical ventilation for more than 7 days under high ventilator settings (FiO2 > 0.9, plateau pressure >30 cmH_2O);
(4) Age: No specific age contraindication, but with increasing age, the risk of mortality rises;
(5) Accompanied by severe multiple organ failure;
(6) If circulatory support with VA-ECMO is required, severe aortic valve insufficiency or acute aortic dissection is also considered a contraindication;
(7) Immunosuppression (neutrophil count less than $0.4 \times 10^9/L$);
(8) There are anatomic malformations or lesions of peripheral large vessels that prevent the establishment of ECMO vascular access.

Most of the COVID-19 critical patients in the early stage have normal cardiac functions. VV-ECMO is the preferred mode of respiratory support, typically utilizing femoral vein and internal jugular vein as vascular access, with the former as the drainage route and the latter as the infusion route.

For the management during ECMO operation, ECMO-related infections, and ECMO weaning, please refer to the Clinical Application Guidelines for Respiratory Support Treatment and Extracorporeal Membrane Oxygenation (ECMO) in Severe COVID-19 Patients (Trial Version) issued by the General Office of National Health Commission.

Based on the characteristics of SARS and MERS and the elaboration of the diagnosis and treatment protocols in China, the mechanical ventilation and ECMO treatment strategies for COVID-19 are also applicable to SARS and MERS.

14.7 TCM diagnosis and treatment

The history of medical records and research on infectious diseases in China is long-standing, with references to epidemic diseases found as early as in the Huangdi Neijing (Inner Canon of the Yellow Emperor). As the Su Wen: Ben Bing Lun (Plain Questions: Treatise on the Origins of Diseases) describes, "During the pure and vibrant season of spring, there is a scarcity of life, and everything is subdued. The dew and frost descend, causing plants and trees to wither". When a common ailment turns into a warm epidemic, the throat becomes dry, the limbs feel swollen, and there is pain in all the joints. As the Liu Yuan Zheng Ji Da Lun (Discussion on the Six Elements and Correct Principles) states, "When the Qi is greatly warm, vegetation flourishes early, and the people are afflicted with pestilence. Warm diseases then arise", "The warm epidemic spreads widely, affecting both nearby and distant areas."

As is recorded in the Su Wen: Ci Fa Lun (Plain Questions: Discourse on the Principles of Needling), "When the five epidemics arrive, they all spread easily, regardless of their size, and their disease manifestations are similar." This clearly explains the contagious and seasonal characteristics of infectious diseases. In traditional Chinese medicine, there are no specific terms such as SARS, MERS, or COVID-19. However, based on the sudden onset, severity, rapid spread, and strong contagiousness, these diseases can be classified under the category of "Wen Yi (pestilence)" in febrile disease of traditional Chinese medicine.

Traditional Chinese medicine has accumulated rich experience in the treatment of infectious diseases. In the Han Dynasty, Zhang Zhongjing, in the preface of his book Shang Han Za Bing Lun (Treatise on Febrile and Miscellaneous Diseases) wrote: "In my extended family, there were originally more than 200 members. Since the Jian'an era, not even 10 remain alive, with three out of every two deaths attributed to the typhoid fever, which accounts for seven out of every 10 illnesses." This passage describes the prevalence of infectious diseases during that period. During the Qing Dynasty, Ye Gui (Tianshi) authored the Wen Re Lun (Treatise on Warm Diseases), advocating the use of the differentiation of the defense, qi, nutrient, and blood aspects to treat warm diseases. Wu Tang.

(Ju Tong) wrote the Wen Bing Tiao Bian (Treatise on Differentiation and Treatment of Epidemic Febrile Disease), proposing the theory of "San Jiao Bian Zheng (Syndrome Differentiation of Triple Energizer)" for the treatment of warm diseases.

SARS, MERS, and COVID-19 are all newly emerging infectious diseases, and both traditional Chinese medicine and Western medicine lack diagnostic and treatment experience. However, it has been proven that the unique diagnostic and treatment system of traditional Chinese medicine (TCM) demonstrates significant advantages when facing unknown emerging infectious diseases. TCM has shown clear efficacy in disease prevention and treatment. Therefore, medical practitioners shall inherit the excellent traditions of TCM and fully promote the advantages of traditional Chinese medicine.

14.7.1 TCM diagnosis and treatment of SARS

SARS occurred during the winter and spring seasons, and it was transmitted by prevailing epidemic pathogenic factors. Just as stated in the Wen Yi Lun (Treatise on Pestilence), "The disease of pestilence is neither caused by wind, cold, heat, nor dampness. It arises from a unique and extraordinary pathogenic factor existing between heaven and earth". The pathogenic factors of pestilence enter through the mouth and nose. In those affected, the disease may lie dormant and manifest later in mild cases, while in severe cases, it may emerge within a day. The clinical manifestations mainly correspond to the classical descriptions of "Warm diseases start with fever, and fever ends with warmth. Warm and fever are inseparable and also known as pestilence" and "When the five pestilences arrive, they all easily infect others, regardless of their size. The symptoms of the diseases are similar." These belong to the category of pestilence.

14.7.1.1 Etiology and pathogenesis

During the SARS epidemic, some individuals got infected while others did not even under the same conditions, crucially depending on the body's ability to resist the pathogenic factors. When the body's balance between cold and warmth is disrupted, accompanied by irregular lifestyle, improper diet, and uncontrolled emotions, causing damage to the body's vital energy and weakening the defensive capabilities, the pathogenic factors of epidemic diseases can take advantage of this vulnerability to invade the body, leading to the occurrence of the disease. The pathogenic toxin serves as the root cause of the disease and runs through the entire course of the illness. The excessive accumulation of pathogenic factors brews toxicity, infiltrating and impairing the functions of the organs and viscera, leading to severe dysregulation. The mutual interaction and intensification of internal and external pathogenic factors determine the progression and outcome of the disease. Heat refers to SARS mainly manifesting as fever with the characteristics of warm-heat diseases. During the course of the disease, there is a pathological manifestation of exuberant heat pathogen burning intensely, presenting a yang-heat appearance, and the pathological features of heat rising and dispersing easily leading to exhaustion of Qi and damage to body fluids. Dampness refers to the Yin pathogen of turbid and lingering nature, which tends to entangle and persist, easily injuring Yang Qi. Stasis refers to the stagnation of the epidemic toxin, with the heat pathogen searing and the blood circulation becoming sluggish, resulting in blood stasis and stagnant blood. It mainly affects the collaterals, starting from the lung collaterals and spreading throughout the whole body, causing the disorder of both the meridians and the diseases related to the collaterals. The microcirculation disorders and pulmonary fibrosis manifested during the course of SARS are all indications of stasis. Stasis can also lead to deficiency. Deficiency refers to the deficiency of vital energy (qi) during the incubation period and early stages of the disease. After the onset of the disease, if the vital energy (qi) remains strong, the prognosis is better; otherwise, it is worse. In severe and moderate to critical stage patients, the manifestation of deficiency syndrome becomes increasingly prominent. Initially, the lung, as the main organ for qi, is affected, leading to respiratory disorders. Subsequently, the kidneys fail to receive qi, and both the heart and spleen experience deficiency. From the perspective of SARS fatal cases, the majority of patients were those with chronic illnesses or elderly individuals with weak constitutions. Both heat toxin and blood stasis are considered as pathogenic factors of excess (shi) in nature. They can coexist and afflict the patient simultaneously, or predominate at different stages of the disease process. "When pathogenic factors are strong, there is excess; when vital energy is depleted, there is deficiency." The pathogenic factors of the epidemic disease enter the body through the mouth and nose, initially attacking the lungs, and can subsequently affect the heart, kidneys, stomach, intestines, and other organs. The lungs govern the surface of the body. When invaded by pathogenic factors, it can cause symptoms of both cold and heat, leading to body aches. The lungs also control the movement of qi and are responsible for respiration. The stagnant pathogenic factors obstruct the lung qi, leading to dry cough, difficulty in breathing, shortness of breath, chest tightness, and wheezing. When pathogenic factors accumulate, they will inevitably weaken the vital energy, leading to extreme fatigue due to the depletion of qi and yin. During the course of the disease, the changes between deficiency and excess are particularly rapid and prominent. In clinical practice, it is essential to grasp the pathogenesis and understand the dynamic changes in the struggle between the body's righteous Qi and the pathogenic factors, so as to formulate appropriate strategies accordingly.

According to the Guidelines for TCM Diagnosis and Treatment of Severe Acute Respiratory Syndrome (SARS), the basic pathogenesis of this disease can be summarized in the following four aspects:

(1) Pathogenic Toxin Blocking the Lungs: It enters through the mouth and nose, primarily affecting the lungs. The lungs govern the surface and control the qi. The struggle between the pathogenic and righteous factors occurs on the surface of the lungs, leading to the invasion of the pathogenic toxin causing fever, body aches, and the blockage of the lungs. The dysfunction of lung's disseminating and descending functions results in high fever with unrelieved sweating, dry cough, and dyspnea. The struggle between the pathogenic and righteous factors is intense, and the pathogenic toxin can penetrate deeply. It is evident that both the vital energy (qi) and nutrient essence (ying) are affected. In some patients, the pathogenic toxin may invade the pericardium, leading to symptoms such as restlessness, delirium, and incoherent speech. When the pathogenic toxin blocks the lungs and high fever persists without remission, the condition becomes severe and prone to developing into various complicated patterns.

(2) Lung-Qi Stagnation: In this disease, the pathogenic toxin accumulates in the lungs, leading to impairment of lung dispersing and descending functions. The mechanism of Lung-Qi stagnation plays

a significant role during the course of the disease, resulting in symptoms of rapid breathing, chest tightness, and wheezing. Lung-Stomach Correlation: When the Lung-Qi fails to descend, it can affect the stomach, leading to symptoms such as distention and fullness in the epigastrium and abdomen, poor appetite, nausea, and vomiting. Lung and Large Intestine are mutually connected in the exterior-interior relationship. When there is a pathological condition in the lung, it can affect the large intestine, leading to symptoms such as constipation or diarrhea. The Lung governs the flow of qi through numerous vessels, and the heart and lung coexist harmoniously in the upper wamer. When the lung qi is stagnant and obstructed, it can lead to disharmony in the hundreds of vessels, resulting in symptoms of wheezing, shortness of breath, and cyanosis.

(3) Damp Phlegm and Stasis Obstruction: The pathogenic factor of the epidemic enters the lungs, causing stagnation and obstruction of lung qi, leading to the inability of qi to transform fluids properly. Consequently, the fluids transform into dampness, and the dampness accumulates to form phlegm. Qi is the commander of blood circulation; when qi fails to circulate properly, blood stasis may occur. Therefore, this leads to the formation of a state of dampness, phlegm, and stasis obstruction in the lungs. Dampness, phlegm, and stasis are both pathological products and pathogenic factors in this condition. The closure of lung qi causes stagnation of the flow of fluids, leading to the accumulation of phlegm and stasis in the lungs, which damages the lung meridians. Therefore, the manifestation includes dry cough, difficulty expectorating phlegm, or the presence of blood streaks in the phlegm.

(4) Qi and Yin Deficiency: The pathogenic influence of the epidemic toxin depletes qi and injures yin, and the deficiency of qi and yin in the lungs can manifest in the early stages of the disease after contracting the pathogen. In the early stages of the disease, one may experience symptoms such as fatigue, lethargy, listlessness, dry mouth, and spontaneous sweating. Moreover, the earlier the deficiency of qi and yin appears, the more severe the condition tends to be. As the disease progresses, the further damage to the qi and yin of the lungs can lead to involvement of the heart, qi, blood, and kidneys. This may result in varying degrees of symptoms such as palpitations, breathlessness, and a sense of impending doom. In

severe cases, there may be a sudden bradycardia, decrease in body temperature and blood pressure, cold limbs, and profuse cold sweating. In the later stages, there may be symptoms indicating further deficiency of both qi and yin, such as dry mouth and thirst, restlessness and heat sensation in the chest, and sweating and breathlessness with minimal exertion.

14.7.1.2 Syndrome differentiation and treatment

14.7.1.2.1 Key points of syndrome differentiation

The key points for syndrome differentiation include the course of the disease, the intensity of fever, the degree of respiratory distress, changes in chest X-rays, and the extent of qi and yin deficiency.

14.7.1.2.2 Treatment principle

(1) Early Treatment: Early diagnosis and prompt administration of traditional Chinese medicine.

(2) Effective Expulsion of Pathogenic Factors: This disease is caused by the invasion of pathogenic factors. Wu Yike, a physician in the Ming Dynasty, emphasized the importance of "expelling pathogenic factors" as the primary principle. Therefore, clearing heat, resolving toxins, and eliminating pathogenic factors are essential throughout the treatment process.

(3) Early Strengthening of the Vital Energy (Qi) and Nourishing Yin: As the pathogenic mechanism of Qi and Yin deficiency persists throughout the disease course, if signs of deficiency appear early in the disease, it is essential to promptly support and nourish the vital energy.

(4) Preventing Disease Progression: As soon as the initial signs of the pathogenic mechanism are observed, measures can be taken in advance, and medication can be administered before the disease progresses, aiming to halt its advancement and prevent damage to other organs.

14.7.1.2.3 Differentiation of syndrome and treatment

(1) The syndrome of epidemic toxin invading the lungs is commonly observed in the early stage

Symptoms: At the onset, there may be fever, or chills, headache, body aches, weakness in the limbs, dry cough, and scanty phlegm; or there may be sore throat, fatigue, shortness of breath, and dry mouth. The tongue coating may be white, yellow, or greasy, and the pulse is slippery and rapid.

Pathogenesis: In this pattern, the pathogenic factor of the epidemic disease initially attacks the superficial layers of the lungs, resulting in a struggle between the pathogenic and righteous factors at the lung surface. Therefore, there are symptoms of alternating chills and fever, headache, and body aches. The lung qi is impaired, leading to a dry cough with little phlegm. The pathogenic factor combines with dampness, leading to limb heaviness and a greasy tongue coating. The pathogenic factor also injures both the qi and yin, resulting in fatigue and dry mouth. This pattern is mostly characterized by deficiency and minor excess.

Treatment Principle: Clear the lungs, resolve toxicity, transform dampness, and expel pathogenic factors.

Basic Prescription and Reference Dosages: Honeysuckle Flower (Yinhua) 15 g, fructus forsythiae (Lianqiao) 15 g, Scutellaria baicalensis (Huangqin) 10 g, Bupleurum chinense (Chaihu) 10 g, Artemisia annua (Qinghao) 15 g, White Cardamom (Baikou) 6 g (crushed), Apricot Kernel (Xingren) 9 g (stir-fried), Raw coix seed (raw Yiren) 15 g, Adenophora stricta (Shashen) 15 g and Reed Rhizome (Lugen) 15 g. Dosage may be adjusted accordingly. If there is no sweating, add mint; for severe heat, add raw gypsum and rhizoma anemarrhenae (Zhimu); for excessive greasy tongue coating, add patchouli and herba eupatorii; for diarrhea, add coptis (Huanglian) and baked ginger (Paojiang); for nausea and vomiting, add Pinellia ternata (Zhibanxia) and bamboo shavings (Zhuru).

(2) The Syndrome of Epidemic Toxin Obstructing the Lungs is commonly observed in the early and progressive stages.

Symptoms: High fever, sweating without relief from heat; cough, little sputum, chest tightness, shortness of breath, diarrhea; nausea and vomiting, or distension and fullness in the epigastrium, or constipation, or loose stools; dry mouth with aversion to drinking, shortness of breath, fatigue, and in severe cases, restlessness and agitation. The tongue may be red or crimson, with a yellow and greasy coating, and the pulse may be slippery and rapid.

Pathogenesis: This syndrome is characterized by the invasion of pathogenic epidemic toxin blocking the lungs, leading to intense heat and fever, with sweating that does not relieve the heat. The heat toxin becomes stagnant in the meridians, causing body aches. If the heat enters the Heart channel, it may lead to restlessness and agitation. The tongue is red, and the lung qi is unable to disperse properly, resulting in dry cough, scanty phlegm, chest tightness, and shortness of breath. The pathogenic toxin of the epidemic disease depletes the qi and yin, leading to fatigue and dry mouth. The combination of pathogenic heat and dampness obstructs the flow of qi, causing irregularity in ascending and descending functions. This results in abdominal distension, reduced appetite for drinking, a red tongue with yellow and greasy coating, and a slippery and rapid pulse.

Treatment Principles: Clear heat, resolve toxins, promote lung function, and transform dampness.

Basic Formula and Recommended Dosages: Raw gypsum (Raw Shigao) 45 g (decoct first), Rhizoma anemarrhenae (Zhimu) 10 g, Stir-fried ephedra (Fried Mahuang) 6 g, Honeysuckle flower (Yinhua) 20 g, Stir-fried apricot kernel (Fried Xingren) 10 g, Raw coix seed (Raw Yiren) 15 g, Thunberg Fritillary Bulb (Zhebeimu) 10 g, Radix pseudostellariae (Taizishen) 10 g, and Raw licorice root (Raw Gancao) 10 g.

Modification: For patients with restlessness, red tongue, and heat invading the heart nutriment, add Radix rehmanniae (Shengdi), Red peony root (Chishao), and Cortex moutan (Danpi); For those with shortness of breath, severe fatigue, and pronounced dry mouth, remove Radix pseudostellariae (Taizishen) and add American ginseng (Xiyangshen); For patients experiencing nausea and vomiting, add Processed Pinellia ternata (Zhibanxia); For constipation, add Whole snakegourd fruit (Quangualou) and Raw rhubarb (Raw Dahuang); for distention and fullness in the abdomen, along with loose and unsatisfactory bowel movements, add Roasted betel nut (Roasted Binlang) and Costus root (Muxiang).

(3) The syndrome of Lung Closure and Dyspnea is commonly observed in the progressive stage and severe cases of SARS

Symptoms: Persistent high fever or decreasing fever, respiratory distress, chest tightness, wheezing, dry cough, scanty phlegm with blood, shortness of breath, fatigue, and weakness. The lips may appear purplish or dark, the tongue can be red or dark red, and the tongue coating is yellow and greasy. The pulse is slippery.

Pathogenesis: This syndrome is caused by the invasion of pathogenic yin-closing the lung, with heat gradually retreating while dampness, phlegm, and stasis obstruct the Lung meridians. The lung's disseminating and descending functions are impaired, leading to respiratory distress, wheezing, chest pain, shortness of breath, dry cough, scanty phlegm with blood, purplish or dark lips, red or dark red tongue, yellow and greasy tongue coating, and a slippery pulse. If the pathogenic yin further depletes the qi and yin, extreme shortness of breath and profound fatigue may occur.

Treatment Principles: Clear heat and drain the lungs, resolve phlegm and eliminate turbidity, while strengthen the body's vital energy.

Basic Formula and Recommended Dosage: Semen lepidii (Tinglizi) 15 g, Cortex mori (Sangbaipi) 15 g, Scutellaria baicalensis (Huangqin) 10 g, Radix curcumae (Yujin) 10 g, Whole snakegourd fruit (Quanguolou) 30 g, Silkworm excrement (Cansha) 10 g (packaged separately), Rhizoma dioscoreae septemlobae (Bixie) 12 g, Salvia miltiorrhiza (Danshen) 15 g, Dahurian patrinia herb (Baijiangcao) 30 g, and American ginseng (Xiyangshen) 15 g.

Modification: For severe shortness of breath, fatigue, and severe wheezing, add Common macrocarpium fruit (Shanyurou); for abdominal distension, poor appetite, add Mangnolia officinalis (Houpu) and Fructus hordei germinatus (Maiya); For cyanosis of the lips, add Pseudo-ginseng (Sanqi) and *Leonurus heterophyllus* (Yimucao).

(4) The syndrome of Internal Closure and External Escape are seen in severe SARS

Symptoms: Breathlessness, labored breathing, shortness of breath, reduced speech volume, restlessness, and agitation. In severe cases, there may be mental confusion, cold extremities with sweating, cyanosis of the lips, dark red tongue, yellow greasy coating, and a faint, thin, almost imperceptible pulse.

Pathogenesis: This syndrome involves the accumulation of dampness, phlegm, and stasis obstructing the lungs, leading to extreme respiratory distress, wheezing, and shortness of breath. The lung impairment affects the heart, and the qi impairment affects the blood circulation. Qi depletion and yang deficiency result in the escape of yang qi to the exterior, manifesting as palpitations and restlessness. In severe cases, there can be a sudden bradycardia, decrease in body temperature and blood pressure, coldness in the extremities, profuse cold sweats, and a weak, deep pulse, indicating critical conditions. This syndrome involves both deficiency and excess patterns, and the condition is critical.

Treatment Principle: Tonify qi and astringe yin, reinforce yang and stabilize the exterior, transform turbid phlegm and promote free flow.

Basic Formula and Reference Dosage: Red Ginseng (Hongshen) 10–30 g (decocted and taken separately), Baked aconite (Baked Fuzi) 10 g, Common macrocarpium fruit (Shanyurou) 30 g, Ophiopogon japonicus (Maidong) 15 g, Radix curcumae (Yujin) 10 g and Pseudo-ginseng (Sanqi) 6 g.

Modification: For patients with delirium, add Cheongsimwon (Angong Niuhuang Wan) for administration; for profuse cold sweating, add Calcined keel (calcined Longgu and Oyster); for cold limbs, add cassia twig (Guizhi) and Dried Ginger; for phlegm rales in the throat, add Hou Tsao powder.

(5) The syndrome of Qi and Yin Deficiency and Phlegm Stasis Obstructing the Meridians are common in the convalescence stage

Symptoms: Chest tightness, shortness of breath, fatigue, and weakness; easily experiencing breathlessness with exertion; or coughing; feeling of fever or low-grade fever, spontaneous sweating, restlessness, anxiety, insomnia, poor appetite, dry mouth, and parched throat. The tongue appears red with little moisture, and the tongue coating is yellow or greasy. The pulse is often deep, thin, and weak.

Treatment Principles: Tonify Qi and nourish Yin, resolve phlegm, and open the meridians.

Basic Formula and Reference Dosage: Codonopsis pilosula (Dangshen) 15 g, Adenophora stricta (Shashen) 15 g, Ophiopogon japonicus (Maidong) 15 g, Radix rehmanniae (Shengdi) 15 g, Red peony root (Chishao) 12 g, Aster (Ziwan) 15 g, Thunbery fritillary bul (Zhebei) 10 g and fructus hordei germinatus (Maiya) 15 g.

Modification: For severe shortness of breath and wheezing, and a dark tongue, add Pseudo-ginseng (Sanqi), Schisandra chinensis (Wuweizi) and Common macrocarpium fruit (Shanyurou); For self-perceived fever or restlessness in the chest with a dark tongue, add Artemisia apiacea (Qinghao), Fructus gardeniae (Shanzhi) and Cortex moutan (Danpi); For loose stools, add Poria cocos (Fuling) and Rhizoma atractylodis macrocephalae (Baizhu); For anxiety and restlessness, add Vinegar-treated bupleurum root (vinegar-treated Chaihu) and Cyperus rotundus (Xiangfu); For insomnia, add Stir-fried date seed (stir-fried Zaoren) and polygala tenuifolia (Yuanzhi); For liver function impairment with elevated liver enzymes, add Capillary artemisia (Yinchen) and Schisandra chinensis (Wuweizi); For bone injury, add Tortoise plastron (Guiban) and Carapax Trionycis (turtle shell).

14.7.1.2.4 Application of Chinese patent medicine

Chinese patent medicines shall be used based on syndrome differentiation and can be used in combination with traditional Chinese herbal decoctions.

(1) Antipyretic medicines are suitable for early and progressive stages of fever, including Guashuang Tiaoreling Capsules, Zixue Powder, Xinxue Granules, Small bupleurum tablets (or granules), Chaiyin Oral Liquid, etc.

(2) Heat clearing and detoxifying medicines are suitable for the syndrome of epidemic toxin invading the lungs, the syndrome of epidemic toxin obstructing the lungs and the syndrome of lung closure and dyspnea at the early and progressive stages. Injectable options include Qingkailing Injection, Houttuynia Injection, Shuanghuanglian Injection and Compound Kushen Injection, etc. Oral options include Qingkailing Oral Solution (or Capsules), Qingre Jiedu Oral Solution (or Granules), Shuanghuanglian Oral Solution, Jinlian Qingre Granules, Kugan Granules, Ge Gen Qinlian Pellets, Plum Blossom Sore-Dispelling Pill, and Zijin Ingot, etc.

(3) Blood-activating and stasis-resolving, dampness-dispelling and phlegm-transforming medicines are suitable for the syndrome of lung closure and dyspnea in the progressive and severe stages of SARS. Injectable options include Danshen Injection, Xiangdan Injection, Chuanxiongqin Injection, and Dengzhanxixin Injection. Oral options include Xuefu Zhuyu Oral Liquid (or Granules), Compound Danshen Dripping Pills, Huoxiang Zhengqi Oral Liquid (or Capsules) and Hou Tsao powder, etc.

(4) Vital energy strengthening medicines are suitable for individuals with deficiency of righteous qi in all stages of the disease. Injectable options include Shengmai Injection, Shenmai Injection, Canmai Injection, and Huangqi Injection, etc. Oral options include Shengmai Decoction, Bailing Capsules, Jinshuibao Capsules, Ningxinbao Capsules, Nuodikang Capsules, Liuwei Dihuang Pills, and Buzhong Yiqi Pills, etc.

For suspected cases, early use of Traditional Chinese Medicine for syndrome differentiation and treatment shall be considered. At the onset of the disease, the main symptoms include fever, aversion to cold, headache, body aches, and limb weakness; dry cough, scanty phlegm, sore throat; shortness of breath, fatigue, and dry mouth. The tongue coating is white or yellow, and the pulse is slippery and rapid. A prescription for clearing the lungs, resolving toxins, transforming dampness, and expelling pathogenic factors can be formulated. Please refer to the following prescription and modifications:

Honeysuckle Flower (Yinhua) 15 g, Fructus forsythiae (Lianqiao) 15 g, Folium isatidis (Daqingye) 15 g, Gypsum (Raw Shigao) 30 g, Rhizoma anemarrhenae (Zhimu) 10 g, Bupleurum chinense (Chaihu) 10 g, Scutellaria baicalensis (Huangqin) 10 g and Apricot Kernel (Xingren) 9 g.

(Fried), Reed Rhizome (Lugen) 15 g and Raw licorice root (Raw Gancao) 9 g.

Modification: For those without sweating, add mint; for those with diarrhea, remove raw gypsum and rhizoma anemarrhenae (Zhimu) and add patchouli, herba eupatorii and huanglian; for those with nausea and vomiting, add Pinellia ternata (Zhibanxia) and bamboo shavings (Zhuru); for those severe coughing, add radix peucedani (Qianhu) and Stir-fried folium eriobotryae; for those with severe wheezing and dyspnea, add Semen lepidii (Tinglizi) and Stir-fried ephedra (Fried Mahuang).

14.7.1.2.5 Medicine prevention

In addition to psychological, behavioral, and lifestyle prevention, medication can also be used for prevention, and taking Chinese herbal medicine proactively has shown good preventive effects. Furthermore, actively treating chronic respiratory and other preexisting diseases helps to enhance the body's resistance against this condition.

Basic Formula and Recommended Dosage:

Prescription One: Radix pseudostellariae (Taizishen) 15g, Herba patriniae (Baijiangcao) 15g, Raw coix seed (raw Yiren) 15g, and Platycodon grandiflorus (Jugeng) 6g. Function: Tonify Qi and resolve Dampness, clear Heat and detoxify. Applicable to individuals with underlying Qi deficiency and concurrent Damp-Heat.

Prescription Two: Houttuynia cordata (Yuxingcao) 15g, Wild Chrysanthemum (Yejuhua) 6g, Trollius chinensis (Jinlianhua) 12g, Capillary artemisia (Yinchen) 15g and Amomum tsao-ko (Caoguo) 3g. Function: Clear heat, detoxify, promote diuresis, and transform turbidity. Suitable for individuals with preexisting internal heat or excessive internal dampness.

Prescription Three: Raw astragalus membranaceus (Raw Huagnqi) 10g, Radix glehniae (Beishashen) 10g, Honeysuckle flower (Yinhua) 10g, Fructus forsythiae (Lianqiao) 10g, atractylodis macrocephalae (Baizhu) 6g, Saposhnikovia divaricata (Fangfeng) 6g, Agastache rugosa (Huoxiang) 10g, and Perilla leaf (Suye) 6g.

Function: Strengthen the spleen, nourish yin, transform dampness, and resolve toxicity. Applicable to cases of both qi and yin deficiency, and individuals with a tendency to catch colds and have dampness in their constitution.

14.7.2 TCM diagnosis and treatment of MERS

MERS has not erupted on a large scale in China, and sporadic imported cases have been reported in relevant literature, and domestic experience in diagnosis and treatment is limited. Based on existing literature, the initiating and fundamental factors for MERS patients are attributed to warm and hot pathogenic factors. Heat pathogen is the key factor in this disease, while toxic and stasis manifestations are not prominent. The transformation of the TCM syndrome in MERS follows the regular pattern of warm diseases affecting the defense qi, nutrient qi, and blood; wherein the defense qi and nutrient qi share the same disease, while the symptoms of blood stasis are not emphasized. MERS shows regular characteristics in its development trend, with the disease location progressing from superficial to deep, the condition shifting from mild to severe, and the disease nature changing from excess to deficiency. Its disease progression follows regular patterns, and it can be treated by referring to the traditional Chinese medicine theories of Defense Qi and Nutrient Blood (Wei Qi Ying Xue) and the syndrome differentiation of triple energizer (San Jiao Bian Zheng) system. The focus of the treatment lies in clearing heat and detoxifying, promoting lung function and relieving cough, while also supporting the body's vital energy.

According to the Diagnosis and Treatment Protocol for Middle East Respiratory Syndrome (MERS) Cases (2015 Edition), combined with the experience of treating diseases such as "Febrile diseases" and "Wind-heat lung heat" in TCM, the treatment can be classified into the following types under the guidance of a traditional Chinese medicine practitioner:

(1) Invasion of the Defensive Qi by Pathogenic Factors in the Lungs

Main Symptoms: Fever, sore throat, headache, body aches, dry cough with little phlegm, fatigue, and poor appetite. Treatment Principles: Detoxify, disperse lung heat, strengthen the body's vital energy, and expel pathogenic factors.

Recommended Formula: Yinqiao Powder combined with Ginseng and Perilla Beverag (Shensuyin).

Commonly Used Medicinal Ingredients: Honeysuckle, Forsythia, Schizonepeta, Peppermint, Perilla Leaf, Common Hogfennel Root, Great Burdock Achene, Platycodon Grandiflorus, American Ginseng, and Licorice, etc. Recommended Patent Chinese Medicines: Lianhua Qingwen Capsules, Qingfei Xiaoyan Pills, Shufeng Jiedu Capsules, Shuanghuanglian Oral Liquid, etc.

(2) Pathogenic Toxins Obstructing the Lungs

Main Symptoms: High fever, sore throat, dry cough with little phlegm, chest tightness, shortness of breath, fatigue, and in severe cases, wheezing, abdominal distension, and constipation, etc. Treatment Principle: Clear heat, drain the lungs, resolve toxicity, and relieve wheezing.

Recommended Formula: Maxing-Shigan Decoction, Xuanbai-Chengqi Decoction combined with Ginseng Baihu Decoction.

Commonly Used Medicinal Ingredients: Ephedra, Apricot Kernel, Raw Gypsum, Anemarrhena Rhizome, Fritillaria Thunbergii, Cortex Mori, American Ginseng, etc. Modification: For those with accumulation in the intestines causing constipation, add Taoren Chengqi Decoction.

Chinese Herbal Injections based on the Conditions: Reduning Injection, Tanreqing Injection, Xuebijing Injection, Qingkailing Injection, etc.

(3) Deficiency of Upright Qi and Invasion of Pathogenic Factors

Main Symptoms: High fever, rapid breathing, profuse sweating, cold extremities, and may be accompanied by delirium, oliguria, or urinary retention. Treatment Principles: Reinforce the primordial Qi, stabilize the collapse, eliminate toxins, and open the orifices.

Recommended Formula: Shengmai Powders combined with Shenfu Decoction, and add Angong Niuhuang Pills.

Commonly Used Medicinal Ingredients: Red ginseng, Ophiopogon Japonicus, Schisandra Chinensis, Baked Aconite, and Common Macrocarpium Fruit, etc.

Chinese Herbal Injections based on the Conditions: Shengmai Injection, Shenfu Injection, and Shenmai Injection.

(4) Deficiency of Zheng Qi and Intrusion of Pathogenic Factors

Main Symptoms: Fatigue, lethargy, loss of appetite, low-grade fever in the afternoon, dry mouth, and occasional cough. Treatment Principle: Tonify qi and invigorate the spleen, nourish yin, and dispel pathogenic factors.

Recommended Formula: Shashen Maidong Decoction combined with Zhuye Shigao Decoction.

Commonly Used Medicinal Ingredients: Radix Adenophorae, Ophiopogon Japonicus, Atractylodes, Poria Cocos, Lophatherum Gracile, Raw Gypsum, Chinese Yam and Tangerine Peel, etc.

14.7.3 TCM diagnosis and treatment of COVID-19

The outbreak of COVID-19 is a significant test for global health and epidemic prevention. China's antiepidemic measures and successful experiences have received full recognition from the World Health Organization. The effectiveness of traditional Chinese medicine in treating COVID-19 is widely recognized, with an overall effective rate exceeding 90%. Traditional Chinese medicine can effectively reduce the mortality rate, alleviate symptoms, and prevent mild and moderate cases from progressing to severe and critical conditions. It also aids in promoting the recovery of patients in the convalescent period.

This disease belongs to the category of "epidemic" in traditional Chinese medicine. The etiology is attributed to the invasion of a pathogenic "epidemic" qi. According to The COVID-19 Diagnosis and Treatment Program (Trial Seventh Edition) issued by the National Health Commission and the National Administration of Traditional Chinese Medicine, the recommended treatment principles are as follows.

14.7.3.1 Medical observation period

(1) Clinical Manifestations: Fatigue accompanied by gastrointestinal discomfort.

Recommended Chinese Patent Medicine: Huo Xiang Zheng Qi Capsules (Pills, Oral Liquid).

(2) Clinical Manifestation: Fatigue accompanied by fever.

Recommended Chinese Patent Medicines: Jinhua Qinggan Granules, Lianhua Qingwen Capsules (Granules), Shufeng Jiedu Capsules (Granules).

14.7.3.2 Clinical treatment period (confirmed cases)

14.7.3.2.1 Lung detoxification decoction

Applicable Scope: Mild, moderate, and severe patients. In the treatment of critical patients, it can be combined with the patient's actual conditions for rational use.

Basic Prescription: Ephedra (Mahuang) 9g, Fried licorice (fried Gancao) 6g, Aprieot seed (Xingren) 9g, Raw Gypsum (Raw Shigao) 15–30g (predecocted), Cassia twig (Guizhi) 9g, Rhizoma Alismatis (Zexie) 9g, Polyporus umbellatus (Zhuling) 9g, Atractylodes (Baizhu) 9g, Poria Cocos (Fuling) 15g, Radix bupleuri (Chaihu) 16g, Scutellaria baicalensis (Huangqin) 6g, Ginger processed pinellia (Jiangbanxia) 9g, Fresh ginger 9g, Aster (Ziwan) 9g, Flos farfarae (Donghua) 9g and Blackberry lily (Shegan) 9g, Asarum (Xixin) 6g, Chinese yam (Shanyao) 12g, Fructus aurantii immaturus (Zhishi) 6g, Tangerine peel (Chenpi) 6g and Agastache rugosus (Huoxiang) 9g.

Dosage and Administration: Traditional Chinese herbal decoction, prepared by boiling in water. Take two doses per day, once in the morning and once in the evening (40 min after meals). Take it warm. Six doses constitute one treatment course. If conditions permit, after taking the medicine, drink half a bowl of rice soup. For those with dry tongue and depleted body fluids, drink up to one bowl. (If the patient is not feverish, the dosage of gypsum shall be reduced. If there is fever or strong fever, the dosage of gypsum can be increased.) If the symptoms have improved but not completely recovered, then the second course of treatment shall be taken. If the patient has special circumstances or underlying conditions, the prescription for the second course can be modified accordingly. Discontinue the medication when the symptoms disappear.

Prescription Source: Notice on Recommending the Use of "Lung Detoxification Decoction" in the Treatment of Novel Coronavirus Pneumonia with Integrated Chinese and Western Medicine (GZYYBYZH No.[2020] 22).

14.7.3.2.2 Mild

(1) Syndrome of Cold-Dampness Stagnating in the Lungs

Clinical Manifestations: Fever, fatigue, generalized body aches, cough, phlegm production, chest tightness, shortness of breath, poor appetite, nausea, vomiting, sticky and unpleasant stool. The tongue appears pale and slightly swollen with tooth marks or a pale red hue. The coating is thick, greasy, and white or white and greasy. The pulse is slippery or smooth.

Recommended Prescription: Raw ephedra (Raw Mahuang) 6g, Raw gypsum (Raw Shigao) 15g, Aprieot seed (Xingren) 9g, Notopterygium root (Qianghuo) 15g, Semen lepidii (Tinglizi) 15g, Cyrtomium fortunei (Guanzhong) 9g, Lumbricus (Dilong) 15g, Paniculate swallowwort root 15g, Agastache rugosa (Huoxiang) 15g, Herba eupatorii (Peilan) 9g, Rhizoma atractylodis (Cangzhu) 15g, Poria (Yuanling) 45g, Raw Atractylodes (Raw Baizhu) 30g, Processed malt, hawthorn and medicated leaven (Jiaosanxian), 9g each, Mangnolia officinalis (Houpu) 15g, Processed betel nut (Processed Binlang) 9g, Stewed amomum tsao-ko (Stewed Caoguo) 9g, and Fresh ginger 15g.

Dosage and Administration: Take 1 dose daily, decoct with 600 mL of water, and divide into 3 doses. Take one dose in the morning, one at noon, and one in the evening, before meals.

(2) Syndrome of Lung with Dampness Heat Retention

Clinical Manifestations: Low-grade fever or no fever, slight aversion to cold, fatigue, heaviness in the head and body, muscle soreness, dry cough with little sputum, sore throat, dry mouth with little desire to drink, or accompanied by chest tightness and epigastric fullness, absence of sweating or insufficient sweating, or nausea and poor appetite, loose stools or sticky and unsmooth bowel movements. The tongue is pale red, the coating is white and thick or thin yellow, and the pulse is slippery and rapid or slippery.

Recommended Prescription: Betel nut (Binlang) 10g, Amomum tsao-ko (Caoguo) 10g, Mangnolia officinalis (Houpu) 10g, Rhizoma anemarrhenae (Zhimu) 10g, Scutellaria baicalensis (Huangqin) 10g, Radix bupleuri (Chaihu) 10g, Red peony root (Chishao) 10g, Fructus forsythiae (Lianqiao).

15 g, Moss (Qingtai) 10 g (added later), Rhizoma atractylodis (Cangzhu) 10g, Folium isatidis (Daqingye) 10g, and Raw licorice (Raw Gancao) 5g.

Dosage and Administration: Take 1 dose daily, decoct with 600 mL of water and divide into 2 doses, one in the morning and one in the evening.

14.7.3.2.3 Moderate

(1) Syndrome of Dampness-Toxin Stagnating in the Lungs

Clinical Manifestations: Fever, dry or productive cough with scanty or yellow sputum, chest tightness, shortness of breath, abdominal distension, and constipation. The tongue appears dark red, with a swollen body, a yellow and greasy or yellow and dry coating. The pulse is slippery and rapid or wiry and slippery.

Recommended Prescription: Raw Ephedra (Raw Mahuang) 6g, Bitter apricot (Kuxingren) 15g, Raw gypsum (Raw Shigao) 30g, Raw coix seed (Raw Yiren) 30g, Atractylis lancea (Maocangzhu) 10g, Pogostemon cablin (Guanghuoxiang) 15g, Artemisia annua (Qinghaocao) 12g, Polygonum cuspidatum (Huzhang) 20g, Vervain (Mabiancao) 30g, Dried rhizoma phragmitis (Ganlugen) 30g, Semen lepidii (Tinglizi) 15g, Exocarpium (Huajuhong) 15g and Raw liquorice (Raw Gancao) 10g.

Dosage and Administration: Take 1 dose daily, decoct with 600 mL of wate and divide into 2 doses, one in the morning and one in the evening.

(2) Syndrome of Cold Dampness Obstructing the Lungs

Clinical Manifestations: Low-grade fever, mild body heat or absence of fever, dry cough with little phlegm, fatigue, chest stuffiness, epigastric discomfort, or nausea, and loose stools. The tongue appears pale or pale red, with a white or white and greasy coating. The pulse is slippery.

Recommended Prescription: Rhizoma atractylodis (Cangzhu) 15g, Tangerine peel (Chenpi) 10g, Mangnolia officinalis (Houpu) 10g, Agastache rugosus (Huoxiang) 10g, Amomum tsao-ko (Caoguo) 6g, Raw Ephedra (Raw Mahuang) 6g, Notopterygium root (Qianghuo) 10g, fresh ginger 9g and Betel nut (Binlang) 10g.

Dosage and Administration: Take 1 dose daily, decoct with 600 mL of wate and divide into 2 doses, one in the morning and one in the evening.

14.7.3.2.4 Severe

(1) Syndrome of Epidemic Toxin Obstructing the Lungs

Clinical Manifestations: Fever with flushed face, cough, yellowish and sticky sputum in small amounts, or sputum with blood, wheezing and shortness of breath, fatigue and lethargy, dry and bitter taste in the mouth, nausea and poor appetite, constipation, and scanty, red urine. The tongue is red, the coating is yellow and greasy, and the pulse is slippery and rapid.

Recommended Prescription: Dampness-Dissipating and Toxin-Dispelling Formula.

Basic Formula: Raw Ephedra (Raw Mahuang) 6g, Apricot seed (Xingren) 9g, Raw gypsum (Raw Shigao) 15g, Licorice (Gancao) 3g, Agastache rugosus (Huoxiang) 10g (added later), Mangnolia officinalis (Houpu) 10g, Rhizoma atractylodis (Cangzhu) 15g, Amomum tsao-ko (Caoguo) 6g, Rhizoma pinellinae praeparata (Fabanxia) 9g, Poria cocos (Fuling) 15g, Raw rhubarb (Raw Dahuang) 5g (added later), Raw astragalus membranaceus (Raw Huagnqi) 10g, Semen lepidii (Tinglizi) 10g and Red peony root (Chishao) 10g. Dosage and Administration: Take 1–2 doses daily, decoct with water, each time 100–200 mL, 2–4 times a day, orally or through nasal feeding.

(2) Syndrome of Qi and Nutritive (Ying) Deficiency with Concurrent Heat

Clinical Manifestations: High fever with restlessness and thirst, wheezing and shortness of breath, delirium and confusion, visual hallucinations or seeing things incorrectly, may present with rashes, vomiting of blood or nosebleeds, or limb convulsions. The tongue is deep red with little or no coating, the pulse is deep, thin, and rapid, or floating, large, and rapid.

Recommended Prescription: Raw Gypsum (Raw Shigao) 30–60g (predecocted), Rhizoma anemarrhenae (Zhimu) 30g, Radix rehmanniae (Shengdi) 30–60g, Buffalo horn (predecocted), Red peony root (Chishao) 30g, Radix scrophulariae (Xuanshen) 30g, Fructus forsythiae (Lianqiao) 15g, Cortex moutan (Danpi) 15g, Coptis chinensis (Huanglian) 6g, Lophatherum gracile (Zhuye) 12g, Semen lepidii (Tinglizi) 15g and Raw liquorice (Raw Gancao) 6g.

Dosage and Administration: Take 1 dose daily, decoct with water, predecoct gypsum and buffalo horn, then add all the other herbs, each time 100–200 mL, 2–4 times a day, orally or through nasal feeding.

Recommended Chinese Patent Medicines: Xiyangping Injection, Xuebijing Injection, Reduning Injection, Tanreqing Injection, and Xingnaojing Injection. Similar drugs with similar effects can be chosen based on individual circumstances, or two drugs can be used in combination based on clinical symptoms. Chinese herbal injections can be used in combination with Chinese herbal decoctions.

14.7.3.2.5 Severe syndrome of internal closure and external prolapse

Clinical Manifestations: Difficulty breathing, easily experiencing wheezing or requiring mechanical ventilation, accompanied by delirium, restlessness, sweating with cold limbs, dark purple tongue, thick and greasy or dry coating, and a floating, large pulse without strength.

Recommended Prescription: Ginseng 15g, Black aconite (Heishunpian) 10g (predecocted), Dogberry (Shanzhuyu) 15g. Administer with Suhexiang Pills or Angong Niuhuang Pills.

For patients requiring mechanical ventilation with abdominal distension and constipation or difficulty in bowel movements, 5–10g of raw Rhubarb (raw Dahuang) can be used. In cases of asynchrony between the person and the mechanical ventilator, while using sedatives and neuromuscular blockers, 5–10g of raw Rhubarb (raw Dahuang) and 5–10g of mirabilite (Mangxiao) can be added.

Recommendation of Chinese Patent Medicines: Xuebijing Injection, Reduning Injection, Tanreqing Injection, Xingnaojing Injection, Shenfu Injection, Shengmai Injection, Shenmai Injection. Similar drugs with similar effects can be chosen based on individual circumstances, and two drugs can also be used in combination based on clinical symptoms. Chinese herbal injections can be used in combination with Chinese herbal decoctions.

Note: Recommended Dosage of Chinese Herbal Injections for Severe and Critical Patients.

The use of Chinese herbal injections follows the principle of starting with a small dosage and gradually adjusting based on individualized diagnosis and treatment, as stated in the drug instructions. The recommended usage is as follows:

Viral Infection or Mild Bacterial Infection: 250 mL 0.9% Sodium Chloride Injection mixed with 100 mg Xiyanping Injection, twice daily; or 250 mL 0.9% Sodium Chloride Injection mixed with 20 mL Reduning Injection; or 250 mL 0.9% Sodium Chloride Injection mixed with 40 mL Tanreqing Injection, twice daily.

High Fever with Consciousness Impairment: 250 mL 0.9% Sodium Chloride Injection mixed with 20 mL Xingnaojing Injection, twice daily.

Systemic Inflammatory Response Syndrome and/or Multiple Organ Dysfunction Syndrome: 250 mL 0.9% Sodium Chloride Injection mixed with 100 mL Xuebijing Injection, twice daily.

Immunosuppression: 250 mL Glucose Injection mixed with 100 mL Shenmai Injection or 20—60 mL Shengmai Injection, twice daily.

14.7.3.2.6 Convalescent stage

(1) Syndrome of Lung and Spleen Qi Deficiency

Clinical Manifestations: Shortness of breath, fatigue and weakness, poor appetite with nausea, feeling of fullness and distension in the abdomen, weak bowel movements, and loose and unsatisfactory stools. The tongue appears pale and swollen, with a white and greasy coating.

Recommended Prescription: Rhizoma pinellinae praeparata (Fabanxia) 9g, Tangerine peel (Chenpi) 9g, Codonopsis root (Dangshen) 15g, Fried astragalus membranaceus (Fried Huangzhi) 30g, Stir-fried atractylodes (fried Baizhu) 10g, Poria cocos (Fuling) 15g, Agastache rugosus (Huoxiang) 10g, Fructus amomi (Sharen) 6g (added later) and Licorice (Gancao) 6g.

Dosage and Administration: Take 1 dose daily, decoct with 600 mL of wate and divide into 2 doses, one in the morning and one in the evening.

(2) Syndrome of Qi and Yin Deficiency

Clinical Manifestations: Fatigue, shortness of breath, dry mouth, thirst, palpitations, excessive sweating, poor appetite, low-grade or no fever, dry cough with scanty phlegm. The tongue is dry and lacks moisture, and the pulse is thin or weak.

Prescription Recommendation: Radix adenophorae and Radix glehniae (Nanshashen and Beishashen) 10g each, Ophiopogon japonicus (Maidong) 15g, American Ginseng (Xiyangshen) 6g, Schisandra chinensis (Wuweizi) 6g, Raw gypsum (Raw Shigao) 15g, Lophatherum gracile (Danzhuye) 10g, Folium mori (Sangye) 10g, Reed Rhizome (Lugen) 15g, Salvia miltiorrhiza (Danshen) 15g and Raw Licorice (Raw Gancao) 6g.

Dosage and Administration: Take 1 dose daily, decoct with 600 mL of wate and divide into 2 doses, one in the morning and one in the evening.

Bibliography

[1] National Health Committee, State Administration of Traditional Chinese Medicine. The COVID-19 diagnosis and treatment program (Trial Eighth Edition). Infect Dis Information 2020;33(4):289—96.

[2] Yuan X, Jinsong M, Guoxin Mo, et al. Respiratory support for severe 2019-nCoV pneumonia suffering from acute respiratory failure: time and strategy. Chin J Tuberc Respir Dis 2020;43(3):177—80.

[3] The Extracorporeal Life Support Professional Committee of the Chinese Medical Association. Recommendations on extracorporeal life support for critically ill patients with Novel coronavirus pneumonia. Chin J Tuberc Respir Dis 2020;43(3):195−8.
[4] Zhong N. Diagnosis and treatment protocol of severe acute respiratory syndrome (SARS). Chinese Med J 2003;83(19):1731−52.
[5] National Health Committee. The diagnosis and treatment protocol for Middle East respiratory syndrome (MERS) cases (2015 Edition). Chin. J Viral Dis 2015;5(5):352−4.
[6] National Health Committee, State Administration of Traditional Chinese Medicine. The COVID-19 diagnosis and treatment program (Trial Eighth Edition). Infect Dis Information 2020;33(4):289−96.
[7] Holshue ML, Debolt C, Lindquist S, et al. First case of 2019 Novel coronavirus in the United States. N Engl J Med 2020;382:929−36.
[8] Beigel JH, Tomashek KM, Dodd LE. Remdesivir for the treatment of Covid-19 preliminary report. N Engl J Med 2020;383:992−4.
[9] Wang Y, Zhang D, Du G, et al. Remdesivir in adults with severe COVID-19: a randomised, double-blind, placebo- controlled, multicentre trial. Lancet 2020;395(10236):1569−78.
[10] Goldman JD, Lye DCB, Hui DS, et al. Remdesivir for 5 or 10 days in patients with severe Covid-19. N Engl J Med 2020;383:1827−37.
[11] Zhong N. Diagnosis and treatment protocol of severe acute respiratory syndrome (SARS). Chinese Med J 2003;83(19):1731−52.
[12] Groneberg DA, Poutanen SM, Low DE, et al. Treatment and vaccines for severe acute respiratory syndrome. Lancet Infect Dis 2005;5(3):147−55.
[13] Timothy P, Sheahan AMYC, et al. Broad-spectrum antiviral GS-5734 inhibits both epidemic and zoonotic coronaviruses. Sci Transl Med 2017;396(9):3653.
[14] National Health Committee. The diagnosis and treatment protocol for Middle East respiratory syndrome (MERS) cases (2015 Edition). Chin J Viral Dis 2015;5(5):352−4.
[15] Al-Tawfiq JA, Momattin H, Dib J, et al. Ribavirin and interferon therapy in patients infected with the Middle East respiratory syndrome coronavirus: an observational study. Int J Infect Dis 2014;20:42−6.
[16] Jianping Z, Yi H, Ronghui Du, et al. Expert Consensus on the use of corticosteroid in patients with 2019-nCoV pneumonia. Chin J Viral Dis 2020;43(3):183−4.
[17] Novel Coronavirus Pneumonia Treatment Cooperation Group of Wuhan Tongji Hospital. Consensus on diagnosis, treatment and management of severe COVID-19. J Inter Intensive Med 2020;26(1):1−5.
[18] The RECOVERY Collaborative Group. Dexamethasone in hospitalized patients with Covid-19 - preliminary report. N Engl J Med 2020. https://doi.org/10.1056/NEJMoa2021436.
[19] Siemieniuk RA, Bartoszko J, Ge L, et al. Drug treatments for Covid-19: living systematic review and network meta-analysis. BMJ 2020;370:2980.
[20] Xu X, Han M, Li T, et al. Effective treatment of severe COVID-19 patients with tocilizumab. Proc Natl Acad Sci 2020;117(20). 202005615.
[21] Somers EC, Eschenauer GA, Troost JP, et al. Tocilizumab for treatment of mechanically ventilated patients with COVID-19. Clin Infect Dis 2020. https://doi.org/10.1093/cid/ciqq954.
[22] Bhimraj A, Morgan RL, Shumaker AH, et al. Infectious diseases society of America guidelines on the treatment and management of patients with COVID-19. Clin Infect Dis 2020. https://doi.org/10.1097/MOG.0000000000000489.
[23] Mair-Jenkins J, Saavedra-Campos M, Baillie JK, et al. The effectiveness of convalescent plasma and hyperimmune immunoglobulin for the treatment of severe acute respiratory infections of viral etiology: a systematic review and exploratory meta-analysis. J Infect Dis 2015;211(1):80−90.

[24] Arabi YM, Arifi AA, Balkhy HH, et al. Clinical course and outcomes of critically ill patients with Middle East respiratory syndrome coronavirus infection. Ann Intern Med 2014;160(6):389–97.
[25] China Association of Chinese Medicine. Guidelines for TCM diagnosis and treatment of SARS. J Tradit Chin Med 2003;44(11):865–71.
[26] National Health Committee. The diagnosis and treatment protocol for Middle East respiratory syndrome (MERS) cases (2015 Edition). Chin J Viral Dis 2015;5(5):352–4.
[27] National Health Committee. The COVID-19 diagnosis and treatment Program (trial seventh edition). Chin Pharm News 2020;15(6):801–5.

CHAPTER 15

Prevention and prognosis

Li Tongzeng[1], Li Zhen[2] and Li Hongyan[3]
[1]Beijing YouAn Hospital, Capital Medical University, Beijing, China; [2]Peking University Health Science Center, Beijing, China; [3]The First Affiliated Hospital of Nanyang Medical College, Nanyang, China

15.1 Prevention and control measures

COVID-19 is a novel respiratory infectious disease. On January 20, 2020, the National Health Commission of China issued Announcement No. 1, incorporating COVID-19 into the Law of the People's Republic of China on Prevention and Treatment of Infectious Diseases as a Class B infectious disease. However, it requires management under Class A infectious diseases. At the same time, pneumonia caused by SARS-CoV-2 infection has been included in the management of quarantinable infectious diseases as stipulated by the Frontier Health and Quarantine Law of the People's Republic of China.

On December 26, 2022, the National Health Commission issued a notice, with the approval of The State Council, lifting the prevention and control measures for Class A infectious diseases stipulated in the Law of the People's Republic of China on the Prevention and Control of Infectious Diseases for the COVID-19 from January 8, 2023; The COVID-19 is no longer included in the management of quarantinable infectious diseases as stipulated in the Frontier Health and Quarantine Law of the People's Republic of China.

Currently, the prevention and control measures for COVID-19 mainly include the following aspects.

(1) General preventive measures for COVID-19: Maintain good personal and environmental hygiene. Have a balanced diet, engage in appropriate exercise, and get sufficient rest to avoid excessive fatigue. Improve health literacy and develop hygienic habits and lifestyles such as maintaining a one-meter social distance, washing hands frequently, wearing masks, and using serving chopsticks. Cover the

The Pathology and Imaging of Coronavirus Pneumonia
ISBN 978-0-443-32930-2
https://doi.org/10.1016/B978-0-443-32930-2.00003-9

nose and mouth when sneezing or coughing. Keep indoor ventilation good and take good personal protection.

(2) Specimen Collection and Laboratory Testing: Relevant specimen collection, transportation, storage, and testing shall be strictly carried out in accordance with the Technical Guidelines for Laboratory Testing of COVID-19 issued by the Chinese Center for Disease Control and Prevention.

(3) Key Locations, Institutions, and Population Control: For example, public places such as stations, airports, docks, shopping malls, public restrooms, elevators, and enclosed transportation vehicles such as buses, trains, and airplanes. Special institutions include schools, kindergartens, nursing homes, welfare facilities for disabled individuals, and correctional facilities.

(4) Nosocomial Infection Control, Specific Location Disinfection, and Personnel Protection: Medical institutions shall strictly implement nosocomial infection control in accordance with the requirements of the Technical Guidelines for the Prevention and Control of SARS-CoV-2 Infection in Medical Institutions. Furthermore, the Technical Specifications for Disinfection in Medical Institutions and the Management Specifications for Hospital Air Purification shall be followed strictly to ensure proper cleaning and disinfection of medical equipment, contaminated items, object surfaces, floors, and air within medical facilities. Proper disposal and management of medical wastes shall be carried out in accordance with the Regulations on the Administration of Medical Wastes and the Measures for the Administration of Medical Waste in Medical and Health Institutions.

(5) Publicity, Education and Risk Communication: Disseminate knowledge about COVID-19 prevention and control, strengthen health education for key populations, provide the public with personal protective guidance through various channels, and reduce potential contacts or exposures within the population.

Before the introduction of vaccines safely and effectively, the prevention and control of COVID-19 are primarily achieved through managing the sources of infection, cutting off transmission routes, and protecting susceptible populations.

According to the characteristics of SARS and MERS, as well as the relevant explanations on the diagnosis and treatment protocol in China, the prevention and control strategies for COVID-19 are equally applicable to SARS and MERS.

15.2 Management source of infection

According to the diagnosis and treatment protocol of COVID-19 in China, the main infection source for COVID-19 is SARS-CoV-2 infected individuals. Some case investigations have found that asymptomatic infected persons may also serve as the infection sources. Therefore, the management of asymptomatic infected persons shall follow similar protocols as confirmed cases. But the infectivity of asymptomatic infected persons is significantly lower than that of symptomatic patients.

(1) Individuals infected with SARS-CoV-2 should be hospitalized or treated at home based on the severity of their illness. During the infectious stage, they should undergo isolation to prevent the spread of the virus to others.

(2) Provide guidance to patients on correctly selecting and wearing masks, and on implementing proper cough etiquette and hand hygiene.

(3) Discharge criteria for hospitalized patients: When the patient's condition has improved significantly, the body temperature has been normal for more than 24 h, the acute exudative lesions shown in lung imaging have been significantly improved, the treatment can be switched to oral medications, and there are no complications that require further treatment, discharge can be considered.

Based on the characteristics of SARS and MERS and the relevant explanations on the diagnosis and treatment protocol in China, the control strategies for the infection sources of COVID-19 are also applicable to SARS and MERS.

15.3 Cut-off of transmission routes

According to the diagnosis and treatment protocol of COVID-19 in China, the transmission routes of COVID-19 are respiratory tract transmission and close contact transmission. There is also a possibility of airborne transmission when exposed to high concentrations of aerosols for a prolonged period in relatively closed environments.

For the prevention and control measures to cut off the transmission routes, the following aspects shall be taken into account.

(1) Block Respiratory Droplet and Contact Transmission: In epidemic areas, avoid using public transportation, and always wear a mask in crowded or poorly ventilated public places such as elevators, public transport, and restrooms. Before eating or after using the restroom, try not to touch your mouth, nose, or eyes; wash hands frequently and adopt correct handwashing method. The managers of public areas need to perform strict daily disinfection of public facilities.

(2) Fecal-oral Transmission: It has not been confirmed up to now, but based on the reported literature of SARS cases in 2003, the possibility of SARS-CoV-2 transmission through the fecal-oral route cannot be ruled out. During the epidemic, strict attention shall be paid to food and water hygiene. A paper published in Science in July 2020 confirmed that SARS-CoV-2 can replicate in the human intestine. Previous studies have found that the duration of SARS-CoV-2 RNA positivity in the feces or anal swabs of COVID-19 patients is longer. Research teams have also isolated live viruses from feces and wastewater, suggesting the possibility of fecal-oral transmission. Although there are no confirmed cases of fecal-oral transmission reported yet, these findings still indicate the importance of proper disinfection of excreta from infected individuals and the need to ensure the safety of drinking water, particularly in areas with poor sanitation conditions.

(3) Take good personal hygiene and protective measures, develop good hygiene habits, wash hands frequently, avoid touching the mouth, nose, and eyes with hands, cover the mouth and nose when coughing or sneezing, wear masks properly, separate and dispose of garbage correctly, maintain social distancing, promote separate meal serving with public chopsticks and adopt online appointment for medical services. Enhance ventilation and hygiene in workplaces and living areas, try to avoid visiting crowded places, especially in enclosed spaces, and maintain a safe social distancing of "1 m apart" upon interacting with others. Wear masks for medical institution staff, as well as employees working in enclosed spaces such as salespersons, security personnel, cleaning staff, transportation operators, station service personnel, police, and patients, etc.

Based on the characteristics of SARS and MERS and the relevant explanations on the diagnosis and treatment protocol in China, the strategies to cut off the transmission routes of COVID-19 are also applicable to SARS and MERS.

15.4 Protection of susceptible population

There are several types of COVID-19 vaccines, including mRNA vaccines (such as the Pfizer-BioNTech and Moderna vaccines), vector vaccines (like the AstraZeneca vaccine), and inactivated vaccines (for example, some vaccines developed in China). The COVID-19 vaccines play a crucial role in reducing the spread of the virus, preventing severe illness, hospitalization, and death. They also help to build herd immunity, where a significant portion of the population is immune to the virus, reducing the overall impact of the pandemic.

Vaccination against the COVID-19 can reduce the incidence and infection of the COVID-19. The effective means of low incidence of severe illness and death should be used by those who meet the vaccination conditions Seed. Those vaccinated who meet the conditions for enhanced immunization should receive enhanced immunization in time Seed.

Protecting vulnerable populations mainly involves conducting health promotion and education through various channels, reminding individuals and relevant entities to take preventive measures.

Based on the characteristics of SARS and MERS and the relevant explanations on the diagnosis and treatment protocol in China, there are currently no safe and effective vaccines available for SARS and MERS. Therefore, the strategies for protecting vulnerable populations for COVID-19 also apply to SARS and MERS.

15.5 Prognosis

15.5.1 Prognosis of COVID-19

The prognosis for the majority of COVID-19 patients is favorable, with a lower mortality rate compared with SARS and MERS. Currently, discharged patients have shown good recovery, and long-term complications and sequelae are uncommon. However, in individuals aged 50 and above, the mortality rate notably increases with age. People with underlying diseases have a higher risk of developing severe and critical cases.

As of 21 July 2024, over 775 million confirmed cases and more than seven million deaths have been reported globally since the beginning of pandemic. The current case fatality rate for COVID-19 was less than 1%.

15.5.2 Prognosis and prevention of SARS and MERS

During the global outbreak of SARS in 2003, a total of 8098 probable SARS cases were reported in 29 countries and regions, with 774 individuals (9.6%) succumbing to the disease. Similarly, there was a significant difference in the fatality rate among different age groups of SARS patients, with individuals aged 60 and above having a considerably higher fatality rate compared with those under 60 years old.

According to data from the World Health Organization, as of January 2020, 27 countries and regions have reported a total of 2519 laboratory-confirmed MERS cases, with 866 deaths, resulting in a fatality rate of 34%. The majority of cases (84.2%) were reported in Saudi Arabia. The statistics indicate that individuals aged 50–59 years are at the highest risk of primary MERS infection, while those aged 30–39 years are most susceptible to secondary MERS infection. The age group with the highest fatality rate of primary MERS infection is also 50–59 years old, while the age group with the highest fatality rate of secondary MERS infection is 70–79 years old.

On May 29, 2015, the former National Health and Family Planning Commission reported the first confirmed imported MERS case in Huizhou City, Guangdong Province, China. The patient, a 44-year-old male and traveled from South Korea, is a close contact of a MERS case in South Korea. After consultation, according to the Diagnosis and Treatment Protocol for Middle East Respiratory Syndrome (MERS, 2014 Edition), the patient was diagnosed as a confirmed MERS case.

Some SARS and MERS patients experience long-term lung damage and pulmonary fibrosis. In an 8-month follow-up of patients in the recovery period of SARS, high-resolution computed tomography scans revealed that most patients had focal or multiple areas of lung fibrosis. Additionally, 7% of patients had restrictive lung ventilation impairment, and 38% of patients had diffusion dysfunction.

Both SARS and MERS currently lack safe and effective vaccines for prevention. Therefore, the principles for their prevention are similar to those for COVID-19. It involves managing and isolating confirmed and suspected cases, tracing and managing close contacts, strengthening health education and risk control, enhancing training for healthcare professionals, and reinforcing laboratory testing and biosafety measures. In addition to preventing human-to-human transmission, it is also necessary to consider the possibility of animal-to-human transmission through intermediate hosts.

Once any suspicious intermediate host animal is discovered, it shall be promptly reported, and multiple departments shall collaborate to conduct investigations to timely cut off the transmission pathway from animals to humans.

Bibliography

[1] Comprehensive Team for Joint Prevention and Control Mechanism for COVID-19 under the State Council. The prevention and control of 2019 novel Coronavirus pneumonia. 7th ed. 2020.
[2] National Health Committee, State Administration of Traditional Chinese Medicine. The COVID-19 diagnosis and treatment program. Infect Dis Informat 2020;33(4):289−96. Trial Eighth Edition.
[3] Zhong N. Diagnosis and treatment protocol of severe acute respiratory syndrome (SARS). Chinese Med J 2003;83(19):1731−52.
[4] National Health Committee. The diagnosis and treatment protocol for Middle East respiratory syndrome (MERS) cases. Chin J Viral Dis 2015;5(5):352−4. 2015 Edition.
[5] National Health Committee. Technical guidelines for the prevention and control of 2019 novel coronavirus infections in medical institutions (1st ed.) [J] Chin J Infect Control, 20, 19 (2): 189−191.
[6] National Health Committee. Interim guidelines for the range of use of common medical protective equipment in prevention and control of 2019 novel Coronavirus pneumonia (trial). 2020.
[7] National Health Committee. Interim work scheme for transshipment of pneumonia cases with 2019 novel Coronavirus infection (trial). 2020.
[8] Lamers MM, Beumer J, Van Der Vaart J, et al. SARS-CoV-2 productively infects human gut enterocytes. Science 2020;369(6499):50−4.
[9] Wu YJ, Guo C, Tang L, et al. Prolonged presence of SARS-CoV-2 viral RNA in faecal samples. Lancet Gastroenterol Hepatol 2020;55(5):434−5.
[10] Zhu FC, Guan XH, Li YH, et al. Immunogenicity and safety of a recombinant adenovirus type-5-vectored COVID-19 vaccine in healthy adults aged 18 years or older: a randomised, double-blind, placebo-controlled, phase 2 trial. Lancet 2020;396(10249):479−88.
[11] Zhu FC, Li YH, Guan XH, et al. Safety, tolerability, and immunogenicity of a recombinant adenovirus type-5 vectored COVID-19 vaccine: a dose-escalation, open-label, non-randomised, first-in-human trial. Lancet 2020;395(10240):1845−54.
[12] Epidemiology Working Group for NCIP Epidemic Response. The epidemiological characteristics of an outbreak of 2019 novel coronavirus diseases (COVID-19) in China. Chin J Epidemiol 2020;41(2):145−51.
[13] Yang XB, Yu Y, Xu JQ, et al. Clinical course and outcomes of critically ill patients with SARS-CoV-2 pneumonia in Wuhan, China: a single-centered, retrospective, observational study. Lancet Respir Med 2020;85(5):475−81.
[14] Wiersinga WJ, Rhodes A, Cheng AC, et al. Pathophysiology, transmission, diagnosis, and treatment of coronavirus disease 2019 (COVID-19): a review. JAMA 2020;324(8):782−93.
[15] Antonio GE, Wong KT, Hui DSC, et al. Thin-section CT in patients with severe acute respiratory syndrome following hospital discharge: preliminary experience. Radiology 2003;228(3):810−5.
[16] Das K, Lee E, Singh R, et al. Follow-up chest radiographic findings in patients with MERS-CoV after recovery. Indian J Radiol Imag 2017;27(3):342−9.

[17] Chiang CH, Shih JF, Su WJ, et al. Eight-month prospective study of 14 patients with hospital-acquired severe acute respiratory syndrome. Mayo Clin Proc 2004;79(11): 1359–66.
[18] Chinese Medical Association, China Association of Chinese Medicine. Diagnosis and treatment protocol of severe acute respiratory syndrome (SARS). Chinese Med J 2003;83(19):1731–52.
[19] General Office of the Notice of National Health and Family Planning Commission on the issuance of the Middle East respiratory syndrome epidemic prevention and control program. Cap J Pub Health 2015;9(4):145–6. 159.

CHAPTER 16

Commonly used cleaning and disinfection methods and medical waste handling in Radiology Departments

Xu Bingren, Zeng Lingzi, Tang Xiaoli, Huang Kuanlong and Zhang Yaming
Shenzhen Qianhai Shekou Free Trade Zone Hospital, Shenzhen, China

The Radiology Department shall establish a dedicated fixed DR (Digital Radiography) room, and hospitals where conditions permit shall give priority to setting up a dedicated CT (Computed Tomography) room, and divide and arrange the areas in accordance with the principle of "Three Areas and Two Passages", namely the Contaminated Area, Potentially Contaminated Area, Cleaning Area, as well as Clean Passage (for medical staff and cleaning items) and Contaminated Passage (for patients and contaminated items). There shall be physical barriers between each zone and all areas and passages shall be prominently marked at their entrances and exits. When designing the room layout, the possibility of rapid conversion of the use during epidemic and nonepidemic situations shall be considered, so as to facilitate the increased utilization efficiency of imaging equipment.

If it is not feasible to establish a dedicated room or equipment, it is necessary to manage the examination sequence of the individuals to be screened scientifically. Patients from regular outpatient departments, fever clinics, and those with higher risk, as well as suspected and confirmed patients in isolation wards, shall be examined in batches and at different time intervals. Strict disinfection measures shall be implemented during the process.

During the COVID-19 epidemic, medical wastes and domestic waste from the dedicated room of the Radiology Department shall be treated as infectious medical wastes and handled strictly in accordance with the relevant regulations, including the Law on the Prevention and Control of

The Pathology and Imaging of Coronavirus Pneumonia
ISBN 978-0-443-32930-2
https://doi.org/10.1016/B978-0-443-32930-2.00013-1

Infectious Diseases, Regulations on the Administration of Medical Wastes and the Measures for the Administration of Medical Waste in Medical and Health Institutions. Some medical wastes from regular rooms and living areas shall be disposed of as routine medical wastes.

16.1 Commonly used cleaning and disinfection methods

16.1.1 Disinfection of DR and CT rooms for COVID-19 in Radiology Department

16.1.1.1 Equipment disinfection

Disinfect the equipment following the disinfection recommendations provided by the manufacturer, use the spraying disinfectant device with caution, as the sprays may penetrate the equipment, causing electrical short-circuits, metal corrosion, or other damage. If a spraying disinfectant device is used, it is required to turn off the equipment and cool it first, then completely cover the equipment with plastic films before starting the spraying process. Remove the plastic film only after all sprays are dispersed, and then wipe and disinfect the equipment. Use the certified cleaning and disinfectant agents to clean the equipment, with particular emphasis on disinfecting areas of the equipment that patients are likely to come into contact with, such as examination beds and sensor areas, at least three times a day; when there are visible contaminations from patient's body fluids, blood, secretions, etc., completely remove them using disposable absorbent material and then clean with a towel or tissue soaked in the cleaning agent. Extra care shall be taken in case of buttons and holes to avoid liquid from infiltrating into the equipment and causing damage.

16.1.1.2 Disinfection of object surface

This includes handrails, handles, windows, walls, switches, tables, chairs, keyboards, mice, and protective equipment in both the dedicated room and the waiting area. Wipe with 1000 mg/L chlorine-containing disinfectant at least 3 times a day; for surfaces of objects that are not susceptible to corrosion, such as metal, use 75% ethanol for wiping at least 3 times a day; when visible contaminations are present, use gauze soaked with 1000 mg/L chlorine-containing disinfectant (except for chlorine hexidine, as above) to remove them carefully; for significant contamination, completely cover with disinfectant powder or bleach powder containing absorbent components, or cover completely with disposable absorbent material and pour sufficient

1000 mg/L chlorine-containing disinfectant on the absorbent material, leave it for at least 30 min, then carefully remove and clean. During the removal process, avoid direct contact with the contaminated materials. Dispose of the cleaned contaminants as medical wastes for centralized disposal, and then use a cloth soaked with 1000 mg/L chlorine-containing disinfectant to wipe the surfaces and the surrounding area of objects that may have come into contact with the contaminants (it is recommended to wipe it twice). Immediately replace the bed sheets after each patient examination. Clean and disinfect the used lead apron using disinfectant wipes containing compound double-chain quaternary ammonium salts (0.18%–0.22%) for wiping and disinfection. Soak the goggles and face masks used by the technicians in 1000 mg/L chlorine-containing disinfectant solution for 30 min.

16.1.1.3 Air disinfection

Turn off all central air conditioning first to avoid cross-infection. Open windows for ventilation and improve air circulation. Keep the air disinfection machine running for continuous disinfection. It is recommended to close the shielding door during air disinfection, and every 4 h, open the door for ventilation for a duration of no less than 30 min each time. During unmanned periods, UV lamps can be used in a suspended or mobile manner for direct irradiation disinfection. The suspension height of the lamp tube from the ground is 1.8–2.2 m. The installation of UV lamps shall have an average of $\geq$1.5 W/m power output, and the disinfection time of $\geq$30 min (starting from 5 min after the lamps are turned on). It is necessary to arrange the lamps reasonably to cover the entire room, with a minimum of three disinfection cycles per day. Additionally, after disinfection, open windows and/or ventilation ducts for at least 30 min. Keep the surface of UV lamps clean, wipe the lamp surface once a week using 70%–80% (volume ratio) ethanol-soaked cotton balls. If any dust or oil stains are found on the lamp surface, promptly clean them. When using UV lamps for air disinfection, ensure that the room is clean and dry, and reduce dust and water mist. If the temperature is $<$20°C or $>$40°C, or the relative humidity is $>$60%, it is recommended to appropriately extend the irradiation time. It is not advisable to use UV lamps when there are people indoors.

16.1.1.4 Ground disinfection

Wipe the floors of the waiting area, dedicated corridors, machine rooms, and operating rooms with 1000 mg/L chlorine-containing disinfectant solution at least 3 times a day. In case of visible contaminations, follow the cleaning process the same as that of surface disinfection, and the cleaning range shall not be less than 2 m.

16.1.1.5 Terminal disinfection

If the dedicated room is designated for confirmed patients, implement the aforementioned disinfection measures and terminal disinfection at least twice a day. If the dedicated room is used for screening febrile patients, implement the aforementioned disinfection measures; When suspected cases are identified, initiate the terminal disinfection immediately.

The specific method for terminal disinfection is as follows: If chemical disinfectants are used for air terminal disinfection, it is recommended to use 1%–3% ultra-low volume hydrogen peroxide or similar aerosol disinfection. Seal the room for at least 30 min and then open the windows for ventilation; next, wipe and disinfect the environmental surfaces, floors, and instrument surfaces, with special emphasis on surfaces that patients and medical staff may come into contact with during work; If the manufacturer explicitly prohibits the use of peroxides or similar substances for terminal disinfection, UV lamps can be used for a 30-min irradiation period. After turning off the UV lamps, proceed with the disinfection of equipment and environmental surfaces.

The above requirements also apply to the disinfection of mobile CT and CT examination rooms of mobile cabin hospital.

16.1.2 Disinfection of general DR and CT rooms in Radiology Department

16.1.2.1 Equipment disinfection

Equipment disinfection shall be based on the disinfection guidelines provided by the manufacturer, using certified cleaning and disinfecting agents. Focus on disinfecting areas of the examination bed and detection regions that are easily accessible to patients, carrying out disinfection at least twice daily. When visible contamination is present, first use disposable absorbent material to completely remove the contaminants, and then use disinfectant wipes containing compound double-chain quaternary ammonium salts (0.18%–0.22%) for wiping and disinfection.

16.1.2.2 Disinfection of object surface

This includes the railings, handles, windows, walls, switches, tables, chairs, keyboards, and other surfaces in both regular rooms and waiting areas. Use disposable disinfectant wipes or a chlorine-containing disinfectant solution with effective chlorine concentration of 500 mg/L to wipe surfaces such as computers, keyboards, at least twice a day. For computers and keyboards, use protective barriers or films, and disinfect them at least twice a day. For surfaces that are not corrosion-resistant, such as metal, use 75% ethanol to wipe them at least twice a day. When visible contamination is present, first use disposable absorbent material to remove the contaminants completely, and then use a cloth soaked with a 1000 mg/L chlorine-containing disinfectant solution for wiping. Immediately replace the bed sheets after each patient examination. To clean and disinfect the object surface, preferably use disinfectant wipes or cloths soaked in disinfectant solution following the proper guidelines. Spray disinfection is not recommended. Clean and disinfect the used lead apron using disinfectant wipes containing compound double-chain quaternary ammonium salts (0.18%–0.22%) for wiping and disinfection.

16.1.2.3 Air disinfection

Natural ventilation is preferred. If natural ventilation is not feasible, mechanical ventilation or air disinfection measures can be considered. For exhaust and return air vents, disinfection shall be conducted once a week, while for air conditioning vents, disinfection shall be conducted once a month. Keep the air disinfection machine running for continuous disinfection. It is recommended to close the shielding door during air disinfection, and every 4 h, open the door for ventilation for a duration of no less than 30 min each time; alternatively, conduct UV radiation for 30 min each time and least 2 times a day, and open the windows for ventilation after the irradiation.

16.1.2.4 Ground disinfection

Wipe the floors of the waiting area, corridors, machine rooms, and operating rooms with 500 mg/L chlorine-containing disinfectant solution twice a day. In case of visible contaminations, first remove them with disposable dehumidifying materials, and then use a cloth soaked with a 1000 mg/L chlorine-containing disinfectant solution for wiping. The cleaning range shall not be less than 2 m.

If there is a suspected patient in a regular room, terminal disinfection shall be initiated immediately after the examination is completed.

16.1.3 Disinfection of mobile DR/CT

When conducting examination with mobile DR/CT, areas that come into contact with patients, such as detectors and shielding protective equipment, shall be prepackaged and isolated with disposable plastic bags or wiped and disinfected after the examination is completed. When moving the equipment to a ward outside the isolation room for use, wipe and disinfect the entire surface of the machine according to the manufacturer's disinfection guidelines, then conduct UV irradiation for at least 30 min before use. After the examination is completed, wipe and disinfect the equipment with a 1000 mg/L chlorine-containing disinfectant solution. If there is a separate room available, implement the disinfection measures for dedicated radiology examination rooms. If any suspected case is discovered, initiate the terminal disinfection immediately.

16.1.4 Disinfection of registry/film pickup room

16.1.4.1 Facility disinfection

Wipe and disinfect the desktop in working areas, windows, countertops that patients may touch, door handles, keyboards, mice, photos, and self-service diagnostic report printers using a 500 mg/L chlorine-containing disinfectant solution at least twice a day; For surfaces that are not corrosion-resistant, such as metal, use 75% ethanol to wipe them at least twice a day. Paper application forms shall be kept at the registration desk and disposed of safety, without being passing on to the next step. When visible contaminations are present, remove them with disposable dehumidifying materials, and then use a cloth soaked with a 1000 mg/L chlorine-containing disinfectant solution for wiping.

16.1.4.2 Ground disinfection

Wipe and disinfect with a 500 mg/L chlorine-containing disinfectant solution at least once a day. In case of contamination, it shall be disinfected immediately.

16.1.4.3 Reduce contact

Hospitals with the necessary conditions can use electronic application forms, images, and reports for data transmission through the internet, reducing direct contact between patients and medical staff.

16.1.5 Regular cleaning and disinfection of daily living areas

Wipe the surfaces of tables, chairs, keyboards, mice, computer screens, pens and mobile phones in the diagnostic room, duty room, rest area, etc. with 75% ethanol once a day; Wipe the floors with a 500 mg/L chlorine-containing disinfectant solution once a day; enhance air circulation and use the air disinfection machine for at least 30 min each time, at least twice a day, or follow the instructions provided by the machine; wipe the toilet and hand sink areas with a 1000 mg/L chlorine-containing disinfectant solution once a day; In case of contamination, it shall be wiped and disinfected immediately.

16.2 Medical waste handling in Imaging Departments

During the COVID-19 epidemic, medical wastes and domestic waste from the dedicated room of the Radiology Department shall be treated as infectious medical wastes and handled strictly in accordance with the relevant regulations, including the Law on the Prevention and Control of Infectious Diseases, Regulations on the Administration of Medical Wastes and the Measures for the Administration of Medical Waste in Medical and Health Institutions.

16.2.1 Classification and collection of medical wastes

16.2.1.1 Clarify the range of classification and collection

The wastes generated by COVID-19 confirmed and suspected patients in dedicated rooms, including medical wastes and household wastes, shall be classified and collected separately according to medical waste regulations.

16.2.1.2 Specify the packaging container

The yellow packing bag specialized for medical wastes and the outer surface of sharp containers shall be marked with warning signs. Before packing medical wastes, a thorough inspection shall be carried out to ensure there is no damage or leakage. The collection bins of medical wastes shall be foot-operated and equipped with a cover. The double-layer packaging bags are used to contain medical wastes. When the medical wastes reach 3/4 of the packaging bag or the sharp container, spray a 1000 mg/L chlorine-containing disinfectant inside the inner layer and adopt a gooseneck-type sealing for inner layer; then spray 1000 mg/L chlorine-containing disinfectant to the surface of the inner bag and adopt a gooseneck-type sealing for outer layer, and label with specialized marks, and finally spray a 1000 mg/L chlorine-containing disinfectant to the surface of the outer bag. Patient gowns, bed sheets, and other items shall be placed in orange waste bags or water-soluble bags, and the bags shall be properly labeled and sent separately to the laundry room for specialized cleaning and disinfection. After removing medical wastes from the trash bin, the trash bin shall disinfected properly.

16.2.1.3 Conduct collection safely

The cleaning staff or dedicated collectors for medical wastes shall wear personal protective equipment (level 2 protection) when collecting infectious medical wastes, to ensure personnel safety and control the infection risk. When the outer surface of the packaging bags or sharp containers containing medical wastes is contaminated with infectious waste, another packaging bag shall be used. Upon classification and collection, do not squeeze or compress disposable isolation gowns, protective clothing and similar items used. Each packaging bag and sharp container shall be attached or affixed with a Chinese label, indicating the generating unit, department, date and category of the medical wastes, as well as the specific note stating "Pneumonia Caused by Coronavirus Infection".

16.2.1.4 Handle by zone

For medical waste generated from confirmed COVID-19 patients and suspected cases in the dedicated room, the surface of the packaging bags shall be disinfected before leaving the contaminated area by evenly spraying 1000 mg/L chlorine-containing disinfectant or by covering additional packaging bag for medical wastes. Medical wastes generated in the cleaning area shall be handled as conventional medical wastes.

16.2.2 Transport of medical wastes

Before transporting medical wastes, it is necessary to check whether the labeling, tags, and sealing of the packaging bags or sharp containers comply with the requirements. During the process of transport, staff shall prevent any damage to the specialized packaging bags and sharp containers for medical waste, avoid direct contact of medical wastes with their bodies and prevent any leakage or spread of medical wastes. At the end of the transport each day, the transport equipment shall be cleaned and disinfected using 1000 mg/L chlorine-containing disinfectant. When the transport equipment is contaminated with infectious medical wastes, it shall be promptly disinfected and handled. Furthermore, during the process of transport, it is essential to ensure that a dedicated vehicle and personnel are used for the collection and transport of medical wastes. Before the transshipment, the transportation route and handover requirements shall be properly determined. Properly conduct the handover registration and sealed transshipment, and store the medical wastes at designated storage locations of the hospital.

Bibliography

[1] National Health Committee. The COVID-19 diagnosis and treatment program. 2020 (trial eighth edition), http://www.gov.cn/zhengce/zhengceku/2020-08/19/content_5535757.htm.
[2] National Health and Family Planning Commission. Regulation for cleaning and disinfection management of environmental surface in healthcare [M]. Beijing: China Standards Publishing House.
[3] Chinese Society of Imaging Technology. Expert consensus on radiological examination scheme and infection prevention and control of COVID-19. Electro J Emerg Infect Dis 2020;5(2):65−73 (trial first edition).
[4] Specialized Committee of Quality Control and Safety Management of Chinese Society of Radiology. Guidelines for CT examination procedures of COVID-19 fever in outpatient and emergency department. J Clin Radiol 2020;39(5):841−5 (1st ed.).
[5] Wu F, Xingxing M, Caiyun W, et al. Prevention and control of COVID-19 epidemic and construction on long-term mechanism of occupational Health in infectious disease hospital. J Emerg Infect Dis 2020;5(3):198−202.
[6] National Health and Family Planning Commission of the People's Republic of China. Technical guidelines for nosocomial infection prevention and control of ebola hemorrhagic fever. Infect Dis Informat 2014;8(5):260−1. 1st ed.
[7] Chinese Society of Imaging Technology. Nosocomial infection prevention and control management with COVID-19 imaging examination: recommendations from the Chinese society of imaging technology. Chin J Radiol 2020;54:009. 1st ed.
[8] Ming T, Xiaoyuan F, Liu S, et al. Guidance on imaging diagnosis and infection control of novel coronavirus pneumonia. Chin Comput Med Imag 2020;26(5):1−19.
[9] Qiuzhen X, Yonggan L. Consensus of experts on radiological examination, diagnosis, prevention and control rules for Coronavirus disease 2019 in Jiangsu Province. J Southeast Univ (Med Sci Ed) 2020;180(2):7−25. 1st ed.
[10] General Office of National Health Committee. Technical guidelines for the prevention and control of 2019 novel Coronavirus infections in medical Institutions. 2019. 2nd ed.

CHAPTER 17

Strategies for prevention and control of nosocomial infection and X-ray protection principles in radiology departments

Huang Kuanlong[1], Fang Wenchun[2], Xu Chuanjun[3], Xu Bingren[1] and Zeng Lingzi[1]
[1]Shenzhen Qianhai Shekou Free Trade Zone Hospital, Shenzhen, China; [2]Comprehensive Clinic of Shenzhen Healthcare Committee Office, Shenzhen, China; [3]Affiliated Nanjing Hospital of Nanjing University of Chinese Medicine, Nanjing, China

The coronavirus is highly infectious and has a high mortality rate. According to relevant laws, regulations, and normative documents in China, and referencing technical documents from the World Health Organization and the Centers for Disease Control and Prevention in the United States, medical staff of radiology department who come into contact with suspected and confirmed COVID-19 patients shall be managed at different zones based on different job positions. Furthermore, emphasis shall be placed on strict adherence to hand hygiene, correct selection and mask wearing, proper ventilation, and other key standard preventive measures. Conduct comprehensive training for all staff and strengthen infection monitoring. Carry out disinfection, isolation, and protection in accordance with the protective norms of medical staff and adopt corresponding preventive measures. Ensure that medical staff have proper personal hygiene protection and X-ray protection in place. Implement relevant management for the professional technicians who have direct contact with patients during examinations according to the prevention and control standards of nosocomial infection.

17.1 Strategies for prevention and control of nosocomial infection

Coronavirus pneumonia has been confirmed to be a highly infectious disease, and the radiology department, as a closely contacted department,

The Pathology and Imaging of Coronavirus Pneumonia
ISBN 978-0-443-32930-2
https://doi.org/10.1016/B978-0-443-32930-2.00006-4

plays a crucial role in the prevention and control of nosocomial infection. Standardized management strategies and workflows can effectively reduce the risk of nosocomial infections.

17.1.1 Diagnosis and treatment environments and post requirements of radiology department

17.1.1.1 Requirements for diagnosis and treatment environment

(1) Provide a separate medical imaging examination room and establish area divisions: To prevent cross-infection, it is necessary to set up a separate medical imaging examination area or dedicated radiology equipment (such as X-ray and CT devices specifically for infected individuals). Clearly demarcate contaminated areas, semicontaminated areas and cleaning areas and implement strict disinfection process in accordance with the requirements of nosocomial infection control. The contaminated and semicontaminated areas of the radiology department are key areas for nosocomial infection control (after the examinations of confirmed patients, thorough disinfection shall be conducted before examining suspected patients).

(2) Establish a dedicated radiographic examination pathway: For general clinic and inpatient cases, implement a comprehensive real-time appointment system (excluding emergency), limiting the number of patients undergoing examinations to prevent gatherings and waiting in crowded areas. Conduct centralized examinations for fever clinics and suspected/confirmed patients in the wards in batches and at different time slots, and carry out disinfection process strictly in accordance with the requirements. Reasonably assign the staff of radiology department with dedicated roles in key areas such as fever clinics, contaminated and semicontaminated areas of the radiology department, and isolation wards of the hospital, so as to ensure specialized and responsible management.

17.1.1.2 Post requirements

(1) Mobile Bedside X-ray Radiographers: When the mobile DR equipment needs to be moved to the emergency room, observation room, infectious isolation units, fever clinics, and other departments, strictly comply with the protection requirements of the department before entering the workspace, and perform two rounds of thorough high-level disinfection (using 1000 mg/L chlorine-containing disinfectant) to the examination equipment before being taken away.

(2) X-ray Radiographers and CT Technologists: For suspected or confirmed COVID-19 patients, establish a dedicated appointment and triage station as well as a dedicated DR/CT scanning station. Upon receiving notification from the fever clinic for examination, inform the staff to prepare for receiving the patient and advise the patient to go to the designated isolation examination area, which shall be physically separated and marked with warning signs to prevent others from approaching; After receiving the examination notice, the staff shall double check their readiness, keep the inner protective door closed, open the outer protective door with an inner switch, and use the intercom system to call the patient in; use the intercom system to verify patient's information, guide the patient into the correct position, and take strict precautions against falling down or off the bed; during the epidemic, clinical sheets that come into direct contact with the skin of patients shall be replaced after each time being used by each patient. If positioning technologists need to have close contact with suspected patients in the examination room, they shall strictly take the Level III protection measures. Operating technologists can take Level I or II protection measures to perform standardized image acquisition on patients; After the examination is completed, they shall open the outer protective door using an inner switch; Once the patient leaves, they shall carry out air and equipment disinfection according to the disinfection requirements, and continue to disinfect the air with an air disinfection machine.

(3) Imaging examination registration personnel: The staff working in key areas shall complete the registration for patients of fever clinic, and no such patient is allowed to go to the conventional appointment and registration service window for registration. The registration personnel for patients from general clinic shall try to obtain patient's appointment information by verbal means, and patients shall keep a distance of at least 1m away from the registration personnel; minimize contact and wash the hands with quick hand sanitizer after touching the appointment form each time. Besides, any forms and documents touched by patients shall be managed separately and disposed of safely. Guide patients in correctly wearing masks and direct them to designated waiting areas to minimize personnel movement. It is recommended to fully utilize the Hospital Information System (HIS), Picture Archiving and

Communication Systems (PACS), and Radiology Information System (RIS) to achieve paperless electronic application forms and medical histories.

(4) Other imaging technologists and diagnostic professionals working in noncontaminated zones: If it is not clear whether to enter the contaminated or semicontaminated zones, they shall wear disposable caps, surgical masks (N95 protective masks are required upon contacting with suspected patients), work clothes (disposable isolation gowns when necessary for precheck triage), and latex gloves when necessary. Perform strict hand hygiene process.

17.1.2 Working mode of radiographers in key areas (trial)

Mobile bedside X-ray technologists, X-ray technologists, and CT technologists working in isolation areas may have close contact with patients. Therefore, it is recommended, if conditions permit, adopting a "2 + 2" working mode.

Assign dedicated personnel to stay and accommodate in specific isolation areas in the hospital. These personnel shall be responsible for and complete mobile bedside X-ray, X-ray, and CT examinations in key areas such as fever clinics, contaminated and semicontaminated areas in the radiology department, and hospital isolation wards for a period of 2 weeks (or as per specific circumstances, the work duration can be shortened). After completing the shift, the personnel need to rest on standby in a specific isolation area specially arranged in the hospital to be prepared for any emergencies. During the work period, no one can leave the designated isolation area and all work and daily needs shall be fulfilled within the key areas and the specified designated isolation area.

After completing their 2-week work, the radiographer will leave the specific isolation area and enter a designated special isolation ward for a 2-week period of medical observation under supervision. During such period, they cannot leave the designated area. After the 2-week period of medical observation under supervision, if there are no abnormalities, they can return to their normal work positions.

17.1.3 Prevention principle for nosocomial infection in radiology department

(1) To prevent cross-infection, the entry and exit passages for department staff shall be closed and access shall be controlled using restricted access measures. Unauthorized personnel will not be allowed to enter or exit the area. Technicians shall wear protective equipment in the cleaning area and take of them in the buffer zone. Positioning technologists working in the contaminated area are not allowed to enter the cleaning area before the scheduled shift. During shift period, the technologists shall take off their protective clothing in the buffer zone and perform proper cleaning process before entering the cleaning area.

(2) When the staff of the radiology department conduct examinations in the emergency room, observation room, infectious isolation units, fever clinics, and other departments, they shall strictly comply with the protection requirements of the department before entering the workspace, and perform two rounds of thorough high-level disinfection (using 1000 mg/L chlorine-containing disinfectant) to the examination equipment before being taken away. Staff shall strictly adhere to proper hand hygiene practices and cannot touch mobile phones, glasses, or other objects before performing hand hygiene process.

(3) Before entering the living area or leaving for home, the staff in general medical technology departments shall change contaminated masks and work clothes and use a disinfectant containing 75% ethanol or disinfectant wipes to disinfect personal glasses, mobile phones, pens, and other items. Besides, they shall also wash their hands with running water.

(4) It is necessary to establish a dedicated system to strictly implement health checks and registration for medical staff, technicians, registration personnel, and cleaning staff in the department, implement a daily health report system for employees, and have emergency plans in place for individuals showing clinical symptoms of COVID-19, with timely reporting as per the established protocols.

(5) Properly manage employee dining by implementing staggered meal times, controlling the number of people dining simultaneously, and maintaining a distance of 1 m between dining tables. If conditions permit, limit each dining table to only one person, locate the dining area shall in the cleaning area, separate from the work area, and employees cannot wear work clothes to the dining area for meals. Staff shall also take personal measures to reduce the risk of infection, such as wearing masks, avoiding touching the mask and eyes, practicing

frequent handwashing (using the seven-step handwashing method), refraining from face-to-face interactions while eating, drinking, or conversing with colleagues, and use disposable plastic bags to cover pagers. Try to avoid holding meetings or shifts with a large number of participants, control the number of attendees and shorten the meeting time if necessary, wear masks and maintain environmental ventilation; whenever possible, conduct meetings through video, online, or other remote means.

(6) Personal Protective Levels: Grade the protective measures based on the exposure risks in different work areas and positions (Table 17.1).

Table 17.1 Graded protection for personnel in different positions.

Position	Protection level	Protective equipment
Diagnostician	Level I	Wear disposable caps, surgical masks (N95 protective masks are required upon contacting with suspected patients), work clothes (disposable isolation gowns when necessary for precheck triage), and latex gloves when necessary. Perform strict hand hygiene process
Technicians and registration personnel in the specialized room	Level II	Applicable to medical staff engaging in diagnostic and therapeutic activities involving close contact with suspected or confirmed patients: Wear disposable caps, protective goggles or face shields (fog-resistant type), N95 protective masks, protective suits or isolation gowns, double-layer disposable latex gloves, disposable shoe covers, and strictly adhere to hand hygiene practices
Fever clinic or mobile DR and CT technicians	Level III	Applicable to medical staff having close contact with suspected or confirmed patients: Wear disposable caps, N95 protective masks, protective goggles, face shields (or full-face respirators or positive-pressure hoods), protective suits, isolation gowns, double-layer latex gloves, shoe covers, and strictly adhere to hand hygiene practices

17.1.4 Patient protection and management

Implement a comprehensive appointment system for general clinic and inpatient patients (excluding emergencies), limiting the number of examination attendees to prevent crowding in waiting areas. The registration personnel shall try to obtain patient's appointment information be verbal means, and patients shall keep a distance of at least 1m away from the registration personnel; minimize contact, and it is recommended to implement a one-stop registration for imaging data at remote fever clinics, allowing patients to upload the information through the internet; If the registration personnel have to contact with the patient appointment application forms, wash the hands with quick hand sanitizer after touching the appointment form each time, guide patients in wearing masks correctly, and direct patients to designated waiting areas to reduce the movement of people. Except for patients, under special circumstances, patients' family members or other companions are prohibited from entering the examination room, and in principle, only one person is allowed. Pregnant women and minors also need to sign a special informed consent form for radiation exposure.

17.1.5 Personnel training

The hospital shall conduct occupational safety training for all staff at least twice a year, advanced learning course of occupational health protection at least once a year and occupational exposure-related trainings in each department at least twice a year, so as to enhance medical staff's awareness of occupational safety precautions.

17.2 X-ray protection against nosocomial infection in radiology department

X-rays are a form of ionizing radiation that is harmful to human tissues. When using medical X-ray equipment, especially during special period of the epidemic, it is important to pay attention to both nosocomial infection prevention and control and radiation protection and safety. The purpose of X-ray radiation protection is to prevent harmful nonstochastic effects and limit the occurrence of stochastic effects to a level considered acceptable. Ensure the health and safety of medical workers, subjects, the public, and their descendants, minimize potential hazards, enhance the effectiveness of X-ray radiation protection, and fully utilize the maximum potential of X-ray technology. Adhere to the three principles of exposure beyond the protection: ① Time Protection: Appropriately reduce CT scanning time without compromising the diagnosis, to minimize the radiation dose received by the subject. ② Distance Protection: The radiation dose received by the human body is inversely proportional to the square of the distance; the farther the distance is, the radiation dose will be decreased exponentially. ③ Shielding Protection: Use materials with high atomic numbers, such as lead, barium sulfate concrete, etc., to effectively absorb or shield radiation, thereby reducing X-ray exposure to the human body. For subjects, personal protective equipment is the simplest and most effective method of protection. Various radiation protection products are designed using lead or lead-containing materials, including lead aprons, lead collars, lead caps, lead neck guards, lead gloves, etc., which can block or shield X-ray exposure.

17.2.1 Basic principles of radiation protection

In order to achieve the purpose of radiation protection, the basic principles of radiation protection have been proposed for radiation exposure resulting from practical activities.

(1) Justification of radiation practices. Any practice that produces ionizing radiation shall be demonstrated or recognized as worthwhile, considering that the harm caused by ionizing radiation is acceptable in relation to the benefits society and individuals gain from it. If the proposed practice cannot yield a net benefit exceeding the cost (including health damage and protective costs), the implementation of such practice shall not be adopted.

(2) Optimization of radiation protection. All unnecessary exposures shall be avoided. Follow the principle of optimizing radiation protection, achieve the maximum net benefit with the minimum cost, and thus keep all necessary exposures at the lowest achievable levels reasonably. Strive for optimization of protection rather than blindly pursuing limitless dose reduction during the process of designing radiation protection. Otherwise, the increased cost of protection would outweigh the benefits, and it cannot be considered reasonable.

(3) Limit of individual dose. While implementing the principles of justification and optimization, it is essential to ensure that the

individual radiation dose received does not exceed the prescribed limit. This ensures that individuals among the radiation workers do not face excessive levels of risk. According to the current standards in China, the annual dose equivalent of uniform whole-body irradiation for radiation workers (including internal irradiation and nonuniform external irradiation) shall not exceed 20 mSv within a 5-year period, in any of which the effective dose shall not exceed 50 mSv. The annual dose equivalent to which individuals in the public are exposed c shall be below 5 mSv (0.1 rem).

17.2.2 Application of mobile DR/CT

Mobile DR/CT can effectively address the issue of patients who are unable to conveniently move to the radiology department but still need imaging. It is often used in places such as the ICU, neonatology department and hospital wards where shielding facilities specialized against X-ray equipment, as found in radiology departments, are not available. The principle of its use is to allow its usage only when it is necessary to perform DR/CT examinations in situations where fixed equipment cannot be used. It cannot be used as a routine examination. Common radiation protection methods include.

17.2.2.1 Site clearing before exposure

Before the exposure, the operators shall notify the surrounding personnel who can evacuate to temporarily leave the area. If it is not possible to clear the site, protective measures shall be taken for adjacent patients, such as using movable lead shields.

17.2.2.2 Use of protective equipment

When operating the equipment, the staff shall wear lead protective equipment, such as lead aprons, lead collars, lead glasses, and lead caps. During X-ray imaging, try to stay as far away from the radiation source as possible. During exposure, select the exposure field and area carefully, minimize the exposure time, and avoid repeated exposures. During patient imaging, use lift-type protective curtains, movable lead screens, and lead protective clothing (lead aprons, lead caps, lead vests, lead glasses, lead collars) for nonexamined areas, and use lead curtains to ensure proper radiation protection around the bed.

Bibliography

[1] National Health Committee. The COVID-19 diagnosis and treatment program. Trial 8th ed. 2020. http://www.gov.cn/zhengce/zhengceku/2020-08/19/content_5535757.htm.

[2] Qiuzhen X, Yonggan L. Consensus of experts on radiological examination, diagnosis, prevention and control rules for coronavirus disease 2019 in Jiangsu Province (1st edition). J Southeast Univ (Med Sci Ed) 2020;180(2):7−25.

[3] National Health and Family Planning Commission. Regulation for cleaning and disinfection management of environmental surface in healthcare. Beijing: China Standards Publishing House.

[4] Chinese Society of Imaging Technology. Nosocomial infection prevention and control management with COVID-19 imaging examination: recommendations from the Chinese society of imaging technology (1st edition). Chin J Radiol 2020;54.

[5] Chinese Society of Imaging Technology. Expert consensus on radiological examination scheme and infection prevention and control of COVID-19 (trial irst edition). Elect J Emerg Infect Dis 2020;5(2):65−73.

[6] Ming T, Xiaoyuan F, Liu S, et al. Guidance on imaging diagnosis and infection control of novel coronavirus pneumonia. Chin Comput Med Imaging 2020:1−19.

[7] Wu F, Xingxing M, Caiyun W, et al. Prevention and control of COVID-19 epidemic and construction on long-term mechanism of occupational health in infectious disease hospital. Elect J Emerg Infect Dis 2020;5(3):198−202.

CHAPTER 18

Coronavirus Infection control during the process of imaging examination

Xu Qiuzhen[1], Xu Dong[2], Xu Chuanjun[3] and Yu Dong[4]
[1]Department of Radiology, Zhongda Hospital Affiliated to Southeast University, Nanjing, China; [2]Beijing Chest Hospital, Capital Medical University, Beijing, China; [3]Department of Radiology, The Second Hospital of Nanjing University of Chinese Medicine, Nanjing, China; [4]Shenzhen Center for Prehospital Care, Shenzhen, China

Imaging examinations are the most crucial means to examine patients with coronavirus infection and play an important role in the diagnosis and evaluation of curative effect. Infection control during the process of imaging examinations is of utmost importance.

18.1 Setup and construction of CT room in fever clinic

18.1.1 Set up CT examination room in fever clinic area

Arrange a relatively independent CT room with easy partition management as a special CT examination room for fever clinic and emergency treatment, and set up clear partition signs. Set up an independent CT examination room and operation room. If possible, set up a self-service film printer (which can be located in the Fever Clinic area). It is recommended to equip a CT scanner with 16 or more detector rows, as it provides higher image resolution.

If it is not feasible to allocate a dedicated CT examination room, examine patients from the fever clinic and for emergency treatment in batches and during different time periods, and give these patients the priority for examination; rigorously disinfect the equipment and air upon the completion of the current patient's examination and then proceed with the next patient's examination.

The Pathology and Imaging of Coronavirus Pneumonia
ISBN 978-0-443-32930-2
https://doi.org/10.1016/B978-0-443-32930-2.00004-0

18.1.2 Set up a dedicated CT examination channel from the fever clinic

Separate the staff passage from the patient passage, and designate different access routes for staff and patients based on designated areas. Staff wearing protective equipment is limited to activities in contaminated or semi-contaminated areas. It is preferable to place directional markers on the floor to prevent patients from taking detours and increasing the risk of transmission.

18.1.3 Set up a dedicated waiting areas for patients from the fever clinic outside the CT room

Use enclosures and other tools to separate this area from other areas. The dedicated waiting area shall be clearly marked with area signs, and unrelated personnel are not allowed to enter this area.

18.1.4 Provide quick-drying hand sanitizer inside and outside the door of different areas

Staff shall perform hand sanitization process when entering and exiting the doors. Provide medical waste bins in all different areas.

18.1.5 Strengthen the management of dedicated CT exhaust vents

Disable central air conditioning systems that do not meet the requirements for prevention and control of nosocomial infection. Seal the exhaust vent of the dedicated CT room to prevent virus transmission through indoor ventilation.

18.2 Infection control and management in imaging department of fever clinic

18.2.1 Establish a sound management system for CT room in the fever clinic

The department head is fully responsible for the operations in the CT room of the Fever Clinic during the epidemic. The chief technologist is responsible for the division of areas and environmental improvement, developing CT examination protocols and quality control plans, as well as conducting technologist training, scheduling, and overall coordination. Appoint 1 coordinator for nosocomial infection prevention and control, responsible for training and supervising staff on the correct usage of protective equipment, understanding examination equipment, floor disinfection methods, managing and replenishing protective equipment, monitoring staff body temperatures, and reporting to department leadership on a daily basis. In the case of special circumstances, such as abnormal body temperatures of staff, they shall also promptly report to the management department and coordinate with the Fever Clinic for timely handling and isolation.

18.2.2 Strengthen the training, assessment and supervision of infection control for imaging staff

(1) Infection control training includes knowledge about nosocomial infections and the proper use of protective equipment, medical isolation techniques and standards, information related to coronavirus infections, hand hygiene, and more. Particular attention shall be given to preshift and postshift protective measures.

(2) Strengthen the supervision and assessment of hand hygiene, adopt the "seven-step handwashing method," and strictly follow the principle of washing hands before contact with patients, before clean or aseptic procedures, after exposure to patients' body fluids, after contact with patients, and after contact with patient's surroundings), ensure to wash hands before and after donning and doffing isolation gowns, protective clothing, and gloves.

(3) Take personal protective measures before and after work

Preparation before work: Due to the need for strict protection during work, enter the cleaning area through a designated staff channel, perform hand hygiene and then sequentially wear a medical protective mask, disposable cap or cloth cap, and change into work shoes. If conditions permit, staff can also change into scrub suits. Wear work clothes before entering the potentially contaminated area (in the control room). Before entering the contaminated area (such as the equipment room or waiting area), remove the work clothes and change into protective clothing or isolation gown, and wear disposable caps, disposable medical surgical masks, protective goggles, gloves, and shoe covers.

Preparation before departure: Prohibit wearing personal protective equipment when leaving the contaminated area to prevent cross-contamination between different areas. Before leaving the contaminated area, disinfect the hands first, then remove protective equipment in the specified order, take off the protective clothing, and perform thorough hand hygiene again. Disinfect the external auditory canal with 75% ethanol, disinfect the nasal cavity with Type III An'er Iodine, and rinse the mouth with saline solution. If possible, perform thorough cleaning for at least 30 min.

(4) Enhance training and assessment on the image recognition of coronavirus infections: The department head or the leader of the chest imaging diagnosis group shall regularly summarize and timely update the various manifestations of coronavirus infections, and train and enhance the technologist's ability to identify CT image features of coronavirus infection through PACS or on-site guidance in the imaging room, etc. After the scan is completed, technologists shall promptly preview the images. If any suspected lesions is observed, they shall report to the attending diagnostic physician in a timely manner. Once confirmed by the attending physician, the findings shall be treated as critical values and reported accordingly, so as to ensure that isolation and protective measures are taken as soon as possible, thereby reducing patients' waiting time and minimizing the spread of the virus.

(5) Only after passing the assessment on nosocomial infection knowledge and proper usage of protective equipment conducted by the infection control training team can the staff be allowed to assume their positions.

(6) The staff of CT room in Fever Clinic are required to report their health conditions on a daily basis. Before starting and after finishing their shifts, they need to measure and record their body temperatures. If any symptoms such as fever or cough occur, they shall visit the Fever Clinic for diagnosis and, if necessary, undergo isolation treatment. The training team of infection prevention and control shall check the body temperature records daily and conduct random inspections on personal protective measures and disinfection from time to time. Identify and summarize any noncompliant points, analyze the identified issues, and conduct targeted training to ensure the correct usage of protective equipment and reduce the risk of infection. The training team of infection prevention and control shall conduct regular inspections of the dedicated channels and waiting areas of the Fever Clinic on a daily basis.

18.2.3 Scientifically and reasonably set up the on-duty post of imaging department

The CT room of the Fever Clinic shall implement a 24-h duty system (and may adopt an on-site duty system when necessary). It is recommended to assign one to two technologists per shift based on actual conditions. If two technologists are assigned, one shall be responsible for operating the equipment in the control room and the other for patient positioning. It is advised that the technologist in the control room follows Level 1 protection (wearing work clothes, medical surgical mask, and latex gloves), while the positioning technologist follows Level 2 protection (wearing work clothes, work cap, medical protective mask, disposable isolation gown, goggles/face shield/face mask, and latex gloves). For designated hospitals for coronavirus infections or during periods of increased epidemic severity, the protection level shall be elevated to Level 3 protection, requiring work clothes, work cap, N95 protective masks, disposable isolation gown, goggles/face shield/face mask, and latex gloves.

Implement a 24-h on-duty system and arrange fixed rest periods based on the actual situation of the hospital (ensuring proper isolation and protection), and adjust the schedule dynamically according to the local epidemic situation and changes in the number of patients.

18.3 CT examination process of fever clinic

18.3.1 Preparation before CT examination

The Fever Clinic shall contact the CT room in advance to make appointments for patients, inform patients in advance and provide necessary information (such as medical insurance card number) and conduct the registration (including PACS triage and registration records), with the registration category of Emergency. The CT room shall make necessary preparations and evacuate other patients through designated channels based on their status, such as confirmed or suspected cases. Scan the handwritten note with the words "Fever Clinic" to replace the requisition form (preferably an electronic requisition form). The requisition form shall clearly indicate whether the patient is a confirmed case, suspected case, or unknown case of coronavirus infection, and then complete the patient's epidemiological history and laboratory test results.

18.3.2 Clinical reception of patients

Patients shall be escorted by a designated nurse or other staff member from the Fever Clinic. After the examination, patients shall avoid wandering around, wear masks properly, minimize conversation with others, and maintain a distance of at least 1 m away from others. After the examination, patients shall immediately return to the examination area of the Fever Clinic to avoid potential nosocomial cross-infection through contact with other patients. Patients shall wear medical surgical masks throughout the examination process. Accompanying doctors, family members, and unrelated personnel shall not enter the control room and examination room to reduce contamination in the control room. Mild patients can get on and off the examination bed under the guidance of the examination personnel. If the patient's condition is critical, the accompanying person shall wear protective equipment or the technologist can assist the patient in getting on and off the examination bed. UV disinfection and other measures can be taken to properly disinfect patient examination requisition forms and other textual materials.

18.3.3 Examination of patients

Complete the examination in a separate room (if possible, in a separate room when there is only one technologist), and setup the equipment with the console on the premise of patient's safety to avoid accidental injuries. If necessary, use isolation gowns (Level 2 protection) upon contact with the patients. After the examination, do not return to the control room, but stay in the waiting area to receive the next patient or assist relevant personnel in environmental disinfection; Follow the operational guidelines to promptly and safely remove the isolation gown before returning to the control room; Gently place it into a medical-grade double-layer yellow garbage bag for proper disposal as medical wastes. It is not recommended for a single technologist to leave the control room upon on duty. Besides, accompanying personnel shall assist in contaminated area operations, such as helping patients get on and off the examination bed. Technologists shall pay attention to hand hygiene and related disinfection procedures.

18.3.4 Image and report processing

Immediately after the examination is completed, the technologist shall promptly perform patient image reconstruction and PACS transmission. It is recommended to routinely perform thin-layer image reconstruction with a thickness of ≤ 1 mm, creating thin (high-resolution) lung window images. If possible, coronal and sagittal plane reconstructions can be performed, focusing on displaying the imaging features of the lesions. Conventional films shall be printed, primarily using 5 mm transverse section images, while thin layers and other body positions shall be primarily read on PACS. Special attention shall be given to enlarging the display of small ground-glass opacity lesions. After being laid out, the films can be printed using the self-service printer located in the cleaning area of the Fever Clinic.

Alternatively, they can be delivered by a designated nurse or other staff member to the cleaning area of the Fever Clinic.

It is recommended to establish a diagnostic room within the radiology department. After the scan is completed, the CT technologist of the Fever Clinic shall promptly notify the emergency CT physician to write the report. The report shall then be reviewed by a verifying physician. It is suggested that the report shall be issued within half an hour for emergency cases. The clinical doctors of the Fever Clinic shall view and print the reports at the workstation. To ensure the timeliness of the reports, it is recommended to have dedicated physicians responsible for directly viewing and reconstructing images on PACS or the backend workstation. For patients highly suspected of viral pneumonia, urgent reports shall be treated as critical values and promptly submitted.

The radiologists shall be familiar with the typical CT manifestations of coronavirus infections. Based on the typical CT manifestations, together with the patient's epidemiological history, clinical symptoms, and laboratory test results, they shall emphasize to the clinicians that these patients are highly likely to have a coronavirus infection, which would be beneficial in improving patient outcomes and reducing the risk of transmission.

18.3.5 Other precautions

(1) Considering that some coronavirus-infected patients may not have a fever in the early stages and might undergo CT scans on other CT machines within the radiology department, all CT technologists and physicians are required to maintain a high level of vigilance. If a patient with a high suspicion of viral pneumonia is found during a CT scan on a nonfever clinic dedicated CT machine, it shall be immediately reported as a critical value, and the examination of other patients shall be stopped. Follow the aforementioned disinfection protocols before receiving the next patient.

(2) Matters requiring overall coordination by the hospital: Fever clinic or nursing department shall arrange dedicated personnel for accompanying patients during examinations. Fever clinic or logistics personnel shall be designated for disinfection tasks. The infection control department shall conduct personnel training. It is recommended that consistent protective equipment be provided and distributed to the Fever Clinic.

(3) Key points for CT technologists: Try to complete CT examinations in separate rooms whenever possible. The concept of "three-zone separation" shall be well understood. Do not walking around in isolation gowns, emphasize hand hygiene and avoid touching things while wearing gloves. Conduct synchronous configurations of medical technology, close collaboration, and prompt completion of examinations and reports.

(4) Conduct in-depth research based on the hospital's own situations to promptly identify infection control loopholes, make timely improvements, and adjust infection control measures to ensure there are no omissions.

Bibliography

[1] Qiuzhen X, Yonggan L. Consensus of experts on radiological examination, diagnosis, prevention and control rules for coronavirus disease 2019 in Jiangsu province (first edition). J Southeast Univ (Med Sci Ed) 2020;180(2):7−25.
[2] Chinese Society of Imaging Technology. Expert consensus on radiological examination scheme and infection prevention and control of COVID-19 (trial first edition). Elect J Emerg Infect Dis 2020;5(2):65−73.
[3] Wu F, Xingxing M, Caiyun W, et al. Prevention and control of COVID-19 epidemic and construction on long-term mechanism of occupational health in infectious disease hospital. Elect J Emer Infect Dis 2020;5(3):198−202.
[4] Specialized Committee of Quality Control and Safety Management of Chinese Society of Radiology. Guidelines for CT examination procedures of COVID-19 fever in outpatient and emergency department (first edition). J Clin Radiol 2020;39(5):841−5.

APPENDIX 1

Guidelines for imaging diagnosis of novel coronavirus pneumonia in both Chinese and English versions (third edition 2021)

The Specialized Committee for Radiology of Infection and Inflammation of Chinese Research Hospital Association, the Imaging Working Committee of Chinese Association of STD and AIDS Prevention and Control, the Radiology of Infectious Diseases of Chinese Society of Radiology, the Professional Committee for Infection Imaging of Chinese Medical Doctor Association Radiologist Branch, the Imaging Management Team for Infectious Diseases of Chinese Hospital Association Management Branch for Infectious Diseases, the Infectious Disease Group of General Radiological Equipment Professional Committee of China Association of Medical Equipment, the Beijing Imaging Diagnosis and Treatment Technology Innovation Alliance.

Since the end of December 2019, there have been a series of cases of pneumonia of unknown cause [1]. The sequencing of respiratory tract samples revealed that the source of infection is a novel coronavirus [2]. The International Committee for Taxonomy of Viruses (ICTV) named it as Severe Acute Respiratory Syndrome Coronavirus 2 (SARS-CoV-2), which can cause severe respiratory system disease similar to Severe Acute Respiratory Syndrome Coronavirus (SARS-CoV) and Middle East Respiratory Syndrome Coronavirus (MERS-CoV), and is officially named Coronavirus Disease 2019 (COVID-19). Chest High-Resolution CT (HRCT) serves as one of the primary methods for screening and diagnosing COVID-19, and its imaging manifestations exhibit certain characteristic features and patterns, which can provide essential evidence-based support for the diagnosis and treatment process. This guideline is formulated based on the guidelines for imaging examination and diagnosis of COVID-19 proposed by clinical staging [3,4] by combining detailed analysis of imaging cases, practical experience, and the latest research achievements in accordance with the first and second editions of the Guidelines for Imaging Diagnosis of Novel Coronavirus Pneumonia in both Chinese and English [3,5], which has significant guiding significance for enhancing the clinical diagnostic capabilities of COVID-19.

1. Clinical diagnosis of COVID-19

1.1 Scope of application

This guideline is applicable to medical institutions of different levels across the country and serves as a reference for imaging examination and diagnosis of COVID-19.

On January 12, 2020, the World Health Organization (WHO) named the novel coronavirus responsible for this outbreak as 2019 novel coronavirus (2019-nCoV).

On February 8, 2020, the Joint Prevention and Control Mechanism of the State Council of China for the novel coronavirus pneumonia outbreak decided to temporarily name the disease caused by the novel coronavirus as Novel Coronavirus Pneumonia (abbreviated as "NCP"). On February 11, 2020, the World Health Organization (WHO) announced that the disease caused by the novel coronavirus would be named COVID-19. At the same time, the International Committee on Taxonomy of Viruses (ICTV) declared that the novel coronavirus would be named SARS-CoV-2. On February 22, 2020, the National Health Commission of China announced that the English name for the disease would follow the WHO designation and be referred to as "COVID-19."

COVID-19 is defined as an acute respiratory infectious disease primarily characterized by pulmonary inflammatory lesions, caused by SARS-CoV-2. It can also lead to damage in the digestive system [4], nervous system, and manifest corresponding symptoms [6].

Coronaviruses are RNA viruses, and based on serotype and genomic characteristics, they are classified into four genera: α, β, γ, and δ. SARS-CoV-2 belongs to the β genus [6]. The possible origin of the virus is believed to be the Chinese horseshoe bat. The exact transmission route and primary host have not been definitively determined yet. However, according to the latest research, there is evidence of zoonotic transmission (from animals to humans) as well as human-to-human transmission. Droplet and contact transmission are considered the main routes of infection; there is a possibility of airborne transmission when exposed to high concentrations of aerosols for a prolonged period in relatively enclosed environments.

Similar to SARS-CoV, and the SARS-CoV-2's pathological mechanism of damaging lung tissue is also through the binding of the spike protein (S protein) of SARS-CoV-2 to the angiotensin-converting enzyme 2 (ACE2) on the alveolar type II epithelial cells in the bronchial mucosa and alveolar epithelium, resulting in the invasion and damage to lung tissue [7]. The pathological manifestations include focal or diffuse acute alveolar inflammation and interstitial inflammation, which are visible under the microscope. In the early stages, there is pulmonary capillary dilation and congestion, with lymphocyte infiltration in the interstitium. Within the alveolar spaces, there is fluid exudation, proliferation, swelling, or degeneration of alveolar epithelial cells, leading to desquamation; monocytes or multinucleated giant cells, lymphocytes, and plasma cells infiltrate, and fibrin exudation may be present. In severe cases, consolidation of the lung can occur, and sometimes pulmonary hyaline membranes can form. In later stages, pulmonary interstitial fibrosis may develop. In some cases, viral inclusion bodies can be observed in the cytoplasm or nuclei of alveolar epithelial cells, which show positive results in immunohistochemistry and nucleic acid testing [8].

1.2 Diagnosis basis

Epidemiological history includes:

① Travel history to or residence in an epidemic area within 14 days before the onset of the disease;

② Contact history with confirmed SARS-CoV-2 infection (positive nucleic acid testing) within 14 days before the onset of the disease;

③ Contact with individuals from an epidemic area or from a community with reported cases who had fever or respiratory symptoms within 14 days before the onset of the disease;

④ Clustering of cases.

1.2.1 Clinical manifestations

Based on current epidemiological investigations, the incubation period of this disease is 1—14 days, with the majority being 3—7 days. The main clinical manifestations include fever, fatigue, and dry cough. In some patients, the initial symptoms may be reduced or loss of smell and taste. A minority of patients may also experience symptoms such as nasal congestion, runny nose, sore throat, conjunctivitis, muscle pain, and diarrhea.

Severe patients often develop dyspnea and/or hypoxemia 1 week after onset, and severe patients can rapidly progress to acute respiratory distress syndrome, septic shock, metabolic acidosis, coagulation dysfunction and multiple organ failure. Very few patients may also have central nervous system involvement, acrovascular necrosis and other manifestations. To be noteworthy, in serious and critical patients infected with 2019-nCoV, middle/low fever can occur during the illness course; and even there is no obvious fever. Patients with mild symptoms may show low fever, mild fatigue, smell and taste disorders, without pneumonia. A few patients have no obvious clinical symptoms after SARS-CoV-2 infection. The majority of patients have a good prognosis, while a small number of patients may suffer from severe conditions, which are more common in elderly individuals, those with underlying chronic diseases, pregnant women in late stages of pregnancy or during the postpartum period, and obese individuals. Pediatric cases generally exhibit milder symptoms. Some children and neonatal cases may have atypical symptoms, such as digestive symptoms such as vomiting and diarrhea, or may present with only poor responsiveness and rapid breathing. In extremely rare cases, children may develop Multisystem Inflammatory Syndrome in Children (MIS-C), characterized by manifestations similar to Kawasaki disease or atypical Kawasaki disease, toxic shock syndrome, or macrophage activation syndrome. MIS-C often occurs during the recovery stage. The main manifestations include fever accompanied by rash, nonpurulent conjunctivitis, mucosal inflammation, hypotension or shock, coagulation disorders, and acute gastrointestinal symptoms. Once it occurs, the conditions can rapidly deteriorate in a short period of time.

According to the No. 680 Notice of the COVID-19 Diagnosis and Treatment Program (trial eighth edition) of GWBYH [2020], the clinical classification of COVID-19 is as follows:

(1) Mild: Clinical symptoms are mild, and no signs of pneumonia are observed on imaging.

(2) Moderate Type: Patients have fever, respiratory symptoms, and imaging shows signs of pneumonia.

(3) Severe

For adults, if any of the following criteria are met:

① Shortness of breath, respiratory rate (RR) $\geq$ 30 breaths per minute;

② Oxygen saturation at rest $\leq$93% while breathing the air;

③ Arterial partial pressure of oxygen (PaO_2)/fraction of inspired oxygen (FiO_2) $\leq$ 300 mmHg (1 mmHg = 0.133 kPa); for high-altitude areas (above 1000 m), the following formula shall be used to correct PaO_2/FiO_2: $PaO_2/FiO_2 \times$ [760/atmospheric pressure (mmHg)];

④ Progressive worsening of clinical symptoms, with pulmonary imaging showing significant lesion progression (>50%) within 24–48 hours.

For children, if any of the following criteria are met:

① Persistent high fever lasting more than 3 days;

② Shortness of breath (under 2 months, RR $\geq$ 60 breaths per minute; 2–12 months, RR $\geq$ 50 breaths per minute; 1–5 years old, RR $\geq$ 40 breaths per minute; older than 5 years old, RR $\geq$ 30 breaths per minute), excluding the influence of fever and crying;

③ Oxygen saturation at rest $\leq$93% while breathing the air;

④ Assisted respiration (nasal ala flap, three concave sign);

⑤ Development of drowsiness or seizures;

⑥ Refusal to eat or difficulty in feeding, with signs of dehydration.

(4) Critical

If any of the following conditions are met:

① Respiratory failure requiring mechanical ventilation;

② Presence of shock;

③ Concurrent organ failure requiring intensive care unit (ICU) monitoring and treatment.

1.2.2 Imaging manifestations

Imaging manifestations from chest X-ray and chest CT examinations are crucial for the detection of pulmonary lesions, assessment of lesion size, density, and progression [4]. Digital radiography (DR) is quick and convenient but has low resolution, making it less effective in detecting early lesions. Additionally, its value in differential diagnosis is limited. Therefore, it is not recommended for initial examination of suspected cases. DR is recommended for use in primary hospitals without CT machines [3,4], as well as for follow-up and reexamination of critical patients and for confirmed cases.

Chest CT, especially High-Resolution CT (HRCT), provides a clear view without overlapping structures, enabling the early detection of small lesions [3,5], which is the preferred imaging modality for current COVID-19 screening. According to the pulmonary imaging manifestations, COVID-19 can be divided into four stages [3]:

① Early Stage: The chest manifestations are often atypical, with lesions showing faint and patchy ground-glass opacities (GGOs), mostly localized and scattered in the lower and peripheral areas of both lungs, primarily seen beneath the pleura.

② Progressive Stage: Multiple lesions are present, characterized by the progression of GGO with consolidation or mixed opacities, which are more commonly distributed in the peripheral areas of both lungs and may be accompanied by a small amount of pleural effusion.

③ Severe Stage (Critical Stage): This stage corresponds to the late phase of the disease, where the density of the lesions in both lungs further increases diffusely and extensively, resulting in a condition known as "white lung". During this stage, the lesions progress rapidly, with an increase of more than 50% within 48 hours. Treatment becomes challenging, and the mortality rate for patients is relatively high.

④ Recovery Stage: During this stage, the lesions start to shrink or absorb, and some cases may show pulmonary interstitial fibrosis changes.

Common features of COVID-19 in all stages include bilateral and multifocal lesions, with ground-glass opacities (GGOs) being commonly observed. Consolidation, bronchial inflation signs, and thickening of interlobular septa may be present. "Paving Stone Sign" can also be seen, along with bronchial inflation signs and vascular passing through the lesions. In the acute stage, the lesion density is relatively uniform. In the absorption stage, uneven dense shadows are more common, and there is rarely pleural effusion or enlarged lymph nodes [3,9].

In the early stages of mild cases, there may be reticular shadows or small ground-glass opacities (GGO) (Fig. A.1). In the progressive stage, the number of lesions increases, and their extent expands. In the consolidation areas, the "air bronchogram sign" can be observed (Fig. A.2). In the severe stage (critical stage), the range and number of lesions increase significantly, with multiple consolidations in both lungs, exhibiting "white lung" (Fig. A.3). In some cases during the recovery stage, pulmonary interstitial fibrosis lines can be observed (Fig. A.4). In cases of MIS-C, patients with heart failure may exhibit cardiomegaly and pulmonary edema on imaging.

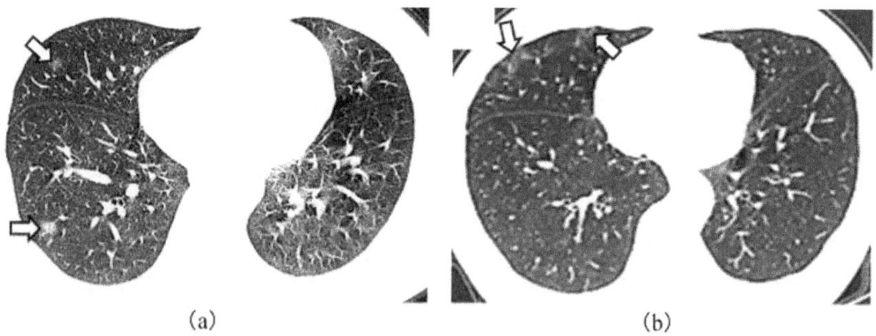

Figure A.1 CT manifestations of lung lesions in early stage. (a) Tiny ground-glass opacities (GGOs) nodules (*arrow*) in the right middle lobe lateral segment and right lower lobe posterior basal segment of the lung; (b) slightly higher density reticular shadows beneath the pleura in the right middle lobe lateral segment (*arrow*).

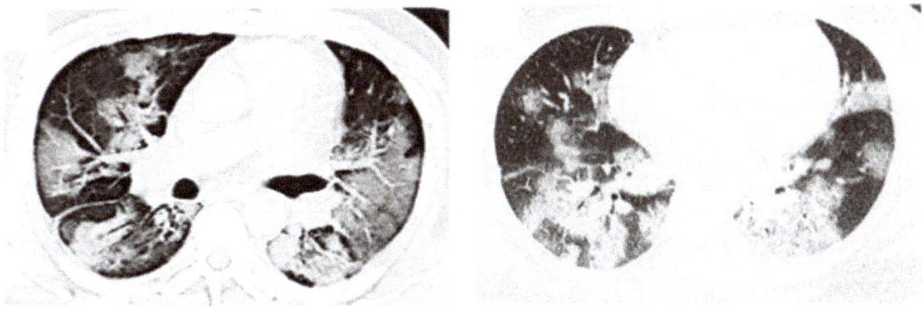

(a) (b)

Figure A.2 CT manifestations of lung lesions in progressive/critical stage. (a) multiple lesions in both lungs; (b) multiple lesions in both lungs, involving the hila.

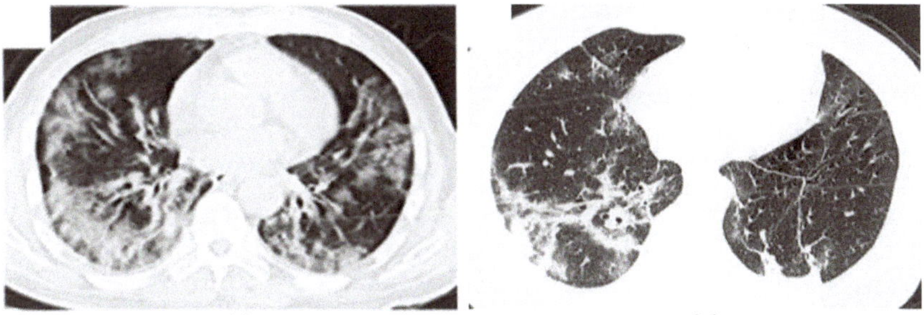

(a) (b)

Figure A.3 CT manifestations of lung lesions in recovery stage. (a) the lesions have partially absorbed compared with before; (b) some lesions appear as fibrous streaks.

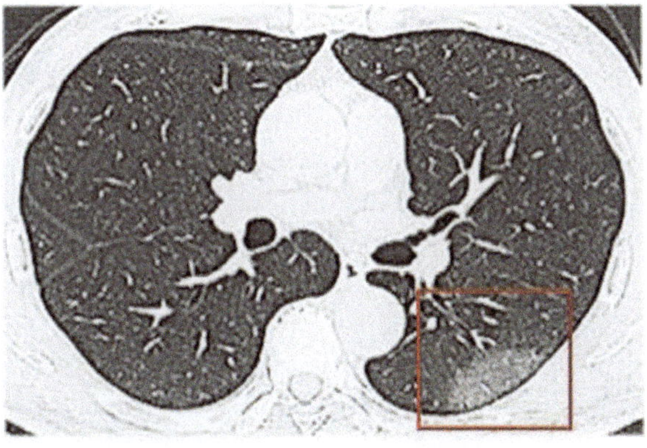

Figure A.4 CT manifestations of COVID-19 in early stage. the patient is a 61-year-old male with a history of fever for 2 days and a faint ground-glass opacity (GGO) shadow in the left lower lung.

1.2.3 Laboratory tests

1.2.3.1 General examinations

In the early stages of the disease, peripheral blood leukocyte count may be normal or decreased, and lymphocyte count may be reduced. Some patients may experience elevated levels of liver enzymes, lactate dehydrogenase (LDH), creatine kinase, myoglobin, troponin, and ferritin. In most patients, C-reactive protein (CRP) and erythrocyte sedimentation rate (ESR) are elevated, while procalcitonin levels are normal. Severe and critical patients may exhibit elevated D-dimer levels, progressive reduction in peripheral blood lymphocyte count, and increased levels of inflammatory cytokines.

1.2.3.2 Microbiological and serological examinations

1.2.3.2.1 Etiological examination SARS-CoV-2 can be detected using RT-PCR and/or NGS methods in specimens such as nasopharyngeal swabs, sputum, other lower respiratory tract secretions, blood, feces, urine, etc. Testing lower respiratory tract specimens (sputum or airway aspirates) is more accurate. Nucleic acid testing may be influenced by factors such as the course of the disease, specimen collection, testing process, and test reagents. To improve the detection positivity rate, specimen collection shall be standardized, and samples shall be sent for testing as soon as possible after collection.

1.2.3.2.2 Serological examination SARS-CoV-2 specific IgM and IgG antibodies are positive, but the positivity rate is relatively low within the first week of the disease. Due to factors such as the positive threshold of the reagent itself or the presence of interfering substances in the body (such as rheumatoid factor, heterophile antibodies, complement, lysozyme, etc.), or specimen-related issues (hemolysis, bacterial contamination, prolonged specimen storage, incomplete coagulation, etc.), antibody testing may result in false positives. Serological testing alone is generally not used as a diagnostic basis. It shall be combined with epidemiological history, clinical manifestations, and the presence of underlying diseases for comprehensive assessment.

The following patients can be diagnosed through antibody testing:

① Patients with clinical suspicion of COVID-19 and negative nucleic acid testing;

② Patients in the convalescence stage of the disease with negative nucleic acid testing.

1.2.4 Diagnostic criteria

Suspected cases can be diagnosed through a comprehensive analysis of the following epidemiological history and clinical manifestations: Patients with any one of the epidemiological history criteria and any two of the clinical presentation criteria; Patients without clear epidemiological history but meeting any two of the clinical presentation criteria, and also testing positive for SARS-CoV-2 specific IgM antibodies, or meeting all three of the clinical presentation criteria.

(1) Epidemiological History:
 ① Travel history or residence in areas with reported cases within 14 days before the onset of symptoms.
 ② Contact with confirmed COVID-19 patients or Asymptomatic infected persons within 14 days before the onset of symptoms.
 ③ Contact with individuals with fever or respiratory symptoms from areas with reported cases within 14 days before the onset of symptoms.
 ④ Clustering of cases (two or more cases with fever and/or respiratory symptoms occurring within 2 weeks in a small area such as a household, office, school class, etc.).

(2) Clinical Manifestations:
 ① Fever and/or respiratory symptoms, and other COVID-19 related clinical manifestations;
 ② Presence of the characteristic COVID-19 imaging features;
 ③ Normal or decreased white blood cell count, and normal or decreased lymphocyte count in the early stage of the disease.

(3) Confirmed Cases:
 ① Suspected cases with either of the following etiological or serological evidence:
  ① Positive SARS-CoV-2 nucleic acid test.
  ② For individuals not vaccinated with SARS-CoV-2 vaccine, positive for both SARS-CoV-2 specific IgM and IgG antibodies [10]

2. Recommended methods for imaging examinations

Standardized chest X-ray and CT examinations are effective guarantees for the screening, early diagnosis, and therapeutic evaluation of SARS-CoV-2 infection cases. They provide high-quality imaging data to support clinical decision-making and help avoid nosocomial cross-infections caused by repeated examinations.

Guidelines for Imaging Diagnosis of Novel Coronavirus Pneumonia (2020 First Edition): The expert group recommends that in medical institutions with the necessary conditions, chest multislice CT scans shall be the preferred imaging modality for COVID-19 patients (Level of evidence: II; Recommended strength: Strong). Chest HRCT scanning and thin-layer reconstructions are more advantageous in revealing early-stage lesions, especially in cases where SARS-CoV-2 nucleic acid testing is negative or unavailable, making them even more valuable. The Diagnosis and Treatment Protocol for Novel Coronavirus Pneumonia (Trial Edition 5) issued by the National Health Commission explicitly states that pulmonary imaging manifestations can serve as an important clinical assessment indicator and be used as one of the clinical diagnostic criteria in key epidemic areas.

2.1 Chest X-ray examination

Chest X-ray examination, including bedside digital radiography (mobile DR) and follow-up imaging, can be performed for critical patients during the severe stage. Although X-ray imaging is convenient and fast, it has lower sensitivity and specificity for displaying and detecting lesions due to image overlap, making it prone to missed diagnoses. General X-ray examination is not recommended as the initial imaging modality but only suitable for primary hospitals (without CT equipment) and for follow-up of critical patients (recommended strength: weak).

2.2 Chest CT examination

Chest CT examination is the most common and essential imaging modality for respiratory system diseases. For suspected COVID-19 patients or patients with viral nucleic acid testing negative, chest CT scan is the preferred imaging method. Chest CT scan is recommended for use in both mild and severe COVID-19 patients, and for severe patients when conditions permit (Level of evidence: II; recommended strength: Strong). Chest HRCT and thin-layer reconstruction techniques are particularly effective in displaying early ground-glass opacities (GGO). It is recommended to use thin-layer reconstruction with a slice thickness of ≤1.0 mm and multiplanar reconstruction. Low-dose CT can be used for follow-up and reexamination of confirmed COVID-19 patients. It is also recommended for use as a screening tool for suspected cases in cluster outbreaks.

3. Imaging manifestations

Different stages of the disease may have overlapping imaging features. The imaging features in the severe stage (critical stage) are relatively distinctive [3,6,9,11−14].

The early imaging features [6,11−14] are most commonly observed within the first week of COVID-19 infection. Single or multiple lesions are commonly observed in both lungs, mostly located in the peripheral regions or subpleural areas of the lungs, with a higher frequency in the back or lateral segments of the middle and lower lungs. The lesions often present as wedge-shaped or fan-shaped opacities pointing towards the hilum of the lungs in a subpleural pattern (Fig. A.4), and they may also appear as patchy or round opacities. According to the time course, the density of the lesions varies, with early-stage lesions showing predominantly faint and thin ground-glass opacities (GGO), and occasionally grid-like shadows. As the disease progresses, the ground-glass opacities (GGO) gradually increase in density. Within the ground-glass opacities (GGO), there may be visible thickening of the bronchovascular bundles or local grid-like thickening of the interlobular septa. Within the ground-glass opacities (GGO), there may be visible thickening of the bronchovascular bundles or local grid-like thickening of the interlobular septa.

As the disease progresses, the early ground-glass opacities (GGO) visible on imaging gradually increase in extent and density (Fig. A.5). They may merge to form subpleural consolidation in a lobular pattern or coalesce to form band-like or extensive areas of increased density. Within these areas, there may be thickened bronchial walls, enlarged bronchovascular bundles, and local tree-in-bud opacities. Grid-like opacities may also be observed. The lesions are distributed asymmetrically in bilateral subpleural regions and appear as wedge-shaped or fan-shaped opacities, with the lung bases and posterior subpleural regions being more commonly affected. Some of the opacities may follow the distribution of the bronchovascular bundles. When there is an increase in exudative fluid within the alveoli, ground-glass opacities (GGO) become more pronounced. In cases of consolidation, the opacities may appear as soft tissue density or dense linear opacities,

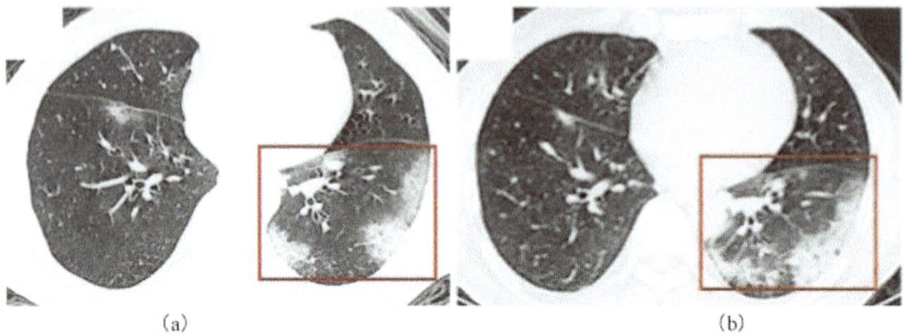

Figure A.5 CT manifestations of COVID-19 in progressive stage. Patient, male and 57 years old, had a fever for 2 days. (a) CT showed multiple patchy GGO lesions located under the pleura of the left lower lobe and near the right lobe fissure; (b) Reexamination 4 days later, CT showed disease progression with a slightly increased range of lesions in the left lower lobe and increased internal density.

distributed in a segmental or lobular pattern. Patients without underlying diseases often do not have pleural effusion, enlargement of mediastinal or hilar lymph nodes.

The imaging manifestations in the severe stage [3,9,14–16] typically occur in patients with underlying diseases or obesity, and the disease rapidly progresses within a few days, leading to respiratory distress and worsening hypoxemia, or respiratory failure, shock, or multiorgan failure, resulting in critical illness, and even death. The chest imaging typically reveals diffuse bilateral lung involvement, with a minority of cases showing a "white lung" appearance. The predominant pattern of the lesions is the consolidation, often accompanied by ground-glass opacities (GGO), bronchial inflation sign (Fig. A.6), and multiple linear opacities. The extent of lesions can increase by 50% within 48 hours, and there may be a small amount of pleural effusion in both lungs.

The imaging manifestations during the recovery stage [11–14] show a decrease in body temperature, reduced cough frequency, and significant improvement in lung function. The imaging manifestations show shrunk lesion range, decreased density, absorbed exudates and gradually resolved lung consolidation. Besides, the lung consolidation lesions may completely disappear or leave residual pulmonary fibrotic bands (Fig. A.7). Existing clinical data shows that the changes in pulmonary imaging manifestations in patients often lag behind the recovery time of the clinical symptom. It is worth noting that during this stage, some patients may still experience disease relapses, with an increase in the number, size, or appearance of new lesions.

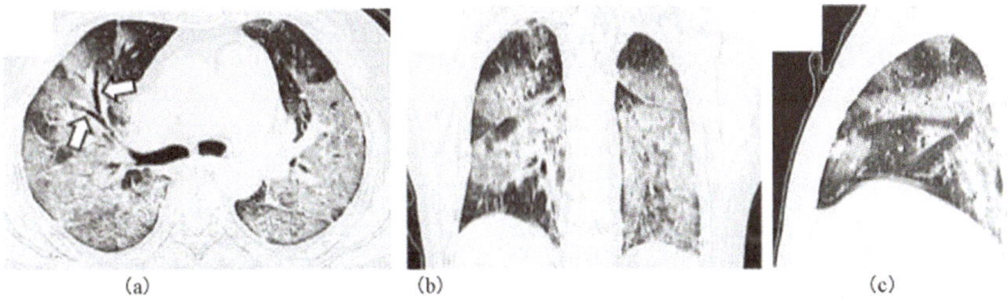

Figure A.6 CT manifestations of COVID-19 in critical stage. the patient was a 57-year-old woman on the eighth day of onset. (a) Axial view, (b) coronal view, (c) sagittal view of CT scan showing extensive ground-glass opacities (GGO) in both lungs, with some areas of consolidation, and visible bronchial inflation sign.

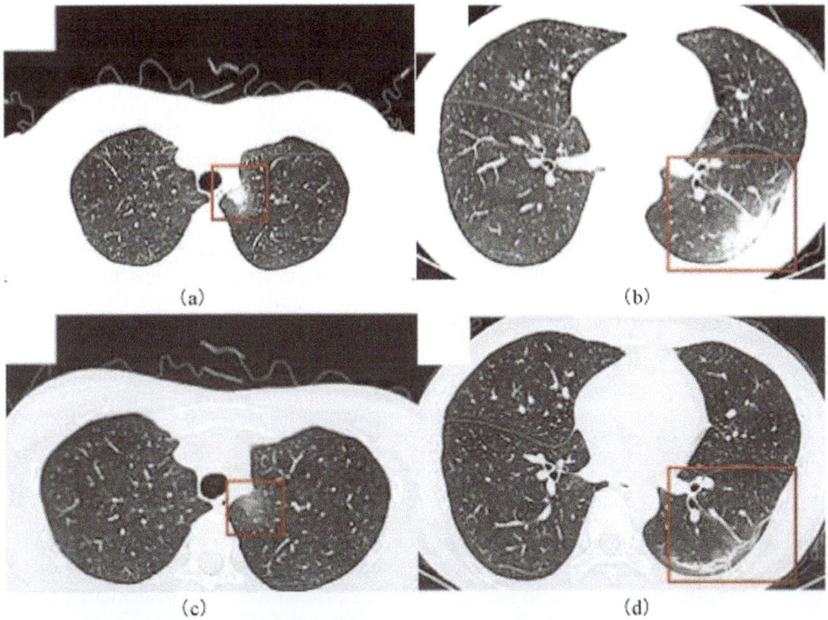

Figure A.7 CT manifestations of COVID-19 in recovery stage. the patient was a 37-year-old male. (a and b) on the sixth day of fever, CT scan showed faint and thin ground-glass opacities (GGO) adjacent to the left upper lobe pleura (a), and fan-shaped high-density shadows across the left lower lobe intersegmental and subpleural areas, with their tips pointing toward the hilum (b); after 2 days of hospitalization, a follow-up CT scan revealed the density of the left upper lung lesion had decreased and its volume had reduced (c), while the majority of the left lower lung lesion had been absorbed, showing fibrous stripe shadows (d).

The imaging manifestations in special populations, including infants, children, adolescents, pregnant women, the elderly, and those with underlying diseases, have their own unique characteristics.

3.1 Infants and young children

3.1.1 Clinical features

Clinical Features Epidemiological data show that currently infants and young children have a lower incidence of COVID-19, and the majority of cases are mild or moderate, with fewer instances of severe or critical symptoms. The youngest confirmed case currently reported is at 30 hours after birth.

3.1.1.1 Imaging manifestations on chest X-ray

3.1.1.1.1 Early stage On plain X-ray, the findings can be negative, and no abnormalities may be detected initially (Fig. A.8a). The missed diagnosis rate is high, and it may appear as bronchitis or bronchiolitis.

3.1.1.1.2 Progressive stage It may manifest as localized or nodular opacities in the lung fields, predominantly in the peripheral regions, without specific characteristics.

3.1.1.1.3 Severe stage (critical stage) Characterized by multiple diffuse consolidations in both lungs, even presenting as "white lung", with or without pleural effusion on one side [1,17].

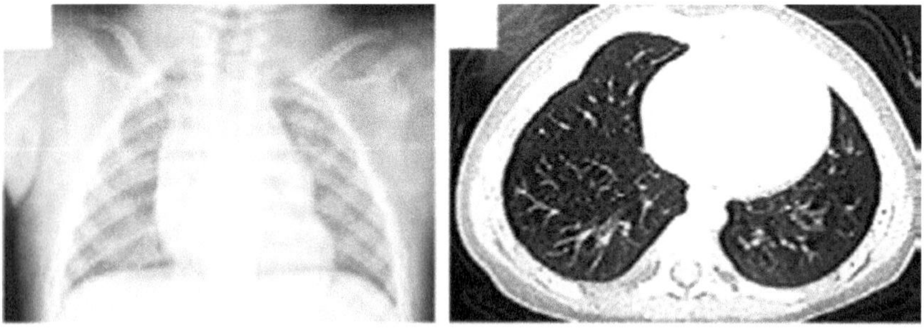

(a) (b)

Figure A.8 Imaging manifestations of infants and young children with mild COVID-19. The patient was a 1-year-old boy, presented with fever of 39°C without apparent cause 10 hours ago, accompanied by runny nose. There is a clear history of contact with a confirmed COVID-19 patient, and the SARS-CoV-2 nucleic acid test is positive. (a) Chest X-ray showed no abnormal changes; (b) CT scan revealed small ground-glass nodules.

3.1.1.1.4 Recovery stage The lesions show a reduction in size and absorption, with residual fibrous stripes. Chest CT Manifestations in Infants and Young Children: For infants and young children, CT scans are recommended to be performed using lower radiation doses according to the equipment, or low-dose CT scans are suggested, while also minimizing the number of follow-up scans whenever possible.

① Early Stage: In the early stage, the lungs may show single or multiple patchy or ground-glass opacities (GGO) located under the pleura with the center around the lobules (Fig. A.8b). The central pulmonary vessels may appear thickened and increased in number. The lung texture may present as a grid-like pattern (paving stone sign). The lesions are often distributed along the bronchovascular bundles or in the posterior and basal subpleural regions, and bronchial inflation signs may be visible.

② Progressive Stage: In the progressive stage, the extent of ground-glass opacities (GGO) increases, and consolidation may appear. The lesions often follow the course of bronchovascular bundles and may be associated with or without thickening of the interlobular septa. Pleural effusion is rare during this stage [16,18,19].

③ Severe Stage (Critical stage): In the severe stage, the lungs exhibit widespread ground-glass opacities (GGO) and multiple large areas of consolidation, often involving both lungs. The interlobular and interlobar pleura may also show thickening [16,18,19], usually with or without pleural effusion on one side.

④ Recovery Stage: The lesion range decreases and is absorbed compared with before, or only residual fibrous stripes may be visible.

3.2 Children and adolescents

3.2.1 Clinical features

Family clustering is the main route of infection in children [1,2]. Clinical manifestations can vary from asymptomatic cases [3] to those presenting with fever, fatigue, dry cough, and in some patients, nasal congestion, runny nose, and sore throat as upper respiratory symptoms, or even digestive symptoms as the initial presentation. In clinical practice, most cases are classified as mild or moderate, with severe and critical cases being rare [4,6,7]. A very small number of children may develop Multisystem Inflammatory Syndrome in Children (MIS-C), and their conditions can rapidly deteriorate in a short period of time [8,9,20].

3.2.2 Imaging manifestations chest X-ray manifestations

① In the early stage, it may appear negative or show increased and blurred lung markings, and decreased lung transparency.

② In the progressive stage, patchy shadows appear in the lungs, mostly located in the periphery of the lungs. Both lungs are often affected, but it can also be unilateral. The changes may also resemble bronchopneumonia.

③ In the severe stage, both lungs show multiple patchy consolidations (as seen in Fig. A.9), or even "white lungs", with unclear cardiac silhouette. Pleural effusion is rare in this stage [6,10].

④ In the recovery stage, the lesions show a reduction in size, clearer boundaries, and a decrease in density. Residual fibrotic streaks may be visible.

When necessary, CT examination is recommended, and it is suggested to lower the radiation dose or perform low-dose CT scanning based on the CT machine model.

3.2.3 Chest CT manifestations

① The lesions in the early stage are mainly located in the lung periphery and subpleural regions, relatively limited in extent, and with small lesion size. Ground-glass opacities (GGO) and patchy shadows are commonly observed. In cases showing features resembling bronchopneumonia,

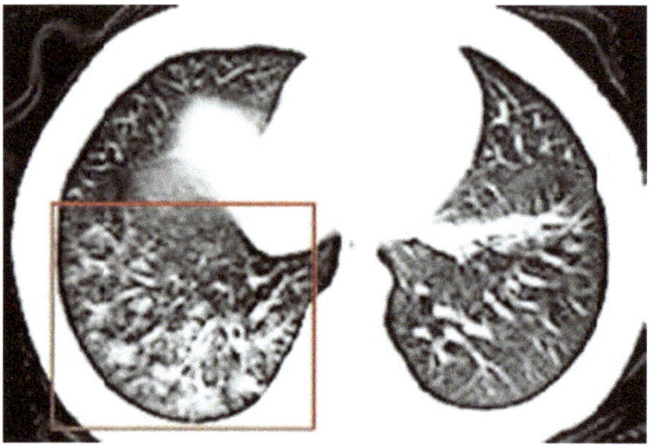

Figure A.9 Imaging manifestations of children with COVID-19. The patient was a 4-year-old male with a history of fever and cough for 5 days, and had a clear history of contact with a COVID-19 patient. White blood cell count 5.30 × $10^{9/L,}$ SARS–CoV–2 nucleic acid test positive CT scan showed multiple ground-glass opacities (GGO) in the right lower lobe and linear opacities in the left lower lobe.

there may be thickening and blurring of the bronchovascular bundles, along with patchy or nodular GGO distributed around the bronchovascular bundles with indistinct margins. Enlarged lymph nodes in the mediastinum and hilum, as well as pleural effusion, are less common in the early stage.

② In the progressive stage, the lung GGO density increases, and consolidation may occur with indistinct borders. Multiple lesions may merge into large areas of consolidation. Within these areas, bronchial inflation signs can be observed. The volume of the lesions may increase, and the number of lesions may also rise. Lesions resembling bronchopneumonia may progress along the bronchovascular bundles. Enlarged lymph nodes in the mediastinum and hilum, as well as pleural effusion, are still less common.

③ In the severe stage, there are multiple scattered patchy consolidations in both lungs, and ground-glass opacities (GGO) may be seen at the periphery of the consolidations. Within the consolidations, bronchial inflation signs can be observed. Almost the entire lung fields may be affected, and pleural effusion is less common.

④ In the recovery stage, the consolidations in the lungs start to resolve, with decreased density. The consolidations may show a decrease in size, with their edges becoming concave and well-defined. Some residual fibrous streaks may be visible. The imaging presentation of mild cases in children can be negative. In patients with MIS-C and heart failure, cardiac enlargement and pulmonary edema may be observed.

3.3 Pregnant women

3.3.1 Clinical features

Pregnant women during viral respiratory infections exhibit significantly increased inflammatory responses and rapid disease progression, which can easily evolve into severe cases, especially in the middle and late stages of pregnancy.

3.3.2 Chest CT manifestations

① Early stage: Nodular and patchy ground-glass opacities (GGO) appear outside the lungs or beneath the pleura, with relatively clear boundaries. Thickened pulmonary vessels and fine reticular patterns are also observed.

② Progressive Stage: Compared with the early stage, the lesions are significantly increased in number and show higher density consolidation with more prominent internal reticular patterns. The boundaries of the lesions are less clear, and some lesions may merge into large consolidated areas. Within the consolidations, signs of bronchial inflation can be observed (see Fig. A.10).

③ Severe Stage (Critical case): The extent of the lesions further expands, involving both lungs with large consolidated areas. Multiple lesions are present with relatively clear boundaries. Signs of bronchial inflation can still be observed within the consolidations, but pleural effusion and enlarged mediastinal lymph nodes are rarely seen.

④ Recovery Stage: As the lesions improve, the ground-glass opacities (GGO) and consolidations gradually get absorbed.

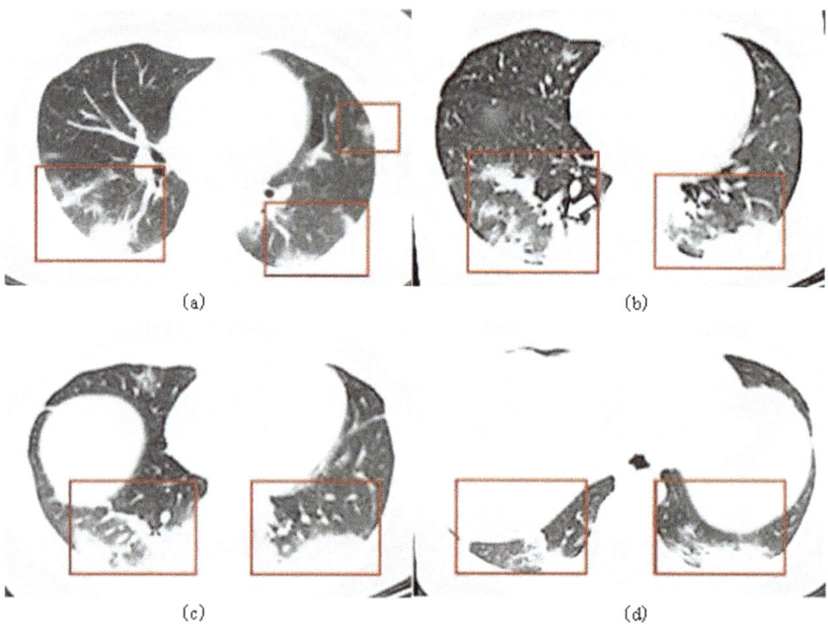

Figure A.10 CT manifestations of COVID-19 in progressive stage. the pregnant woman, 26 years old, at 29 weeks of gestation, presented with fever, with the highest body temperature reaching 38.5°C, and tested positive for SARS-CoV-2 nucleic acid. (a)–(d) CT showed multiple patchy consolidations in both lungs, with signs of bronchial inflation observed within. Additionally, there is ground-glass opacity (GGO) in the right middle lobe.

3.4 The elderly

3.4.1 Clinical features

Elderly patients experience rapid disease progression, severe conditions, and high mortality rates after infection. As of now, more than 80% of the reported fatalities are among individuals aged 60 and above, with few cases being classified as mild.

3.4.2 Imaging manifestations chest X-ray manifestations

① Early stage: There are generally no abnormal changes observed in both lungs, or it may show increased and blurred lung markings or localized patchy shadows in the lung fields.

② Progressive Stage: Within a short period (1−3 days), rapid progression can be observed, characterized by increased, thickened, and disorganized lung markings. There will be single or multiple patchy areas of increased density in the lung fields, commonly seen in the periphery of both lower lung fields.

③ Severe Stage (Critical case): There is a significant increase in lung lesions, with multiple patchy and large consolidated shadows. Alternatively, it may manifest as a widespread decrease in lung transparency, with diffuse multiple areas of consolidation.

④ Recovery Stage: The extent of the lesions shrinks, leaving behind residual fibrous stripes on the images.

3.4.3 Chest CT manifestations

① Early Stage: In unilateral or both lungs, there are patchy ground-glass opacities (GGO) (Fig. A.11a) or small nodular shadows. In some cases, there may be localized fine reticular patterns within a relatively faint GGO background. The lesions are predominantly located in the peripheral subpleural regions of the lungs [15].

② Progressive Stage: The lesions progress and change rapidly (within 1−3 days). The extent of lung involvement increases, affecting multiple lung lobes, spreading from the periphery to the central areas. The density of ground-glass opacities (GGO) intensifies (Fig. A.11b), with signs of bronchial inflation and thickened blood vessels observed within some lesions. Significant thickening of interlobular septa is evident, and there is a notable increase in lung consolidations, often accompanied by fibrous stripes. A "halo sign" may appear around the nodules. Newly developed lesions are mainly distributed in the subpleural regions of the middle and lower lobes of both lungs, and typically, there is no pleural effusion.

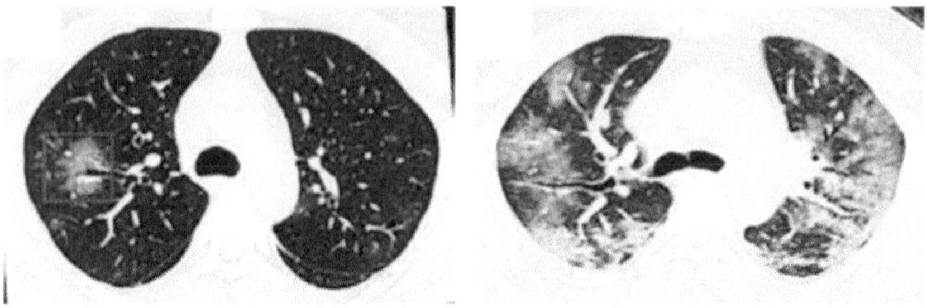

Figure A.11 CT manifestations of the elderly with COVID-19. The patient was a 62-year-old male and presented with fever, fatigue, and dry cough, which appeared 1 day ago without any apparent inducement and body temperature of 39.1°C. Chest imaging revealed patchy ground-glass opacities (GGOs) with thickened blood vessels in the right upper lobe (a); Five days later, the lesions significantly increased in number and showed higher GGO density. The lesions were primarily distributed in the peripheral subpleural regions of the lungs, with progression from the lung periphery toward the central areas [Figure (b)].

③ Severe Stage (Critical case): Within 48 hours, the lesions increase by more than 50%. The lungs show widespread or diffuse consolidations as the predominant feature, with signs of bronchial inflation. In some cases, it may exhibit a "white lung" appearance. A small proportion of cases may exhibit a small amount of pleural effusion on one side or both sides. Recovery Stage: The lung lesions gradually absorb, and the extent of the lesions shrinks. Residual fibrous stripes may be visible, and in some cases, there may be signs of bronchiectasis [2].

3.5 Patients with underlying diseases

3.5.1 Clinical features

Approximately 25% of COVID-19 patients have at least one underlying disease [1]. The most common comorbidities are hypertension and diabetes, followed by cardiovascular and respiratory diseases. A smaller proportion may have chronic liver disease, chronic kidney disease, malignant tumors, and other conditions. Individuals with underlying diseases often have compromised immune systems, leading to rapid disease progression, poor prognosis, and a higher mortality rate after infection [3]. In the clinical classification, the common types are moderate, severe, and critical.

3.5.2 Imaging manifestations

3.5.2.1 Chest X-ray manifestations

① Early Stage: The imaging changes may be negative or caused by underlying medical conditions, such as interstitial changes in the lungs, pulmonary cysts, etc. [3].

② Progressive Stage: Within a short period (1–3 days), rapid progression can be observed, characterized by increased, thickened, and disorganized lung markings, forming a reticular or honeycomb pattern. The lower lobes of both lungs are commonly affected [3].

③ Severe Stage: On the basis of diffuse interstitial changes in both lungs, there are also patchy and consolidative opacities interspersed [3,9].

④ Recovery Stage: The extent of the lesions shrinks, and their density decreases. Residual fibrous stripes may be present [9].

3.5.2.2 Chest CT manifestations

① Early Stage: Manifested by thickening of the interlobular and intralobular septa, with the latter being more predominant. This may be accompanied by ground-glass opacities (GGO) and consolidative opacities with a mixed appearance of ground-glass and consolidation. The lesions are commonly distributed in the peripheral regions and lower lobes of both lungs [3,9]. It may be accompanied by fibrous stripes. Some lesions may have atypical presentations, such as reticular shadows or single, tiny ground-glass opacities (GGO) [9]. A few small bubbles may be present within the GGO, showing a "bubble sign".

② Progressive Stage: Interstitial changes become evident. The extent of lung lesions increases, advancing from the periphery towards the central areas [3,6–8,20], involving more than three lung lobes. It may be accompanied by a small amount of pleural effusion.

③ Severe Stage: Diffuse reticular thickening of the interlobular septa in both lungs, with a honeycomb appearance when consolidations are present within the lobules. There are often scattered patchy and consolidative opacities, and signs of bronchial inflation can be seen within the consolidations. It may be accompanied by pleural effusion on one side or both sides [3,9], and in some cases, there may be enlargement of mediastinal lymph nodes.

Recovery Stage: The lesions gradually get absorbed [3,9], and residual fibrous stripes may be visible [3,9]. Some fine bronchi may undergo traction and dilation [20].

3.6 Asymptomatic infected persons

3.6.1 Definition of asymptomatic infected persons

Asymptomatic infected persons refer to those who have no related clinical symptoms, such as fever, cough, sore throat and other self-perceptible or clinically recognizable symptoms and signs, but with positive nucleic acid test or Serum-specific IgM antibody test of in respiratory tract and other samples [1].

3.6.2 Clinical characteristics of asymptomatic infected persons

Asymptomatic infected persons can be classified into two situations: First. The infected individual tests positive for nucleic acid but remains asymptomatic throughout the 14-day incubation period, showing no self-perceived or clinically recognizable symptoms or signs. They continue to be in an asymptomatic infected state; Second. The infected individual tests positive for nucleic acid without any self-perceived or clinically recognizable symptoms or signs at the time of sampling. However, they subsequently develop certain clinical manifestations, indicating a state of "asymptomatic infected persons" during the incubation period. Research has shown that asymptomatic infected persons account for approximately 17.9% of SARS-CoV-2 infections and have a certain level of contagiousness [2,5]. Among adolescents and pregnant women, the proportion of asymptomatic infected persons is relatively high, while it is lower among the elderly. For individuals initially diagnosed as asymptomatic infected persons, approximately 77% of them develop into clinically confirmed cases during subsequent observation [6].

3.6.3 Imaging manifestations

(1) Imaging manifestations can be negative, accounting for 29.2%—34.6% of cases, while abnormal CT manifestations are present in 60%—62.2% of cases [21].

(2) Asymptomatic infected persons, on chest CT, may show patchy ground-glass opacities (GGO) or GGO with consolidative opacities, distributed in the lung periphery and subpleural regions. The lesions can be absorbed within a short period, and in some patients, the lesions may be completely absorbed.

(3) Patients transitioning from asymptomatic infected persons to clinically confirmed cases mostly exhibit imaging features that follow the developmental pattern of COVID-19, characterized by early stage - progressive stage - recovery stage imaging manifestations:

① Early Stage: The imaging may show patchy or nodular ground-glass opacities (GGO), mainly distributed in the lung periphery or subpleural regions, involving fewer lung lobes. Within the lesions, thickened blood vessels may be visible, with or without thickening of the interlobular septa;

② Progressive Stage: The extent of lung involvement increases, advancing from the lung periphery towards the center of lung lobes. The density of the lesions can change from ground-glass opacities to consolidations, and consolidations may become more predominant. Signs of bronchial inflation can be observed within the consolidations. Concurrently, there is thickening of the interlobular and intralobular septa, resembling a "paving stone" appearance;

③ Recovery Stage: The extent of the lesions shrinks, and their density decreases. The number of lesions decreases, and in some cases, only a small amount of residual fibrous stripes may be present. Some lesions may be completely absorbed.

In a minority of patients, the chest CT manifestations do not follow the dynamic developmental pattern. In these cases, the initial chest CT may show the most severe manifestations, but in subsequent follow-up scans, the chest CT manifestations gradually improve, and the lung lesions are absorbed; In some patients, the progression and absorption of lung lesions coexist, meaning that while some lesions are resolving, new lesions may appear in other areas or the extent of existing lesions may increase [5].

3.7 Re-positive cases of SARS-CoV-2 nucleic acid test

3.7.1 Clinical manifestations

Clinical Manifestations Recovered COVID-19 patients may experience repositive results of nucleic acid test upon discharge. The duration of the positive NAT results can vary, and it may occur either during the patient's stay at a designated isolation point or after they have returned home from isolation. In most cases, repositive cases do not exhibit significant discomfort, or they may only experience mild symptoms such as cough, dry throat, or sore throat. Laboratory tests show that WBC (white blood cell count), CRP (C-reactive protein), PCT (procalcitonin), LAC (lactic acid), and LDH (lactate dehydrogenase), often return to normal levels, indicating that patients with re-positive NAT are still in the stage of continuous recovery [1]. The duration of hospitalization for patients with re-positive NAT is typically shorter than their initial hospitalization period.

3.7.2 Imaging manifestations

In terms of imaging manifestations, the majority of patients with re-positive NAT show lesion absorption, improvement, or no significant changes on CT scans. Instances of worsening CT manifestations are rare. Chest CT can show the coexistence of multiple types of lesions or may present as a single lesion. Ground-glass opacities (GGO) can be observed in the lungs, appearing as small patchy or nodular lesions. The small bronchi exhibit traction dilation, with increased and thickened bronchovascular bundles. There is also thickening of intralobular septa, along with the presence of fibrous stripes. Thickening of the pleura and pleural effusion, etc. For cases with worsening lesions after re-positive NAT, the CT manifestations show an increase in the number and thickening of the bronchovascular bundles in both lungs, accompanied by decreased lung transparency, demonstrating mild ground-glass opacities.

3.8 Imaging differential diagnosis

3.8.1 Distinguishing from other viral pneumonias

The imaging manifestations of viral pneumonia predominantly show interstitial changes with accompanying alveolar wall edema. On CT scans, the manifestations include ground-glass opacities (GGO), consolidations, thickening of the interlobular septa, reticular shadows, centrilobular nodules, tree-in-bud sign, air trapping, and fibrous stripes, etc. The imaging manifestations of different viral pneumonias are similar. The final diagnosis requires the integration of clinical data, epidemiology, and laboratory results. Definitive diagnosis relies on pathogen detection.

In cases of H1N1 influenza virus pneumonia, the chest CT may show unilateral or bilateral focal or multiple ground-glass opacities (GGO), with or without consolidations. These opacities are distributed along the bronchovascular bundles or subpleurally [17,22]. In early stages of COVID-19, it may present as small GGO or patchy GGO with visible thickened blood vessels, which may aid in distinguishing it from early-stage H1N1 lesions. However, definitive differentiation requires pathogen detection.

Avian influenza virus pneumonia (H7N9) may show bilateral ground-glass opacities (GGOs) or consolidations in both lungs, along with signs of bronchial inflation. In the early stages of the disease, lesions in avian influenza virus pneumonia (H7N9) can occur simultaneously in the central and peripheral regions of the lungs, and pleural effusion is relatively common. This feature can help differentiate it to some extent from COVID-19 [23].

In severe acute respiratory syndrome (SARS), chest CT may show unilateral or bilateral ground-glass opacities (GGO), localized unilateral or bilateral consolidations, or a combination of both. Ground-glass opacities (GGO) is accompanied by thickening of the interlobular septa and the "paving stone sign". The lung lesions are predominantly distributed in the middle and lower lung fields. The disease progresses rapidly, quickly advancing to multifocal consolidations, with "white lung" being common in severe cases. Rarely seen are cavities, calcifications, reticular or nodular opacities, and enlargement of lymph nodes and pleural effusion [17,22]. The causative viruses belong to the coronavirus family, and both have highly similar pathogenic mechanisms and imaging characteristics. It is challenging to distinguish them solely based on imaging manifestations.

Human parainfluenza virus pneumonia is a common cause of seasonal respiratory tract infections. The imaging manifestations of human parainfluenza virus pneumonia are diverse and may include multiple peribronchial small nodules, ground-glass opacities (GGO), and consolidations containing inflated bronchi [17,22]. The lesions can have a central distribution, which is different from the characteristic subpleural distribution seen in COVID-19.

Adenovirus pneumonia is common in children. It presents as multifocal ground-glass opacities (GGO) with patchy consolidations in both lungs, and may show a tendency for segmental distribution within lung lobes. In children, it can lead to lung atelectasis. Sometimes it is difficult to distinguish from bacterial pneumonia [17,22].

Respiratory syncytial virus pneumonia commonly occurs in infants, young children, individuals with congenital defects, immunocompromised individuals, and those with chronic lung diseases. The most characteristic manifestation is centrilobular nodules, which can be observed in up to 50% of cases, making it helpful for distinguishing from COVID-19; Additionally, other manifestations may include consolidations with air bronchograms (35%), ground-glass opacities (30%), and thickened bronchial walls (30%), distributed in the central or peripheral areas of the lungs with a bilateral asymmetric distribution [17,22].

3.8.2 Distinguishing from nonviral infectious pneumonia

Mycoplasma pneumoniae pneumonia is common in children and adolescents, which presents with centrilobular nodules, GGO, and consolidations. Other manifestations may include thickened bronchial walls, tree-in-bud sign in the small bronchioles, and enlargement of hilar and mediastinal lymph nodes [24]. Laboratory tests show positive mycoplasma antibodies.

Bacterial pneumonia often lacks preceding symptoms of upper respiratory tract infection. Common clinical features include productive cough with purulent, bloody, or rust-colored sputum. Laboratory tests typically show increased white blood cell count. Imaging manifestations mostly consist of homogeneous consolidation in a single lobe or subsegment. Antibiotic treatment is generally effective.

3.8.3 Differentiation from noninfectious lung lesions

Cryptogenic Organizing Pneumonia The typical manifestation of cryptogenic organizing pneumonia (COP) includes bilateral subpleural patchy ground-glass opacities (GGO) or consolidations, with bronchial inflation signs within. Some lesions may show central GGO, peripheral ring-like or crescentic consolidations, exhibiting a "reversed halo sign" appearance. The disease may have migratory features, and in some cases, there can be enlargement of hilar and mediastinal lymph nodes and pleural effusion [23].

Acute Eosinophilic Pneumonia Diffuse ground-glass opacities (GGOs) and micronodular infiltrations (with scattered or localized distribution in mild cases), with minimal pleural effusion. There is a significant increase in eosinophils in peripheral blood or bronchoalveolar lavage fluid.

Hypersensitivity Pneumonitis Bilateral patchy or diffuse ground-glass opacities (GGOs), hazy centrilobular nodules, mosaic perfusion, and expiratory air trapping. In the chronic stage, fine reticular opacities and traction bronchiectasis are observed in the lung fields. Be a bird breeder or have a history of occupational exposure [25].

Vasculitis often presents with multiple nodules accompanied by cavitation, nodules connected to the pulmonary vessels (nutritional vessel sign), halo or reverse halo signs, multiple consolidations, reticular opacities, and diffuse distribution of ground-glass opacities (GGO). Pleural involvement is uncommon and usually seen in the middle and inner lung zones. Diffuse alveolar hemorrhage can occur, leading to clinical manifestations of hemoptysis, and pleural effusion is common [26]. A positive cANCA (antineutrophil cytoplasmic antibody) in laboratory tests is helpful for the diagnosis.

3.8.4 Value of imaging examinations in diagnosis and treatment of COVID-19

The Diagnosis and Treatment Protocol for Novel Coronavirus Pneumonia (Trial Fifth Edition) proposes using the detection of pneumonia features on chest imaging as a criterion for clinical diagnosis of COVID-19 in Hubei Province. Subsequent releases of the Diagnosis and Treatment Protocol for Novel Coronavirus Pneumonia (Trial Sixth to Eighth Edition) eliminate the differentiation between Hubei Province and other provinces, and unify the classification into two categories: "suspected cases" and "confirmed cases".

Chest imaging is primarily performed using CT scans, with chest X-rays as a supplementary method. The misdiagnosis rate of chest X-rays is over 50%. On thin-layer CT scans, ground-glass opacity (GGO) is the most common manifestation in COVID-19, and as the disease progresses, other changes may appear. The respective occurrence rates are as follows: GGO 85.49%, GGO with mixed consolidation 58.42%, thickening of adjacent pleura 52.46%, thickening of interlobular septa 48.46%, air bronchogram sign 46.46%, paving stone sign 14.81%, pleural effusion 5.88%, bronchial dilation 5.42%, pericardial effusion 4.55%, and lymphadenopathy 3.38%. The distribution of lesions is mainly multilobar, with the following occurrence rates: bilateral lungs 81.8%, peripheral regions of the lungs 76.95%, and peribronchovascular regions 10.81%; right lower lobe 87.21%, left lower lobe 81.41%, bilateral lower lobes 65.22%, right upper lobe 65.22%, right middle lobe 54.95%, and left upper lobe 69.43%; bilateral upper lobes 60.87%, three or more lung lobes 70.81%, four lung lobes 20.51%, and all lung lobes 39.54%.

CT, especially HRCT, can detect pulmonary abnormalities in the early stage, and its characteristic CT manifestations are one of the three clinical features of suspected cases. It is of great significance for screening suspected cases and can sensitively evaluate changes in the disease after treatment, offering the advantages of being timely, convenient, and efficient.

In the clinical classification of COVID-19, mild cases show no pulmonary abnormalities on imaging, while common cases exhibit pneumonia manifestations; Progressive worsening of clinical symptoms with significant lesion progression (>50%) on lung imaging within 24–48 hours is one of the diagnostic criteria for severe pneumonia. In adults, significant progression of pulmonary lesions on chest imaging, and in children, bilateral or multilobar infiltrates, pleural effusion, or rapid lesion progression within a short period are early warning indicators for severe/critical pneumonia.

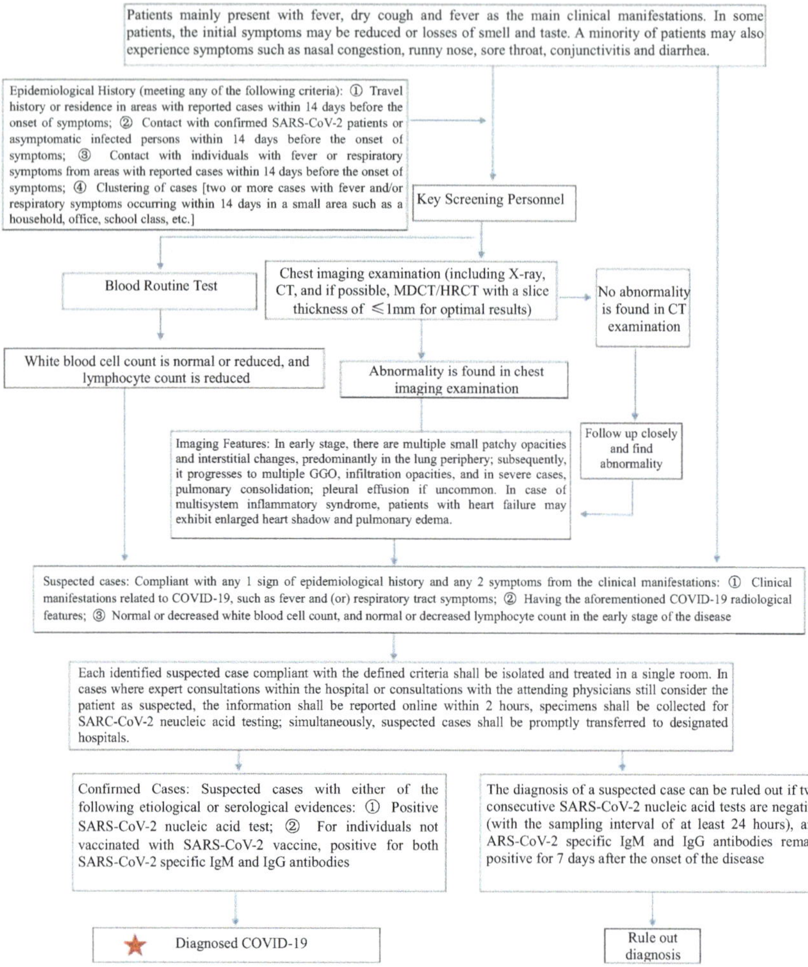

Figure A.12 Flowchart for diagnosing COVID-19.

Patients who have a body temperature returning to normal for more than 3 days, significantly improved respiratory symptoms, evident improvement of acute exudative lesions in lung imaging, and two consecutive negative respiratory specimens for nucleic acid testing (with at least 24 hours apart in sampling time) are eligible for discharge (Fig. A.12).

3.9 AI-assisted COVID-19 diagnosis

Artificial Intelligence (AI) enables CT to achieve automatic human body three-dimensional modeling, scan anatomy positioning, bed and transfer movements, and intelligent determination of scanning dosage. Technicians can perform the entire scanning process without entering the operating room.

Precise and early screening AI technology can quickly accomplish automatic lesion detection, segmentation, accurate quantification, and automated generation of standardized reports for CT or DR. This technology helps assist radiologists in promptly identifying hidden lesions in patients with inconspicuous symptoms or negative nucleic acid tests, thereby improving diagnostic accuracy [1].

Quantitative assessment AI technology enables comprehensive quantification of individual lesions, lobar lesions, and overall lung lesions [2].

Intelligent follow-up AI technology enables quantitative comparative analysis of lesion count, lesion volume changes, and lesion density changes in multiphase images. It helps assess the severity of pneumonia, predict disease progression, and particularly forecast severe and critical cases.

The differential diagnosis AI technology utilizes the distribution, morphology, density, and other features of lesions to provide auxiliary and differential diagnosis for the detected abnormalities [5]

4. Implementation, promotion and updating of COVID-19 imaging diagnostic guidelines

4.1 Implementation and promotion of the COVID-19 imaging diagnostic guidelines involve the estimation of favorable and unfavorable factors

4.1.1 Favorable factors

① The widespread and in-depth adoption of evidence-based medicine principles among Chinese healthcare professionals has increased the objective demand for high-quality evidence-based guidelines

② Pneumonia is the most common reason for patient visits, significantly impacting their quality of life and even posing life-threatening risks. It also adds to the social and economic burdens. Currently, there is a strong clinical demand for diagnostic guidelines to curb the spread of COVID-19.

4.1.2 Adverse factors

① COVID-19 is a newly emerging infectious disease with limited understanding. Due to differences in the interpretation of the importance and recommendations of the first edition of the imaging diagnostic guidelines among clinical doctors in different medical institutions, comprehensive promotion, publicity, and implementation of this guideline may require time;

② Some units have not yet started HRCT examinations, real-time fluorescent RT-PCR testing of respiratory samples or blood samples, virus gene sequencing, and other tests. These limitations in resources may have certain impacts on the application and promotion of this guideline.

Update Plan: The plan is to update the COVID-19 imaging diagnostic guidelines based on clinical practice conditions.

5. Appendix A: Supplement of diagnostic guidelines for COVID-19 in children, infants and young children

A1 Epidemiology

Referring to the Diagnosis and Treatment Protocol for Novel Coronavirus Pneumonia (Trial Sixth Edition) released by the National Health Commission [27], the diagnosis of COVID-19 in children and infants focuses on whether they have a history of close contact with family members who have been diagnosed with the disease. There have been individual cases where family members have not been confirmed as infected.

A2 Clinical Manifestations

Most children and infants show mild clinical manifestations, including symptoms such as fever, fatigue, dry cough, nausea, vomiting, cough, and sneezing, which are typical of respiratory tract infections.[28] In some cases, no respiratory tract infection symptoms were found.

A3 Diagnostic Criteria

A3.1 Refer to the relevant criteria in the Diagnosis and Treatment Protocol for Novel Coronavirus Pneumonia (Trial Fifth Edition) issued by the National Health Commission.

A3.2 In the diagnostic criteria for suspected cases, it is recommended to emphasize the epidemiological aspect, particularly the close contact of affected children or infants with family members who are confirmed cases. These family members may include parents, siblings, and there have also been reports of familial cluster infections. In the diagnostic criteria for suspected cases, clinical manifestations need to meet any two of the three diagnostic conditions. Among them, imaging is one of the important examination methods. Due to the radiation protection requirements for lung examination in children and infants, X-ray chest radiography is recommended as the primary imaging method. If interstitial pneumonia specific to children is detected in the lungs, this diagnostic indicator shall be considered positive. X-ray chest radiography has lower sensitivity for diagnosing COVID-19, which may lead to missed diagnoses. Therefore, when necessary for definitive diagnosis, low-dose chest CT examination is recommended, and there are reports of ground-glass opacities (GGO) appearing in lung CT scans [15].

A3.3 The clinical classification of affected children and infants is mostly common type, and currently there are no relevant reports of severe and critical types.

A4 Imaging Manifestations and Clinical Significance

Chest X-ray in children with COVID-19 may show increased and thickened lung markings, disordered lung parenchymal structures, and some distortion, mainly in the perihilar regions. Thickening of the bronchial walls around the hilum, referred to as the "cuff sign," can be observed, but there is no enlargement of the pulmonary hilum. The lung transparency is uneven, and uneven lung emphysema can lead to irregular depression of the diaphragmatic surface. The above X-ray manifestations may overlap with other common viral infections in children, such as respiratory syncytial virus (RSV) infection and mycoplasma infection.

In the fourth edition of the diagnosis and treatment protocol, the contents related to the clinical diagnosis of severe cases based on the progression of imaging manifestations have been removed. At the same time, the criteria for the removal of isolation or discharge of patients, based on significant absorption of pulmonary imaging manifestations, have been removed. There is currently insufficient evidence in pediatric and infant patients to demonstrate the value of imaging manifestations in determining the release from isolation or discharge of patients.

According to the introduction of the sixth edition of the diagnosis and treatment protocol, for patients with an epidemiological association with COVID-19, even if common respiratory pathogen tests are positive, it is recommended to promptly conduct novel coronavirus pathogen testing. Common pathogens causing community-acquired pneumonia in children include respiratory syncytial virus, adenovirus, Mycoplasma pneumoniae, Streptococcus pneumoniae, etc. [29]. Common viral infections in children can manifest as lobular pneumonia or interstitial pneumonia. Mycoplasma pneumoniae infection can also present as interstitial pneumonia or other types of pneumonia. Community-acquired pneumonia in children can be caused by mixed infections of pathogens, and the younger the age is, the more likely the mixed infections may occur. If there is a clear epidemiological history, even if the clinical and radiological manifestations are suggestive or laboratory tests confirm infection by other pathogens, the possibility of COVID-19 mixed infection cannot be completely ruled out [30].

6. Appendix B: Imaging examination protection and technical parameters

B1 X-ray Examinations

B1.1 General X-ray Examinations

B1.1.1 Preparation for Independent DR Examination in Fever Clinic.

B1.1.1.1 Preparation before Reception by Radiographer

Priority shall be given to selecting a model with a console equipped with a lift detector, and using an amplifier to communicate with the patient for proper positioning. If the above conditions cannot be met, the technologist shall enter the control room (maintaining a distance of at least 1m away from the patient) to adjust the patient in a posteroanterior position; To avoid cross-infection, a disposable cover or film shall be placed on the detector, ensuring one person uses it once and then replaces it with a new one for the next person. Electric radiation protection equipment racks shall be used preferentially (which can be adjusted in conjunction with the detector to position the radiation protection equipment), followed by fixed protection racks. The personal protective measures for the operating technologist shall follow the corresponding level in Appendix C.

B1.1.1.2 Preparation by Patients

The patient shall wear an N95 mask throughout the examination process. Before entering the examination room, the patient shall disinfect their hands with hand sanitizer or wear disposable gloves, remove clothing with zippers, buttons, paint patterns, or any other elements that may affect image quality and, if necessary, wear specialized patient gowns. Accompanying personnel shall wear N95 masks and disinfect their hands with hand sanitizer or wear disposable gloves before entering the examination room, and wait outside the examination room once the patient is ready. If the patient requires assistance during the examination and the accompanying personnel need to help, they shall be provided with radiation protection equipment (lead apron, lead cap, and lead collar).

B1.1.1.3 Disinfection after Examination

Air disinfection shall be performed at fixed times in the morning, afternoon, and evening using an air disinfection machine (in accordance with Standards for the Medical Air Purifier) or ultraviolet disinfection for 60 minutes. Additionally, windows shall be opened for ventilation for at least 30 minutes. Use a 2000 mg/L concentration of chlorine-containing disinfectant to wipe the floor at least twice a day. When there is a large number of patients, the disinfection frequency can be increased. Wipe the examination bed, operation table keyboard, and mouse with 75% ethanol after each patient examination.

B1.1.2 DR Technical Parameters

B1.1.2.1 Photographic Distance: 180 cm.

B1.1.2.2 Filter Grid: Available (grid ratio of at least 10:1).

B1.1.2.3 Exposure Conditions: High kilovolt photography at 120–125 kW, using automatic exposure techniques.

B1.1.2.4 Positioning: Posteroanterior (PA) position, the patient stands (or sits) facing the imaging rack, with the central axis of the body aligned with the central line of the detector. Spread the legs apart, relax the shoulders downwards, and grip the handles on both sides or hold onto the detector. Make head positioned straight, with the chin slightly lifted. Let the front chest as close as possible to the detector, with the upper edge of the detector extending 5 cm beyond both shoulders, including the outer sides of the acromioclavicular joints. Align the central line to the sixth thoracic vertebra for vertical projection.

B1.1.2.5 Respiratory Requirement: Take a deep breath and hold it for exposure.

B1.2 Bedside DR Technical Parameters

B1.2.1 Photographic Distance: about 100 cm.

B1.2.2 Filter Grid: Fixed filter grids can be used for output power greater than 23 kW; no filter grid is used for output power less than 23 kW.

B1.2.3 Exposure Conditions: Use a filter grid: 75–85 kW; Not use a filter grid: 65–70 kW. The milliampere-seconds are automatically adjusted by the machine according to the patient's body size.

B1.2.4 Positioning: If the patient's condition permits, the preferred position is standing or sitting for the posterior-anterior (PA) projection, with both hands holding the detector. If the patient's condition is critical, a supine position is adopted, and the detector is placed close to the patient's back, with the centerline aligned vertically to the midpoint between the nipples. The detector angle and the incident angle are adjusted according to the specific situation.

B1.2.5 Respiratory Requirement: Take a deep breath and hold it for exposure.

B1.2.6 Protection requirements: Lead apron for gonadal protection, lead thyroid shield for thyroid protection.

B1.2.7 Precautions for Using Fixed Filter Grid: The correct orientation of the filter grid shall be ensured, and the distance of exposure shall be controlled within ±20% of the grid focal distance. The central line shall be perpendicular to the imaging receptor, as it may otherwise affect the image quality.

B2 CT Examination

B2.1 Preparation before Reception by Radiographer

B2.1.1 Select 1 CT scanner exclusively for suspected or confirmed cases. It is preferable to choose a CT machine with a console that allows remote scanning bed movement. If this functionality is not available, two technologists can be arranged to work together, with one operating the machine and the other positioning in the control room (following the corresponding protection level as described in Appendix C).

B2.1.1 Whenever possible, select machines with independent operating rooms to avoid sharing operating rooms with other machines. If it is not feasible to meet the aforementioned condition, after examining each patient, standard air disinfection shall be performed for other connected machine rooms to the operating room.

B2.1.3 In computer rooms equipped with a fresh air system and central air conditioning, the supply and exhaust air volume of the air conditioning shall be set to the maximum. For computer rooms equipped with regular central air conditioning, it is recommended to turn off the central air conditioning in the computer room and operating area, and switch to standby independent air conditioning. If there is no standby independent air conditioning, the central air conditioning shall be turned on after completing the inspection and disinfection.

B2.1.4 Cover the examination beds with disposable bed sheets, isolate the protective equipment from the patient using disposable bed sheets, change the bed sheet for each patient, and after the examination is completed, wipe the examination bed with 75% ethanol. But it is not recommended to use 75% ethanol to wipe important machine components to prevent water from seeping in and damaging the equipment.

B2.2 Patient preparation same as regular X-ray examination.

B2.3 Position Scanning and Breath-holding Exercises

Generally, take the supine position, with the body lying flat in the middle of the examination bed and the arms raised. For individuals who have difficulty raising their arms, they can be positioned by placing their arms on both sides of the body. Before the scan, conduct respiratory training with the patient and instruct them to cooperate with the examination according to the respiratory instructions. Typically, perform the end-inspiratory hold scan. For severe and critical patients, priority is given to ensuring breath-holding.

B2.4 Scanning Range and Direction

Scanning Range: Scan from the lung apices to the posterior costophrenic angles, ensuring that the entire lung tissue is within the scanning range.

Scanning Direction: For general patients, scan from the lung apices to the lung bases. For severe and critical patients, scan from the lung bases to the lung apices to reduce respiratory motion artifacts caused by breath-holding difficulties in the lower lung fields and ensure image quality.

B2.5 Selection of Scanning Parameters

The helical scanning mode is used with automatic tube voltage or tube voltage set at 120 kV, employing intelligent current ranging from 50 to 350 mAs. Beam collimator: 0.5−5.0 mm * rows of X-ray detector; Slice thickness: ≤1.0 mm, tube rotation speed: 0.6−0.8 s/rotation, and thread pitch: 1.0 to 1.3. It is recommended to use iterative reconstruction technique. For severe and critical patients, priority can be given to reducing the scan time by using a large pitch (1.5 or 1.7), shortening the tube rotation speed, and increasing the collimation width to adjust and minimize the patient's respiratory motion artifacts and ensure the image quality.

B2.6 Selection of Image Reconstruction Parameters

Use the raw data to reconstruct two sets of images using the lung algorithm, with the slice thickness of 5 mm and ≤1.0 mm (equivalent to the conventional HRCT), respectively; the reconstruction interval: 75%−100%; the window width: 1000 to 2000 Hu, and the window level: -700 to -500 Hu.

Use the soft tissue algorithm to reconstruct two sets of mediastinal window images, with the slice thickness of 5 mm and not greater than 1.0 mm, respectively. The reconstruction interval: 75%−100%, the window width: 300 to 400 Hu, and the window level: 30 to 50 Hu.

B2.7 CT Image Postprocessing

B2.7.1 Multiplanar Recombination (MPR)

Conventional CT images are only cross-sectional images. MPR is a postprocessing technique that recombines the calculated pixel data into coronal, sagittal and arbitrary orientation images. MPR allows for more accurate visualization of the relationship between lesions and blood vessels, pleura and chest wall, aiding in precise lesion localization. MPR can also provide a better display of the relationship between lesions and bronchi, which is conducive to the imaging diagnosis of COVID-19.

B2.7.2 Volume Rendering (VR)
The displayed anatomical structures are realistic, providing clear visualization of organ morphology and spatial relationships. It is primarily used for three-dimensional bronchial imaging, vascular imaging, and displaying complex anatomical relationships, helping to assess the patency of bronchi in COVID-19 patients and understand the spatial distribution of lung lesions.

B2.7.3 Maximum Intensity Projection (MIP)
MIP images can better display the long axis of blood vessels compared with cross-sectional images, thereby improving the three-dimensional orientation effect. This aids in distinguishing blood vessels from nodules, and enhances the detection rate of nodules. MIP can not only distinguish nodules from adjacent vascular branches, but also accurately display nodules around the lobular center and bronchovascular structures, which is of great significance for precise differential diagnosis in the later stages.

B2.7.4 Minimum Intensity Projection (MinIP)
Its principle is similar to MIP, but the reconstructed images highlight low-density structures, providing significant advantages in displaying airway density. MinIP is beneficial in displaying the pulmonary emphysema and mosaic sign resulting from the alveoli hyperinflation and the destruction of alveolar walls.

B2.7.5 CT Virtual Endoscopy Imaging (CTVE)
Reconstruct the structure relationship between the tracheal lumen and the wall based on the volume data obtained from the scan and observe the tracheal wall and lumen inside using a simulated endoscope method. CTVE can assist in accurately locating the bronchoscope before the procedure, greatly enhancing the diagnostic rate of transbronchial needle aspiration and biopsy (TBNA) performed during the bronchoscopy.

7. Appendix C: Guidelines for prevention and control of nosocomial infections during radiological examination of COVID-19

To strengthen the prevention and control of nosocomial infections (commonly referred to as "Nosocomial Infection") related to COVID-19, this prevention and control guideline is specifically developed to provide guidance to radiology staff on implementing proper disinfection, isolation, and examination procedures, aiming to ensure the protection of medical staff from nosocomial infections and effectively reduce the risk of transmission within the hospital.

C1 Requirements for Prevention and Control of Nosocomial Infection in Radiology Department

C1.1 Registration staff, nurses, technologists, and physicians shall clearly identify the responsible person and contact person for nosocomial infection prevention and control, along with their respective duties.

C1.2 Each responsible person for nosocomial infection prevention and control shall closely monitor the occurrence of infections among medical staff within the hospital, promptly contact the nosocomial infection control department, public health department, or disease control department and compile a daily summary of nosocomial infection cases for reporting to the respective authorities.

C2 Basic Requirements of Prevention and Control and Control Areas of Nosocomial Infection in Radiology Department

All radiology staff are required to implement standard preventive measures and take precautions such as droplet isolation, contact isolation, and airborne isolation to protect against infections. Area management shall be implemented based on different workplaces, such as the radiology registration room, examination room, control room, diagnostic room, rest area, meeting room, and duty room, especially emphasizing on maintaining good ventilation, strictly adhering to hand hygiene, and correctly selecting and wearing masks, etc. It is required to develop emergency plans and workflows, conduct comprehensive staff training, ensure proper protection for medical staff, strengthen infection surveillance, implement effective cleaning and disinfection management, enhance patient management and education, reinforce infection outbreak management, and improve medical waste management.

C2.1 Set up contaminated areas, potentially contaminated areas, and cleaning areas in the radiology department:

C2.1.1 Contaminated Areas: Registration room, Imaging (CT and X-ray examination) rooms.

C2.1.2 Potentially Contaminated Areas: Control room, Patient Passage.

C2.1.3 Cleaning Areas: Diagnostic room, Rest area, Duty room, Meeting room.

C3 Protection Level and Grouping Rules

C3.1 Protection Level

C3.1.1 Level 1 protection: Applicable to medical staff in pre-check triage, fever clinics, and outpatient of infectious disease department. Medical staff are required to wear disposable caps, disposable surgical masks (N95 protective masks for close contact with individuals with an epidemiological history), work clothes, isolation gowns (disposable isolation gowns if necessary during precheck triage), and disposable latex gloves when necessary, and strictly adhere to hand hygiene.

C3.1.2 Level 2 protection: Applicable to medical staff engaged in diagnostic and therapeutic activities involving close contact with suspected or confirmed patients. Medical staff are required to wear disposable caps, protective goggles or face shields (antifog type), N95 protective masks, protective suits or isolation gowns, disposable latex gloves, disposable shoe covers, and strictly adhere to hand hygiene.

C3.1.3 Level 3 protection: Applicable to medical staff performing aerosol-generating procedures on suspected or confirmed patients, such as sputum suction, respiratory sampling, tracheal intubation, and tracheostomy, which may involve respiratory secretions, body fluids, or splashes. Medical staff are required to wear disposable caps, N95 protective masks, protective face shield (or full-face respirators or positive-pressure hoods), protective suits, disposable latex gloves, shoe covers, and strictly adhere to hand hygiene practices.

C3.2 Protection Grouping

C3.2.1 Fever clinic technicians: Follow Level 2 or Level 3 protection and wear double-layer gloves.

C3.2.2 Bedside technicians: Follow Level 2 or Level 3 protection and wear double-layer gloves.

C3.2.3 Regular outpatient technicians: Technicians operating in the machine room shall follow Level 1 protection; technicians in the examination room shall follow Level 2 protection and wear double-layer gloves.

C3.2.4 Diagnostic physicians: During routine work, wear disposable surgical masks and white coats. For regular gastrointestinal examinations and allergy emergencies, follow Level 2 or Level 3 protection and wear double-layer gloves.

C3.2.5 Registration and other staff: Follow Level 2 protection.

C4 Disinfection of Radiation Rooms

C4.1 Imaging Equipment Cleaning and Disinfection Plan: First thoroughly clean the fever clinic and machine room equipment and only use soap and water to clean the machine surfaces. Do not use detergents or organic solvents to clean CT equipment because strong cleaning agents, ethanol, and organic cleaning agents may damage the surface gloss of the equipment and weaken its structural integrity. Perform the disinfection in accordance with the guidelines for intermediate and low-level disinfection as defined by the Centers for Disease Control and Prevention (CDC) (http://www.cdc.gov/hicpac/pdf/guidelines/disinfection_nov_2008.pdf).

C4.1.1 The following materials can be used for cleaning and disinfection:

① Spray cleaners or wipes with equivalent effectiveness to bleach (with concentration below 10%);
② Low-level or intermediate-level disinfectant wipes or liquids;
③ 3% hydrogen peroxide;
④ Ethanol;
⑤ Desulfurized liquefied petroleum gas;
⑥ Quaternary ammonium compounds;
⑦ Benzyl-C12-18-alkyldimethyl ammonium chloride;
⑧ Salt with 1,2-benzisothiazolin-3(2H)-keto-1, and 1-dioxide (1:1);
⑨ Distilled water;
⑩ Industrial alcohol.

C4.1.2 During the cleaning process of the equipment, pay attention to

① Shield the microphone when cleaning the front and back covers of the scanner to prevent cleaning agents from leaking in.

② Be very careful when cleaning the buttons and rack holes to prevent the cleaning liquid from penetrating into the interior of the rack.

③ Due to the potential health risks associated with blood and contrast agents, appropriate safety measures shall be taken when removing blood or residual contrast agents, including wearing appropriate protective equipment.

④ Spray-type disinfection tools are not recommended for use.

⑤ The use of flammable or explosive disinfectant sprays are strictly prohibited.

⑥ Upon disinfecting the machine room, it is not recommended to use high-concentration chlorine-containing disinfectants, especially at night, as they may form a film on the slip ring surface and interrupt scanning. Instead, consider using alternative disinfection methods or perform the disinfection during the daytime when the equipment is not in use.

C4.2 Disinfection of the Operating Console: It can be performed using a chlorine-containing disinfectant with a concentration of 250–500 mg/L.

C4.3 Floor Disinfection: The floors in the fever clinic and examination rooms shall be disinfected using a chlorine-containing disinfectant with a concentration of 2000 mg/L. For regular machine rooms, disinfection can be performed using a chlorine-containing disinfectant with a concentration of 250–500 mg/L. When visible contaminants are present, they shall be completely removed using disposable absorbent materials before disinfection. Disinfection shall be carried out at least twice daily. Disinfection shall be carried out immediately when contamination occurs.

C4.4 Terminal Disinfection Method: For examination rooms that have been used by suspected or confirmed patients, terminal disinfection shall be performed daily. Dispose of disposable bed sheets, and use disinfectant wipes containing chlorine or alcohol to wipe down disinfection machines, examination beds, and related items (with a disinfection time of 4 minutes). For the floor disinfection, use a chlorine-containing disinfectant with a concentration of 2500 mg/L. Close the machine room for 30–60 minutes during air disinfection. During the operation, continuous disinfection can be achieved by using a circulating air disinfection machine.

C4.5 Medical waste management: All waste from patients shall be considered as infectious medical waste and strictly managed according to the Regulations on the Administration of Medical Waste and the Measures for the Administration of Medical Waste in Medical and Health Institutions. The personal protective equipment used by staff who have been in contact with suspected or confirmed patients shall be directly discarded into medical waste bins after use, with double-layer sealing, clear labeling, and sealed transportation.

C5 Management of Patents to be Examined

C5.1 Before receiving the requisition form (electronic requisition form is recommended), ensure that the patient is wearing a mask, minimize conversation with the patient, and maintain a distance of at least 1.5 meters away from the patient.

C5.2 Set up a separate report collection area for fever patients (self-service printers are recommended), with clear signage and instructions, to avoid multiple inquiries and movements by patients.

C5.3 When examining multiple patients consecutively, adopt a reasonable examination order:

① Perform terminal disinfection after suspected patients have been examined.

② Perform terminal disinfection after confirmed patients have been examined.

③ When examining one suspected patient and one confirmed patient, perform the examination for the suspected patient first, then the confirmed patient, and finally conduct terminal disinfection.

C6 Training, Inspection, and Supervision

C6.1 Infection control personnel, quality control personnel, and the overall responsible person for the department are responsible for daily infection control inspections within the department. The department head, diagnostic team leader, and technical team leader are responsible for supervision.

C6.2 Infection control personnel shall conduct infection control training as needed and communicate the latest infection control arrangements.

8. Appendix D: Procedures for medical staff to don and doff protective equipment

D1 Procedures for Medical Staff to Wear Protective Equipment before Entering the Isolation Area

D1.1 Medical staff enter the cleaning area through the staff dedicated passage. They shall thoroughly wash their hands and then sequentially wear medical protective masks, disposable caps or cloth caps, and change into work shoes and socks. If conditions permit, they may also change into scrub suits.

D1.2 Before entering the potentially contaminated area, wear work clothes, and if there are any skin lesions or suspected injuries on the hands, wear gloves when entering the potentially contaminated area.

D1.3 Before entering the contaminated area, remove work clothes and change into protective clothing or isolation gown. Additionally, wear disposable caps and disposable surgical masks (two layers of caps and masks), protective goggles, gloves, and shoe covers.

D2 Procedures for Medical Staff to Remove Protective Equipment before Leaving the Isolation Area

D2.1 Before leaving the contaminated area, medical staff shall first disinfect their hands, then sequentially remove protective goggles, outer layer of disposable surgical masks, outer layer of disposable caps, protective clothing or isolation gowns, shoe covers, gloves, and other items. These items shall be placed in dedicated containers. After disinfecting their hands again, they can enter the potentially contaminated area and change into work clothes.

D2.2 Before leaving the potentially contaminated area and entering the cleaning areas, wash hands and perform hand disinfection. Remove work clothes, wash hands, and perform hand disinfection again.

D2.3 Before leaving the cleaning areas, wash hands and perform hand disinfection, remove the inner disposable cap or cloth cap, inner medical protective mask, take a shower, change clothes, and clean the oral cavity, nasal cavity, and external auditory canal.

D2.4 After each contact with the patient, immediately wash and disinfect your hands.

D2.5 Disposable medical surgical masks, medical protective masks, protective clothing, or isolation gowns and other protective equipment shall be replaced immediately if they are contaminated with patient's blood, body fluids, secretions, or other potentially infectious materials.

D2.6 Before leaving work, properly carry out personal hygiene and pay attention to protect the respiratory tract and mucous membranes.

9. Appendix E: Radiological protection in imaging examination of novel coronavirus–infected pneumonia

Imaging examinations play a crucial role in the screening of current cases of novel coronavirus-infected pneumonia. The radiology department is making every effort to conduct chest imaging examinations for patients with novel coronavirus-infected pneumonia while minimizing radiation exposure to both patients and radiology personnel. In accordance with the relevant provisions of the Basic Requirements for Radiation Protection in Medical Radiation and the Radiation Health Protection Standards for X-ray Diagnosis, the following radiation protection principles shall be adhered to.

E1 Radiographers shall strengthen their sense of responsibility and enhance their awareness of radiation protection.

E2 During X-ray radiography and CT examinations, it is advised to avoid or minimize the presence of accompanying individuals.

E3 Chest X-ray examination shall select the appropriate position of the subject based on the direction of X-ray projection, aiming to avoid or minimize the exposure of nonexamined areas to the X-ray beam.

E4 During CT examinations, it is essential to control the scanning range for chest scans, limiting the range from the thoracic inlet to the level of the upper margin of the adrenal glands, thereby avoiding excessive scanning coverage.

E5 It is essential to cover the nonexamined areas of the subject, especially the X-ray sensitive regions, with lead aprons and pay special attention to protect radiation-sensitive organs such as the gonads, thyroid gland and fetus, etc.

E6 Digital radiography shall be applied as much as possible, utilizing "high voltage, low current, and small field" techniques that do not compromise the acquisition of optimal diagnostic information. Subsequently, postprocessing functions can be used to enhance the images and reduce unnecessary radiation exposure.

E7 Upon bedside radiography, it is crucial to utilize the room's corners and the bathroom in the ward for protection and take proper X-ray protection measures for radiologic technologists to reduce the harm caused by unnecessary radiation exposure.

E8 Appoint the dedicated personnel for X-ray radiography to ensure control over the radiographic parameters, avoid setting excessively high parameters or retaking images due to inadequate diagnostic quality, thus resulting in unnecessary radiation exposure for the patients.

References

[1] General Office of National Health Committee, State Administration of Traditional Chinese Medicine Office. The COVID-19 Diagnosis and Treatment Program. Trial Sixth Edition 2020.

[2] Huang C, Wang Y, Li X, et al. Clinical features of patients infected with 2019 novel coronavirus in Wuhan, China. Lancet 2020;395(10223):497–506.

[3] Special Committee on Radiology of Infection and Inflammation of Chinese Research Hospital Association. Committee of the Infectious Diseases Radiology Group of Chinese Society of Radiology; Committee of the Infectious Diseases Radiology Section of Chinese Medical Doctor Association; Infectious Disease Imaging Group. Guideline for Medical Imaging in Auxiliary Diagnosis of Coronavirus Disease 2019. Chin J Med Imag Technol 2020;36(3):1–11.

[4] Chinese Society of Imaging Technology. Expert consensus on radiological examination scheme and infection prevention and control of COVID-19 (trial first edition). Electron J Emerg Infect Dis 2020;5(2):65–73.

[5] Committee of the Infectious Diseases Radiology Group of Chinese Society of Radiology. Committee of the Infectious Diseases Radiology Section of Chinese Medical Doctor Association; Infectious Disease Imaging Group, Infectious Disease Branch, Chinese Research Hospital Association, etc. Guideline for Imaging Diagnosis of Novel Coronavirus (2019-nCoV) Infected Pneumonia (1st Edition 2020). New Med 2020;30(1):22–34.

[6] Zhu N, Zhang DY, Wang WL, et al. A novel coronavirus from patients with pneumonia in China, 2019. N Engl J Med 2020;382(8):727–33. https://doi.org/10.1056/NEJMoa2001017.
[7] Drosten C, Giinther S, Preiser W, et al. Identification of a novel coronavirus in patients with severe acute respiratory syndrome. N Engl J Med 2003;348:1967–76.
[8] Xu Z, Shi L, Wang Y, et al. Pathological fifindings of COVID-19 associated with acute respiratory distress syndrome. Lancet Respir Med 2020;8(4):420–2.
[9] Zheng Q, Lu Y, Lilian T, et al. Advances in clinical and imaging studies of novel coronavirus pneumonia. Electron J Emerg Infect Dis 2020;5(2):140–4.
[10] General Office of National Health Committee, State Administration of Traditional Chinese Medicine Office. The COVID-19 Diagnosis and Treatment Program. Trial Eighth Edition 2021. Revised).
[11] Liu P, Tan X. Novel Coronavirus (2019-nCoV) pneumonia. Radiology 2019;295(1):19.
[12] Chung M, Bernheim A, Mei X, et al. CT imaging features of 2019 novel coronavirus (2019-nCoV). Radiology 2020;295(1):202–7.
[13] Song F, Shi N, Shan F, et al. Emerging coronavirus 2019-nCoV pneumonia. Radiology 2020;295(1):210–7.
[14] Wang D, Hu B, Hu C, et al. Clinical characteristics of 138 hospitalized patients with 2019 novel coronavirus-infected pneumonia in Wuhan, China. JAMA 2020. https://doi.org/10.1001/jama.2020.1585.
[15] Chan JF, Yuan S, Kok KH, et al. A familial cluster of pneumonia associated with the 2019 novel coronavirus indicating person-to-person transmission: a study of a family cluster. Lancet 2020;395(10223):15–21. https://doi.org/10.1016/S0140-6736(20)30154-9.
[16] Lei JQ, Li JF, Li X. CT imaging of 2019 novel coronavirus(2019-nCoV) pneumonia. Radiology 2020. https://doi.org/10.1184/radiol.2020200236.
[17] Franquet T. Imaging of pulmonary viral pneumonia. Radiology 2011;260(1):18–39.
[18] Dongchi Z, Jin R, Liu Z, et al. Recommendation for the diagnosis and treatment of novel coronavirus infection in children in Hubei (Trial Edition 1). Chin J Contempor Pediat 2020;22(2):96–9.
[19] The Children's Hospital Zhejiang University School of Medicine. Guidelines for Diagnosis and Treatment of COVID-19 Infection in Children. Trial First Edition. Pediatric Time; 2020. p. 31.
[20] General Office of National Health Committee, State Administration of Traditional Chinese Medicine Office. The COVID-19 Diagnosis and Treatment Program. Trial Eighth Edition 2020 [S].
[21] Holshue ML, Chas DB, Scott L, et al. First case of 2019 novel coronavirus in the United States. N Engl J Med 2020;382:929–36.
[22] Koo HJ, Lim S, Choe J, et al. Radiographic and CT features of viral pneumonia. Radiographics 2018;38(3):719–39.
[23] Feng F, Jiang YB, Yuan M, et al. Association of radiologic findings with mortality in patients with avian influenza H7N9 pneumonia. PLoS One 2014;9(4):e93885.
[24] Bajantri B, Venkatram S, Diaz FG. Mycoplasma pneumoniae: a potentially severe infection. Clin Med Res 2018;10(7):535–44.
[25] Silva CI, Churg A, Muller NL. Hypersensitivity pneumonitis: spectrum of high-resolution CT and pathologic findings. AJR 2007;188:334–44.
[26] Castaner E, Alguersuari A, Gallardo X, et al. When to suspect pulmonary vasculitis: radiologic and clinical clues. Radiographics 2010;30(1):33–53.
[27] National Health Committee. The COVID-19 Diagnosis and Treatment Program. Trial Sixth Edition 2020.

[28] Pediatrics Branch of Hubei Medical Association, Pediatrics Branch of Wuhan Medical Association, Hubei Pediatric Medical Quality Control Center. Recommendation for the diagnosis and treatment of novel coronavirus infection in children in Hubei (Trial Edition 1). Chin J Contempor Pediat 2020;22(2):96–9.
[29] National Health Commission, National Administration of Traditional Chinese Medicine. Guideline for diagnosis and treatment of community-acquired pneumonia in children (2019 version). Chin J Clin Infect Dis 2019;12(1):6–13.
[30] Zhang H, Kang ZJ, Gong HY, et al. The digestive system is a potential route of 2019-nCov infection: a bioinformatics analysis based on single-cell transcriptomes. bioRxiv 2020. https://doi.org/10.1101/2020.01.30.927806.

APPENDIX 2

Expert consensus on radiological examination scheme and infection prevention and control of COVID-19 (trial first edition)

Chinese Society of Imaging Technology[1,2]

Expert consensus of radiological examination scheme and infection prevention of the 2019 Novel Coronavirus related pneumonia (First version)

Radiological Technology Committee of Chinese Medical Association Expert Group of Specialized Committee for Technologists on Infectious Diseases[3]

In December 2019, cases of pneumonia with unknown causes were discovered. On January 7, 2020, a new type of coronavirus was detected through laboratory testing. The viral nucleic acid testing was completed on January 10, 2020. On January 12, 2020, the World Health Organization (WHO) officially named the novel coronavirus responsible for this pneumonia outbreak as the 2019 novel coronavirus [1–4]. In just over a month, the epidemic has spread throughout the country. As of 5:00 PM on February 1, 2020, the total number of confirmed cases nationwide has reached 11,889, and the number of suspected cases has reached 17,988, with a continuous upward trend. Currently, in accordance with the

[1] Foundation Projects: 1. Key Project of National Natural Science Foundation (61,936,013); 2. Project of Shanghai Science and Technology Commission (19,411,965,800)

[2] Citation Format: Chinese Society of Imaging Technology. Expert Consensus on Radiological Examination Scheme and Infection Prevention and Control of COVID-19 (Trial First Edition) [J]. Electronic Journal of Emerging Infectious Diseases, 2020, 5, 65–73

[3] Radiological technology committee of Chinese medical. Expert consensus of radiological examination scheme and infection prevention of the 2019 Novel Coronavirus related pneumonia (First version) [J].Electronic Journal of Emerging Infectious Diseases, 2020, 5(2):65–73.

provisions of the Law of the People's Republic of China on the Prevention and Control of Infectious Diseases, the National Health Commission of the People's Republic of China has classified the novel coronavirus pneumonia (COVID-19) as a Class B infectious disease [5]. Prevention and control measures for Class A infectious diseases are being implemented, and it is also included in the management of quarantine infectious diseases as stipulated in the Law of the People's Republic of China on Frontier Health and Quarantine [6].

According to the Diagnosis and Treatment Protocol for Novel Coronavirus Pneumonia (Trial Fifth Edition) released by the Office of the National Health Commission of the People's Republic of China [7], imaging examinations are one of the crucial methods for diagnosing novel coronavirus pneumonia. Standardized CT examination techniques serve as an effective guarantee for screening, diagnosis, therapeutic evaluation, and information sharing related to novel coronavirus infections. Radiologic technologists may have direct or indirect contact with infected patients during X-ray radiography or CT examination processes [8]. In order to prevent the continued spread and dissemination of COVID-19, as well as to improve the level of radiology imaging examination techniques and effectively reduce the risk of nosocomial infections among radiology personnel, the Radiological Technology Branch of the Chinese Medical Association, in collaboration with renowned national infectious disease imaging experts, medical imaging technology experts, and infection prevention and control experts, has established an expert consensus collaboration team and developed the Expert Consensus on Radiological Examination Scheme and Infection Prevention and Control of COVID-19 (Trial First Edition) through the joint efforts of radiology professionals in China. The aim is to standardize radiological examination techniques, share information resources, prevent cross-infections within hospitals, enhance the quality of radiology imaging examinations, guide infection prevention and control measures in radiology departments, and strive for the goal of "zero infections" among radiographers. Due to the severe epidemic situation, limited time, and heavy workload, there might be imperfections and questions about the rationality of the technical processes during the writing process, which need to be further improved in the updated version.

1. Characteristics, clinical manifestations, and reporting procedures of COVID-19 [7]

1.1 Etiology characteristics

The novel coronavirus belongs to the β genus of coronaviruses. It is enveloped and its particles are typically round or oval-shaped, often displaying pleomorphism, with a diameter ranging from 60 to 140 nm. Its genetic characteristics are significantly different from SARS-CoV and MERS-CoV. Current research shows that there is a homology of over 85% with bat-SARS-like coronavirus (bat-SL-CoVZC45). During in vitro isolation and culture process, the novel coronavirus can be detected in human respiratory epithelial cells after approximately 96 hours. However, it takes about 6 days for isolation and culture in VeroE6 and Huh-7 cell lines. Our understanding of the physicochemical characteristics of coronaviruses is largely derived from research on SARS-CoV and MERS-CoV. The virus is sensitive to ultraviolet light and heat. It can be effectively inactivated by exposure to 56°C for 30 minutes, as well as by using ether, 75% ethanol, chlorine-based disinfectants, peracetic acid, and chloroform, which are all lipid solvents. However, chlorhexidine is not effective in inactivating the virus.

1.2 Epidemiologic characteristics

1.2.1 Source of infection

Currently, it is believed that the main source of transmission is patients with novel coronavirus pneumonia. Asymptomatic infected persons can also become sources of transmission.

1.2.2 Route of transmission

Respiratory droplets and contact transmission are the main routes of transmission. The transmission routes through aerosols and the digestive tract are still to be clarified.

1.2.3 Susceptible population

The general population is generally susceptible.

1.3 Clinical manifestations

According to the current epidemiological investigation, the incubation period is 1—14 days, with the majority being 3—7 days. The main symptoms include fever, fatigue, and dry cough. A small number of patients may also experience symptoms such as nasal congestion, runny nose, and diarrhea. Severe cases often develop respiratory distress around 7 days after the onset, and in severe cases, it may rapidly progress to acute respiratory distress syndrome, septic shock, difficult-to-correct metabolic acidosis, and coagulation dysfunction. During the course of severe and critical patients, they may have low-grade fever or even no obvious fever. Some patients may only present with mild symptoms such as low-grade fever and slight fatigue, without any signs of pneumonia. Most of them recover within 7 days. From the current cases under treatment, most patients have a good prognosis, while a small number of patients are in critical condition. Pediatric cases generally exhibit relatively mild symptoms. Deaths are more common in elderly individuals and those with underlying chronic health conditions.

1.4 Clinical classification

1. Mild: Clinical symptoms are mild, and there are no signs of pneumonia on imaging.
2. Moderate: It presents with symptoms such as fever and respiratory tract issues, and imaging examinations show signs of pneumonia.
3. Severe: Any of the following criteria is satisfied.
   ① Respiratory distress, with a respiratory rate (RR) of ≥30 breaths per minute;
   ② Oxygen saturation at rest (SpO2) ≤93%;
   ③ Arterial partial pressure of oxygen (PaO2)/fraction of inspired oxygen (FiO2) ratio ≤300 mmHg (1 mmHg = 0.133 kPa).

   Critical patients meet any of the following conditions.

① Respiratory failure requiring mechanical ventilation;
② Presence of shock;
③ Organ dysfunction requiring intensive care unit (ICU) monitoring and treatment.

Case reports

When medical staff at all levels of medical institutions discover suspected cases that meet the case definition, they shall immediately isolate and treat the patients. In cases where expert consultations within the hospital or consultations with the attending physicians still consider the patients as suspected cases, the information shall be reported online within 2 hours. Respiratory tract or blood specimens shall be collected for novel coronavirus nucleic acid testing. Simultaneously, suspected patients shall be promptly transferred to designated hospitals. For individuals with an epidemiological relation to patients with novel coronavirus pneumonia, even if the respiratory pathogen testing is positive, it is still recommended to conduct timely testing for the novel coronavirus pathogen. For suspected cases, they can be ruled out only when two consecutive respiratory virus nucleic acid tests are negative (with a minimum interval of 1 day between samplings). If there is a high clinical suspicion, continuous monitoring of nucleic acid testing shall be maintained.

2. Strategies for prevention and control of nosocomial infection in radiology department

2.1 Diagnosis and treatment environments and post requirements of radiology department

It is equipped with an independent medical imaging examination room and divided into regions

① To prevent cross-infections, it is necessary to set up a separate medical imaging examination area or dedicated radiology equipment (e.g., X-ray and CT equipment for infected individuals) and film printers. According to the requirements of nosocomial infection prevention and control, it is essential to clearly demarcate contaminated areas, semicontaminated areas, and cleaning areas, and strict disinfection shall be implemented in all designated areas. If it is not possible to allocate a separate dedicated examination room (e.g., CT examination room), strict equipment and air disinfection shall be conducted after each patient's scan is completed before proceeding with the examination of the next routine patient. Fever clinics, contaminated and semicontaminated areas in the radiology department, and hospital isolation wards are considered key areas for nosocomial infection prevention and control. After conducting examinations on confirmed patients, terminal disinfection must be performed before conducting examinations on suspected patients.

② A dedicated radiology examination pathway shall be established.

③ Fever clinic patients and suspected or confirmed patients in isolation wards shall be examined in batches and during designated time slots, and strict disinfection protocols shall be followed.

Radiology personnel shall be reasonably divided into specialized roles to manage the key areas, such as fever clinics, contaminated and semi-contaminated areas in the radiology department, and hospital isolation wards, with dedicated responsibilities [9,10].

Department Administrator of Nosocomial Infection: In the radiology department, a designated person shall be appointed as the department's nosocomial infection administrator, responsible for guiding and supervising all staff members in their disinfection and protective measures, as well as preparing disinfectants. This ensures consistency and clear division of responsibilities and prevents the transmission of the virus within the hospital's radiology department.

Mobile Bedside X-ray Technologists: Assign a radiographer to specialized positions to perform mobile bedside X-ray imaging in key areas, ensuring standardized imaging procedures. Technologists are required to strictly adhere to Level 2 protection. However, in situations where procedures such as sputum suction, respiratory tract sampling, endotracheal intubation, and tracheostomy are performed and there is a possibility of patient respiratory secretions or bodily substances being sprayed or splashed, Level 3 protection shall be adopted. After completing the X-ray imaging, the equipment shall be disinfected (by wiping with 75% ethanol).

X-ray Technologists and CT Technologists: Positioning technologists working in the contaminated area shall strictly adhere to Level 2 protection. However, in situations where procedures such as sputum suction, respiratory tract sampling, endotracheal intubation, and tracheostomy are performed and there is a possibility of patient respiratory secretions or bodily substances being sprayed or splashed, Level 3 protection shall be adopted. Operating technologists can use Level 1 or Level 2 protection and if standardized image acquisition for patients is required, after completing the scanning, both the equipment and room shall be promptly disinfected according to the nosocomial infection control requirements.

Image Examination Registration Personnel: Patients in key areas shall be prohibited from going to regular appointment registration service windows for business transactions. It is recommended to fully utilize the Hospital Information System (HIS), Picture Archiving and Communication System (PACS), and Radiology Information System (RIS) to achieve paperless electronic application forms and medical history. In key areas, radiographers shall complete the registration process, and any documents or forms handled by patients shall be managed separately and securely.

Other imaging technologists and diagnostic professionals working in nonkey areas: If it is not explicitly required to enter contaminated or semicontaminated areas, they may wear medical protective masks or N95 medical protective masks, hats, work clothes, and gloves. After entering contaminated areas, Level 2 protection shall be followed strictly. In situations where procedures such as sputum suction, respiratory tract sampling, endotracheal intubation, and tracheostomy are performed and there is a possibility of patient respiratory secretions or bodily substances being sprayed or splashed, Level 3 protection shall be adopted.

The working mode (pilot) for radiographers in key areas: For mobile bedside X-ray technologists, X-ray technologists, and CT technologists working in isolation areas, due to the possibility of close contact with patients, it is recommended to adopt the "2+2" working mode when conditions permit.

Assign specific personnel to stay and lodge in designated isolation areas within the hospital, and be responsible for carrying out mobile bedside X-ray, X-ray, and CT examinations in key areas such as fever clinics, radiology-contaminated areas, semicontaminated areas, and hospital isolation wards for a period of 14 days (or a shorter duration based on specific circumstances). After completing the photography work in shift, the personnel need to stay and lodge in a specific isolation area specially arranged in the hospital to be prepared for any emergencies. During the work period, no one can leave the designated isolation area and all work and daily needs shall be fulfilled within the key areas and the specified designated isolation area.

After completing their 14-day work, the radiographer shall leave the specific isolation area and enter a designated special isolation ward for a 14-day period of medical observation under supervision. During such period, they cannot leave the designated area. After the 14-day period of medical observation under supervision, if there are no abnormalities, they can return to their normal work positions.

Based on past experiences in the prevention and control of severe infectious diseases and the latest epidemiological investigation data on the novel coronavirus, a 14-day medical observation period is considered appropriate.

2.2 Rules for prevention and control of nosocomial infection in radiology department

2.2.1 Requirements for department staff

To prevent cross-infection, the access channel for department staff shall be closed. Access will be restricted using an entry system and unauthorized personnel are not allowed to enter or exit. The radiographer shall wear protective equipment in the cleaning area and take of them in the buffer zone [11,12]. Positioning technologists working in the contaminated area are not allowed to enter the cleaning area before the scheduled shift. During shift period, the technologists shall take off their protective clothing in the buffer zone and perform proper cleaning process before entering the cleaning area.

2.2.2 Personal protection level

General protection includes wearing work clothes, disposable medical masks, and a work cap. Gloves shall be worn when necessary.

Level 1 protection is applicable to medical staff working in preexamination triage, fever clinics, and infectious disease outpatient departments. It involves wearing disposable caps, disposable surgical masks (N95 protective masks for close contact with individuals with an epidemiological history), work clothes, isolation gowns (disposable isolation gowns if necessary during precheck triage), and disposable latex gloves when necessary, and strictly adhering to hand hygiene.

Level 2 protection is applicable to medical staff involved in diagnosis and treatment activities with close contact with suspected or confirmed patients. It involves wearing disposable caps, protective goggles or face shields (antifog type), medical protective face mask, protective suits or isolation gowns, disposable latex gloves, disposable shoe covers, and strictly adhering to hand hygiene.

Level III Level 3 protection is applicable to personnel performing aerosol-generating procedures on suspected or confirmed patients, such as sputum suction, respiratory tract sampling, tracheal intubation, and tracheostomy, where there is a possibility of patient respiratory secretions or body substances being aerosolized or splashed. It involves wearing disposable work caps, medical protective masks, protective face shields (or full-face respiratory protective devices or positive-pressure hoods), protective suits, disposable latex gloves, disposable shoe covers, and strictly adhering to hand hygiene practices.

2.2.3 Procedures for radiology staff to don and doff protective equipment [13]

2.2.3.1 The process of donning protective equipment

Perform the seven-step handwashing process - Wear a cap - Wear a medical protective mask (perform an air leakage test) - Put on the protective suit (after removing shoes) - Wear disposable latex gloves (inner layer) - Put on a disposable isolation gown - Wear disposable latex gloves (outer layer) - Put on rubber boots - Wear boot covers - Wear protective goggles/face shield - Check wearing tightness.

2.2.3.2 The process of doffing protective equipment

In contaminated area, remove visible contaminants - Perform hand hygiene - Take off the shoe covers (outer layer) - Perform hand hygiene - Remove the isolation gown along with the outer gloves - Perform hand hygiene) - (In semicontaminated areas, remove protective goggles/face shield - Perform hand hygiene - Take off the protective suit along with the inner gloves and boot covers - Perform hand hygiene - Remove medical protective mask - Remove the cap - Perform the seven-step handwashing process.

2.2.3.3 Precautions

In the cleaning area, be sure to perform air leakage test to ensure that the medical mask is tightly worn; ensure that the front tape of the zipper on the protective suit is securely fastened; There shall be a mirror in the area for donning protective equipment; After donning the complete set of protective gear and before entering the contaminated area, it is essential to check the tightness and flexibility of the equipment. This process shall be completed under the guidance and assistance of a supervisor to ensure safety.

In contaminated and semicontaminated areas, the process of doffing protective equipment shall be done gently and carefully to avoid generating aerosols and potential exposure; When removing the protective suit, be cautious not to touch the contaminated surface of the suit to prevent skin exposure; The doffing process shall follow a sequence from the most contaminated areas to the least contaminated areas and cannot be reversed. If conditions permit, a supervisor can observe and provide guidance in the cleaning area.

2.3 Equipment cleaning and disinfection of radiology department

2.3.1 Daily cleaning

Metal surfaces and surfaces with paints shall be wiped with a mild cleaning agent and then dried with a dry towel. Do not use corrosive cleaning agents, corrosive detergents, or corrosive polishing agents. If it is unable to confirm the characteristics of the cleaning agent, stop using it.

Chrome-plated components shall only be wiped with a dry towel. Do not use abrasive polishing agents. To protect the surface coating, please use nonabrasive wax. Plastic surfaces shall only be cleaned with soap and water. If other cleaning agents, such as high-concentration ethanol, are used, the plastic material may lose its luster and become prone to cracking.

Any standard glass cleaner can be used to clean the touch screen, but avoid using products containing ammonia. Spray the glass cleaner onto a cloth or towel and then wipe the touch screen. Be sure to promptly remove any liquid droplets to prevent them from seeping into the device's gaps. Dust and fingerprints generally do not affect the use of a sealed touch screen.

2.3.2 Equipment disinfection

For digital X-ray (DR), CT, MRI, and other equipment, after completing the examination for each patient, use 75% ethanol for disinfection by wiping. If there is dirt or visible stains, first use disposable absorbent material to completely remove the stains before proceeding with disinfection.

Do not use corrosive disinfectants or sterilizing agents.

Use disinfectant spray devices with caution as these sprays may penetrate the equipment, causing electrical short circuits, metal corrosion, or other damage. If a spraying disinfectant device is used, it is required to turn off the equipment and cool it first, then completely cover the equipment with plastic films before starting the spraying process. Remove the plastic film only after all sprays are dispersed, and then wipe and disinfect the equipment.

2.3.3 Floor disinfection

For the floor in the equipment room, use a 2000 mg/L chlorine-containing disinfectant (except chlorhexidine) for wiping and disinfection. If there are visible contaminants, first use disposable absorbent material to completely remove the stains before proceeding with disinfection.

For the floor in waiting areas and corridors, use a 2000 mg/L chlorine-containing disinfectant (except chlorhexidine) to disinfect waiting area and corridor passages (including handrails, doorknobs, windows, wall switches, etc.). If there are contaminants, follow the same disinfection procedure as used for the equipment room.

Perform the air disinfection in an unoccupied state using ultraviolet (UV) radiation, with continuous exposure for at least 30 minutes. After UV exposure, ensure proper ventilation by opening windows and/or using ventilation ducts for at least 30 minutes. UV disinfection requires a proper configuration to cover the entire equipment room. Alternatively, continuous disinfection can be achieved using air disinfection machines.

Supplementary instructions after completing the examination of suspected or confirmed patients, the equipment, floors, and air shall be disinfected once.

2.4 Handling measures of medical wastes in radiology department

2.4.1 Management regulations

All wastes generated by patients shall be treated as infectious medical wastes and managed strictly in accordance with the Regulations on the Administration of Medical Wastes and the Measures for the Administration of Medical Waste in Medical and Health Institutions.

2.4.2 Medical waste collection process

Infectious wastes (including items contaminated with patient's blood or body fluids; household wastes generated by isolated patients; used disposable medical equipment and supplies such as syringes, needles, etc.), shall be placed in a yellow medical waste collection bag, and filled up to 3/4 of its capacity. After spraying the inside of the bag with a 5000 mg/L chlorine-containing disinfectant (except chlorhexidine), the inner bag shall be sealed with a goose-neck closure. The outer surface of the inner bag shall be sprayed with a 5000 mg/L chlorine-containing disinfectant (except chlorhexidine). The outer bag shall also be sealed with a gooseneck-type sealing, and a specialized label shall be affixed. The outer surface of the outer bag shall be sprayed again with a 5000 mg/L chlorine-containing disinfectant (except chlorhexidine). The bag shall be stored in the medical waste temporary storage area of the department.

2.4.3 Personnel protection

The cleaning staff or dedicated medical waste collectors shall wear personal protective equipment (Level 2 protection) during the collection of infectious medical waste. Properly conduct the handover registration and sealed transshipment, and store the medical wastes at designated storage locations of the hospital.

Precautions:

When entering or leaving each room or isolation ward, perform quick hand sanitization to avoid surface contamination. When donning and doffing protective equipment, it is essential to follow the designated procedures and strictly adhere to hand hygiene practices.

If it is not possible to assign a dedicated technician exclusively to the isolation ward, then following the principles of infection control, a fixed semicontaminated area and a cleaning area shall be set up in the radiology department. Strictly adhere to the procedures for donning and doffing protective clothing as described in this document. Additionally, set up designated collection points to facilitate the subsequent handling of infectious medical waste and the disinfection of reusable items.

After using, place protective goggles in the designated disinfection area and soak them in a closed container with either a 1000 mg/L chlorine-containing disinfectant (except chlorhexidine) or 75% ethanol for at least 1 hour.

3. Radiological examination scheme for novel coronavirus pneumonia

The expert group recommends that, for medical institutions with the necessary conditions, lung CT scans shall be the preferred imaging method when conducting examinations for patients with COVID-19.

3.1 DR examination scheme

3.1.1 Examination precautions

Install one dedicated digital radiography (DR) machine as a dedicated unit for patients with novel pneumonia; Strictly follow the above-mentioned disinfection measures for equipment and equipment room management; For critical patients, bedside radiography is recommended.

3.1.2 Radiological examination scheme

3.1.2.1 Preparation before examination

3.1.2.1.1 Staff preparation If there is the availability of intelligent positioning systems, the radiology technologist shall adopt Level 1 or Level 2 protection. If such conditions are not available, the radiology technologist shall strictly follow Level 2 protection. However, in situations where procedures such as sputum suction, respiratory tract sampling, endotracheal intubation, and tracheostomy are performed and there is a possibility of patient respiratory secretions or bodily substances being sprayed or splashed, Level 3 protection shall be adopted.

3.1.2.1.2 Patient preparation Carefully verify the examination and consultation form as well as the notes on precautions, and clarify the purpose and requirements of the examination. During the examination, patients shall wear a medical surgical mask or an N95 medical protective mask throughout the process. Remove necklaces, chest accessories, and other high-density items (such as bras, zippers, buttons, etc.).

Note: In special cases where patients are children or critical, and they require accompaniment from family members or healthcare personnel during the examination, the protective requirements for accompanying individuals shall not be lower than Level 2 protection standards. Additionally, proper radiation protection shall be provided for both the patient and the accompanying individuals.

3.1.2.2 Radiographic examination scheme for adults

Photographic distance: 180 cm

Filter Grid: Minimum ratio of 10:1.

Exposure Conditions: Typically, high kilovolt (kV) radiography at 120 kV is used with automatic exposure technology.

Photographic Posture: Posteroanterior (PA) position. The patient stands in front of the detector with the back toward the X-ray tube. The anterior chest is in close contact with the detector, and both hands are placed on the iliac crest or holding the detector. The shoulders are drooping, the upper arms are internally rotated (pulling scapula apart), the head is slightly tilted backward, and the chin is placed on the upper edge of the detector. The centerline is aligned with the level of the sixth thoracic vertebra and perpendicular to the detector's entrance.

Protection Requirements: Use lead aprons or similar equipment to cover other parts of the body as much as possible. Breathing Requirement: Collect the image at the end of deep inspiration and breath-holding.

3.1.2.3 Radiological examination scheme for children

(1) For cooperative subjects aged 3 years and older, the photographic distance is 150 cm. Filter Gird: Unavailable.

Exposure Conditions: Do not use automatic exposure technology and avoid high kilovolt (kV) radiography. Recommended range: 55–65 kV, 2–5 mAs. Adjust the settings appropriately based on the age and body thickness of the subject.

Photographic Posture: Posteroanterior (PA) position. The subject faces the upright detector with the feet apart and stands firmly. The body's mid-sagittal plane is aligned with the detector's central axis and perpendicular to the detector. The chin is slightly lifted to prevent the overlap of lung apices. Both elbows are bent, and the backs of the hands are placed on the hips. The shoulders are kept flat and internally rotated, leaning against the detector to pull the scapulae away from the lung fields. The centerline is aligned with the level of the sixth thoracic vertebra and perpendicular to the detector's entrance. If younger subjects cannot achieve the above position, they can be instructed to tightly hug the detector.

Protection Requirements: Use lead aprons or similar equipment to cover other parts of the body as much as possible.

Breathing Requirement: Collect the image at the end of deep inspiration and breath-holding. If the subject does not cooperate, capture the exposure at the end of their natural inhalation based on their breathing pattern.

(2) For noncooperative subjects aged 0–3 years and aged 3 years and older, the photographic distance is 100 cm.

Filter Gird: Unavailable.

Exposure Conditions: No automatic exposure technology is used, the reference range is 50–60 kV and 0.6–2.0 mAs.

Photographic Posture: The subject lies supine on the imaging table, with the body's mid-sagittal plane perpendicular to the tabletop and aligned with the centerline of the table. Both arms are raised above the head, and sandbags, straps, and/or assistance from family members are applied to secure the legs, arms, and head. The lower limbs of the subject shall be kept as straight as possible, the upper limbs shall be clamped close to the head, and the back shall be pressed against the tabletop to reduce motion blur. The head is aligned, and the chin is slightly lifted to prevent the overlap of lung apices. The centerline is directed perpendicularly to the center of the line connecting both nipples.

Protection Requirements: Use lead aprons or similar equipment to cover other parts of the body as much as possible.

Breathing Requirement: Capture the exposure at the end of the subject's natural inhalation according to their breathing pattern.

3.2 Examination scheme of mobile bedside X-ray photograph

3.2.1 Examination precautions

① One mobile DR (Digital Radiography) unit is dedicated and placed in key areas such as the isolation ward for COVID-19 patients to perform bedside imaging for critical patients.

② Before entering the COVID-19 isolation ward, the technologist shall strictly adhere to Level 2 protection. In situations where procedures such as sputum suction, respiratory tract sampling, endotracheal intubation, and tracheostomy are performed and there is a possibility of patient respiratory secretions or bodily substances being sprayed or splashed, Level 3 protection shall be adopted.

③ After completing the bedside imaging, the technologist shall follow the procedures outlined in this document to remove the protective clothing. The used protective equipment shall be placed in yellow trash bags and handled as required. When conditions permit, it is recommended that the imaging technologists work on standby within the isolation ward for 14 days. After exiting the isolation ward, the imaging technologists shall undergo medical observation in the designated area of the hospital for 14 days. Once it is confirmed that there is no infection, they can return to their normal work positions.

④ For bedside imaging, it is recommended to ensure proper protection of the flat-panel detector or IP (Image Plate). It is advised to cover the flat-panel detector with plastic bags and perform disinfection after use.

⑤ When performing mobile bedside X-ray examinations on critically diagnosed patients in key areas, it is essential to ensure that the equipment is dedicated to this purpose. The mobile device shall not leave the isolation ward or other key areas until the end of the epidemic. After the epidemic is over, the mobile DR unit can only be returned to its original workplace after comprehensive cleaning and disinfection.

3.2.2 Radiological examination scheme

Preparation before the examination is the same as for DR examination

Radiographic examination scheme for adults [14] Photographic distance: 100 cm.

Exposure Conditions: 75–85 kV is recommended if the filter grid is used; and 65–70 kV is suggested when no filter grid is used.

Photographic Posture: The detector is placed close to the patient's back, and the centerline is directed perpendicularly to the center of the line connecting both nipples. If the patient cannot lie flat, the position and incident angle can be adjusted according to the specific situation.

Protection Requirements: Use lead aprons or similar equipment to cover other parts of the body as much as possible. Breathing Requirement: Collect the image at the end of deep inspiration and breath-holding.

Radiographic examination scheme for children Photographic distance: 100 cm.

Filter Gird: Unavailable.

Exposure Conditions: No automatic exposure technology is used, the reference range is 50–60 kV and 0.6–2.0 mAs.

Photographic Posture: The subject lies supine on the bed, with the detector placed close to the back. The arms are raised, and sandbags, straps, and/or assistance from family members are used to secure the legs, arms, and head. The legs of the subject shall be kept as straight as possible, and the upper limbs shall be pressed close to the head. The head is aligned, and the chin is slightly lifted to prevent the overlap of lung apices. The centerline is directed perpendicularly to the center of the line connecting both nipples. If the subject is unable to lie flat, the subject's angle and the angle of the central axis of the X-ray beam can be adjusted according to the specific situation to ensure that the central axis is perpendicular to the detector.

Protection Requirements: Use lead aprons or similar equipment to cover other parts of the body as much as possible. Breathing Requirements: Capture the exposure at the end of the subject's natural inhalation according to their breathing pattern.

3.3 CT examination scheme

3.3.1 Examination precautions

3.3.1.1 Isolation ward equipped with dedicated CT

① The patient shall wear a medical surgical mask or N95 medical protective mask and enter the CT room accompanied by medical staff or positioning technicians. Medical staff or positioning technicians shall strictly adhere to Level 2 protection. However, in situations where procedures such as sputum suction, respiratory tract sampling, endotracheal intubation, and tracheostomy are performed and there is a possibility of patient respiratory secretions or bodily substances being sprayed or splashed, Level 3 protection shall be adopted.

② Guide the patient to lie on the examination bed and operate the equipment rack to complete the positioning process.

③ After the CT operator completes the examination, perform rapid hand disinfection.

④ Open the shielded door, and the accompanying medical staff will escort the patient back to the ward. After the examination is completed, the radiology operator shall inform the cleaning personnel to disinfect the CT room.

3.3.1.2 Isolation ward not equipped with dedicated CT

① Set up designated CT scanning for diagnosed patients with specific personnel at the appointed time.

② Cooperate with nosocomial infection, medical and radiology departments, as well as wards to establish a closed-loop process and passages for diagnosed patients' transportation, examination, and return to the ward.

③ Before the examination, cooperate with hospital security and logistics personnel to ensure the clearance of other patients and all unnecessary personnel from around the designated CT area and the established passage.

④ The patient shall wear a medical surgical mask or N95 medical protective mask.

Enter the CT room accompanied by medical staff (or nursing personnel). Medical staff (or nursing personnel) shall strictly adhere to Level 2 protection. However, in situations where procedures such as sputum suction, respiratory tract sampling, endotracheal intubation, and tracheostomy are performed and there is a possibility of patient respiratory secretions or bodily substances being sprayed or splashed, Level 3 protection shall be adopted. The transportation personnel shall use 75% ethanol or 1000 mg/L chlorine-containing disinfectant (except chlorhexidine) to spray and disinfect the surrounding air at any time.

⑤ The accompanying medical staff (or nursing personnel) shall assist the patient in lying on the examination bed, while the radiology technologist operates the equipment rack to complete the positioning process.

⑥ After the patient's scan is completed, open the shielded door, lower the patient from the examination bed, and have the accompanying medical staff (or nursing personnel) escort them back to the ward. After the examination is completed, inform the cleaning personnel to disinfect the CT room. During the return process to the ward, the transportation personnel shall also use 75% ethanol or 1000 mg/L chlorine-containing disinfectant (except chlorhexidine) to spray and disinfect the surrounding air at any time.

⑦ In case of insufficient hospital staff, the designated radiology technologist can also perform Step ④ to ⑥. The technologist shall strictly adhere to Level 2 protection. However, in situations where procedures such as sputum suction, respiratory tract sampling, endotracheal intubation, and tracheostomy are performed and there is a possibility of patient respiratory secretions or bodily substances being sprayed or splashed, Level 3 protection shall be adopted.

⑧ The CT operator shall operate the equipment in the control room to complete the examination and perform rapid hand hygiene disinfection after the examination is completed.

Important Note 1: If the same CT scanner is to be used for both diagnosed cases and suspected cases from the fever clinic, it is essential to strictly separate the examination times of the both, with a minimum interval of 1 hour, to allow for thorough cleaning and disinfection of the CT room and surrounding environment.

Important Note 2: The pulmonary manifestations of COVID-19 often appear as multiple small patchy opacities, ground-glass opacities, and consolidations, with a significant involvement of the subpleural, lower lungs, and posterior regions. Therefore, except for pediatric patients, CT scanning is recommended as the preferred imaging method for diagnostic evaluation.

Important Note 3: For pediatric patients, when hardware conditions permit, low-dose wide-detector volume scanning on high-end CT scanners can be used as an alternative to conventional X-ray imaging.

3.3.2 Commonly-used examination scheme

Due to the diverse CT equipment manufacturers, models, and types in different hospitals, this consensus only provides the recommended examination scheme.

3.3.2.1 Preparation before examination

(1) Staff Preparation: It is recommended to assign two technologists with one for operating the CT equipment and the other for positioning the patients in the CT room and instructing them on breathing and breath-holding techniques. If conditions do not permit, family members or clinical staff accompanying the patient may be asked to provide assistance, but it is essential to ensure that they wear personal protective equipment. If the operator does not come into direct contact with the patient, Level 1 or Level 2 protection can be implemented. Positioning technicians shall strictly adhere to Level 2 protection. However, in situations where procedures such as sputum suction, respiratory tract sampling, endotracheal intubation, and tracheostomy are performed and there is a possibility of patient respiratory secretions or bodily substances being sprayed or splashed, Level 3 protection shall be adopted. After completing the examination for a diagnosed patient, all the above-mentioned disposable items shall be replaced before the technologist can participate in the examination of the next patient.

(2) Preparation of Auxiliary Equipment:

① For scanning suspected or diagnosed cases, prioritize using a CT machine with an adjustable examination bed at the console. If there is no such CT machine, place an auxiliary device with a step-like footboard in front of the examination bed to assist the patients in getting on and off the bed.

② The CT room shall be a separate operating space and, in principle, not shared with other machines. If it is not possible to meet the above conditions, during disinfection after the examination, the connected rooms shall undergo air disinfection to ensure thorough sanitization.

③ For CT rooms equipped with a fresh air system and central air conditioning, set the supply air volume and exhaust air volume to the maximum. For CT rooms with regular central air conditioning, the central air conditioning in the CT room and control room shall be turned off, and the backup independent air conditioning shall be turned on. If there is no independent air conditioning, the central air conditioning can be turned on after completing the examination and disinfection.

④ The examination bed shall be covered with a disposable bed sheet without folding, covering the entire examination bed surface. Protective equipment shall include lead caps and rectangular lead aprons. When using them, they shall be separated from the patient's body and clothing by a disposable bed sheet.

(3) Patient Preparation:

① Carefully verify the CT examination consultation form, understand the medical conditions, and clarify the purpose and requirements of the examination. Patients (including accompanying caregivers) shall wear a medical surgical mask or N95 medical protective mask throughout the examination process. Before entering the CT room, they shall use hand sanitizer to disinfect their hands or wear disposable gloves. Remove necklaces, chest jewelry, and other high-density items (such as bras, clothing with zippers, buttons, paint, etc.).

② Before the scan, conduct respiratory training with the patient and instruct them to cooperate with the examination according to the respiratory instructions. Usually, a deep breath followed by breath-holding at the end of inhalation (end-inspiratory hold) is employed. For critical patients, priority is given to ensuring breath-holding. For patients unable to hold their breath, they shall be instructed to breathe calmly and avoid coughing.

(4) Other Isolation Measures

① When explaining the examination precautions to patients, use intercom communication whenever possible. When it is necessary for the technologists to come into contact with the patient, maintain a distance of at least 1 meter whenever possible. For patients who can cooperate, the technologist can guide the positioning using voice control in the control room while ensuring the patient's safety. Alternatively, accompanying personnel can assist the patient in getting onto the examination bed. When it is necessary for the technologists to personally position the patient, they shall keep their head as far away from the patient's respiratory tract as possible. Before and after touching the patient, the technologists shall promptly perform hand disinfection. Throughout the patient's entry into the examination area and the entire examination process, both the technologist and the patient shall wear masks. If a patient refuses to wear a mask, the technologist can inform the attending physician to assist in ensuring the patient wears a mask before proceeding with the examination.

3.3.2.2 CT scanning scheme for adults [14,15]

(1) Scanning Position: The standard position is supine, with the body placed in the middle of the examination bed, and both arms raised, hugging the head. For individuals who have difficulty raising their arms, they can be positioned by placing their arms on both sides of the body.

(2) Scanning Method: Transverse helical scanning.

(3) Positioning Scan: Determine the scanning range, slice thickness, and interslice gap.

(4) Scanning Range: From the lung apices to the diaphragm (including bilateral costophrenic angles). For critical patients who cannot hold their breath for an extended period, scan from the diaphragm to the lung apices (where the respiratory motion amplitude is greater at the lung bases than at the lung apices). This helps reduce respiratory motion artifacts caused by the inability to hold breath, ensuring image quality.

(5) Display Field of View (DFOV): For individuals with a general body size, it is typically set between 35 and 45 cm. It is recommended to use a fixed DFOV to facilitate follow-up and comparison. For larger body sizes, a DFOV of 45 cm is suggested, while for smaller body sizes, a DFOV of 35 cm is recommended, depending on individual situations.

(6) Scan Parameters: For the positioning image scan, axial scan is generally selected, and it is recommended to use 80 kV and 25 mAs. Transverse section scanning is generally performed using helical scan. Automatic tube voltage or fixed tube voltage is recommended based on the patient's body mass index (BMI): For patients with a BMI <19 kg/m^2, 100 kV is suggested; for patients with a BMI between 19 and 24 kg/m^2, 120 kV is suggested; and for patients with a BMI >24 kg/m^2, 140 kV is suggested. The intelligent radiation dose tracking technology is used with a range of 50–350 mAs. The slice thickness shall be set between 0.5 and 1.0 mm, tube rotation speed between 0.27 and 0.80 r/s, pitch between 1.0 and 1.3, and iterative reconstruction technology be enabled. For severe and critical patients, it is preferable to prioritize reducing the scan time by adjusting the pitch to a larger value of 1.5–1.7, increasing the tube rotation speed, and widening the collimator width. This approach helps to minimize respiratory motion artifacts caused by patient breathing.

3.3.2.3 CT scanning scheme for children

The main differences between pediatric and adult scans lie in the setting of the display field of view, scan dose, and radiation protection level.

(1) Display Field of View (DFOV): Adjust the appropriate display field of view according to the size of the subject.

(2) Scan Parameters: For the positioning image scan, axial scan is generally selected, and it is recommended to use 80 kV and 25 mAs. Transverse section scanning is generally performed using helical scan with low dose and a tube voltage of 100 kV. The intelligent radiation dose tracking technology is used. The slice thickness for data acquisition is set between 2 and 5 mm, and the reconstruction slice thickness/interval is set between 0.5 and 1.0 mm. The tube rotation speed is between 0.27 and 0.80 r/s, and the pitch is between 0.5 and 1.0. Iterative reconstruction technology is enabled.

(3) Radiation Protection: Strictly adhere to radiation protection regulations and shield other parts of the child's body, especially radiation-sensitive areas such as the reproductive organs. For infants and young children, ensure that accompanying personnel are also adequately protected during the examination.

Important Note: For pediatric patients who cannot cooperate with breath-holding or require sedation before the examination, the scan shall be completed while the child is breathing calmly.

3.3.2.4 CT image reconstruction [14,15]

(1) Routine Image Ceconstruction: Typically, images are reconstructed with a slice thickness of 5 mm to create both lung window images (window width 1000 to 1500 HU), window level (−650 to −500 HU) and mediastinal window images (window width 250 to 300 HU, window level 30 to 50 HU).

(2) Thin-layer Image Reconstruction: Typically, images are reconstructed with a slice thickness below 1 mm to create thin-layer lung window images (lung window algorithm, window width 1000 to 1500 HU), window level −650 to −500 HU).

3.3.2.5 CT image postprocessing technology

(1) Maximum Intensity Projection (MIP): MIP images can better display the long axis of blood vessels compared with thin-slice scanning, thereby improving the stereoscopic orientation and facilitating the differentiation between blood vessels and solid lesions, leading to an increased detection rate of solid lesions. It is recommended to choose Slab-MIP with a slice thickness of 10 mm, as it is more advantageous for lesion detection and feature display.

(2) Minimum Intensity Projection (MinIP): MinIP can highlight organs with significantly low density, such as air-containing structures (e.g., bronchi), providing a significant advantage in displaying airways with densities close to that of air. MinIP is advantageous in displaying conditions such as emphysema, bronchiectasis, and mosaic attenuation and detecting early ground-glass opacities, with recommended slice thickness of 3−5 mm.

(3) Volume Rendering (VR): It is recommended to use Maximum Intensity Projection Volume Rendering (MP-VR), which provides a VR display of a certain thickness of tissue. It is suggested to use a slice thickness of 20 mm. MP-VR can vividly demonstrate three-dimensional anatomical structures and provide clear visualization of the morphology and spatial relationships of organs. Volume Rendering (VR) has advantages in spatial three-dimensional morphology and density differentiation of inflammatory lesions.

(4) Multiplanar Reconstruction (MPR) is an image postprocessing technique that recombines the already computed pixel data into images of arbitrary angles (coronal, sagittal, or any other angle) to generate sectional views. MPR allows for more accurate visualization of the relationship between lesions and blood vessels, pleura and chest wall, aiding in precise lesion localization. MPR can also provide a better display of the position relationship between lesions and bronchi and detailed and characteristic information about the lesions.

3.4 Other radiological examinations

Currently, in medical radiological examinations, the screening and diagnosis of COVID-19 primarily rely on X-ray imaging and CT scans. Magnetic Resonance Imaging (MRI) is the preferred imaging modality when patients present with neurological disorders. Radiology staff shall strictly adhere to infection prevention and control regulations to prevent cross-infection between patients and between patients and medical staff. During patient examinations, it is important to avoid wearing masks with ferromagnetic metal materials (such as iron wires or clips) into the magnetic field. After the examination is completed, necessary equipment cleaning, disinfection, and air disinfection shall be strictly implemented. If necessary, radiology technologists in close contact with confirmed COVID-19 cases can be isolated for 14 days according to the incubation period of COVID-19, and then resume normal work in the working environment.

Explanation of ethical and conflict of interest statements

This article does not involve ethical issues. All contents are sourced from guidelines and documents provided by the National Health Commission and the practical experience of frontline hospital staff. There are no other conflicts of interest.

References

[1] Hui DS, Azhar EI, Madani T, et al. The continuing 2019-nCoV epidemic threat of novel coronaviruses to global health-the latest 2019 novel coronavirus outbreak in Wuhan, China. Int J Infect Dis 2020;91:264.
[2] Huang C, Wang Y, Li X, et al. Clinical features of patients infected with 2019 novel coronavirus in Wuhan, China. Lancet 2020 (Epub ahead of print).
[3] Zhu N, Zhang D, Wang W, et al. A novel coronavirus from patients with pneumonia in China, 2019. N Eng J Med 2020. https://doi.org/10.1056/NEJMoa2001017.
[4] Ketai L, Paul NS, Ka-tak TW. Radiology of severe acute respiratory syndrome (SARS): the emerging pathologicradiologic correlates of an emerging disease. J Thoracic Imag 2006;21(4):276–83.
[5] Law of the People's Republic of China on the Prevention and Control of Infectious Diseases [Z]. 2013-06-29.
[6] Law of the People's Republic of China on Frontier Health and Quarantine [Z]. 2007-12-29.

[7] National Health Committee, State Administration of Traditional Chinese Medicine. The COVID-19 diagnosis and treatment program (trial fifth edition) [EB/OL]. [2020-02-04]. http://www.nhc.gov.cn/yzygj/s7653p/202002/3b09b894ac9b4204a79db5b8912d4440/files/7260301a393845fc87fcf6dd52965e cb.pdf.
[8] King AD, Ching ASC, Chan PL, et al. Severe acute respiratory syndrome: avoiding the spread of infection in a radiology department. AJR Am J Roentgenol 2003;181(1):25−7.
[9] Ministry of Health of the People's Republic of China. WS/T3112009 technical specification for hospital isolation [S/OL]. http://www.nhc.gov.cn/cmsresources/mohyzs/cmsrs- document/doc5841.pdf.
[10] Ministry of Health of the People's Republic of China. WS/T3132019 Codes for Hand Hygiene for medical staff [S/OL].http://www.nhc.gov.cn/wjw/s9496/200904/40118/files/5fe4afce5b874512a9780c724a4d5be0.pdf.
[11] Hou Y. Strategies and effects of prevention of major emerging infectious diseases. Electron J Emerg Infect Dis 2019;4(3):129−32.
[12] Wang X, Zhang G. Management of Radiology Department in SARS Prevention and Control. Chine J Nosocomiol 2004;14(5):558.
[13] Chen J. Interpretation of the process of donning and doffing protective clothing. Electron J Emerg Infect Dis 2016;1(1):63.
[14] Jianming Yu, Zeng Y. Medical Imaging Examination Technology [M]. Beijing: People's Medical Publishing House; 2016. p. 139−42.
[15] Hongjun Li. Radiology of Infectious Diseases [M]. Beijing: People's Medical Publishing House; 2014. p. 1−32.

Index

Note: 'Page numbers followed by *f* indicate figures, *t* indicate tables, and *b* indicate boxes.'

T

U

V

X